T0137104

Intelligent Systems Reference Library

Volume 135

Series editors

Janusz Kacprzyk, Polish Academy of Sciences, Warsaw, Poland
e-mail: kacprzyk@ibspan.waw.pl

Lakhmi C. Jain, University of Canberra, Canberra, Australia;
Bournemouth University, UK;
KES International, UK
e-mail: jainlc2002@yahoo.co.uk; jainlakhmi@gmail.com;
URL: http://www.kesinternational.org/organisation.php

The aim of this series is to publish a Reference Library, including novel advances and developments in all aspects of Intelligent Systems in an easily accessible and well structured form. The series includes reference works, handbooks, compendia, textbooks, well-structured monographs, dictionaries, and encyclopedias. It contains well integrated knowledge and current information in the field of Intelligent Systems. The series covers the theory, applications, and design methods of Intelligent Systems. Virtually all disciplines such as engineering, computer science, avionics, business, e-commerce, environment, healthcare, physics and life science are included. The list of topics spans all the areas of modern intelligent systems such as: Ambient intelligence, Computational intelligence, Social intelligence, Computational neuroscience, Artificial life, Virtual society, Cognitive systems, DNA and immunity-based systems, e-Learning and teaching,Human-centred computing and Machine ethics, Intelligent control, Intelligent data analysis, Knowledge-based paradigms, Knowledge management, Intelligent agents, Intelligent decision making, Intelligent network security, Interactive entertainment, Learning paradigms, Recommender systems, Robotics and Mechatronics including human-machine teaming, Self-organizing and adaptive systems, Soft computing including Neural systems, Fuzzy systems, Evolutionary computing and the Fusion of these paradigms, Perception and Vision, Web intelligence and Multimedia.

More information about this series at http://www.springer.com/series/8578

Margarita N. Favorskaya
Lakhmi C. Jain
Editors

Computer Vision in Control Systems-3

Aerial and Satellite Image Processing

 Springer

Editors
Margarita N. Favorskaya
Institute of Informatics and
 Telecommunications
Reshetnev Siberian State University
 of Science and Technology
Krasnoyarsk
Russian Federation

Lakhmi C. Jain
Faculty of Education, Science, Technology
 and Mathematics
University of Canberra
Canberra
Australia

and

Bournemouth University
UK

and

KES International
UK

ISSN 1868-4394   ISSN 1868-4408   (electronic)
Intelligent Systems Reference Library
ISBN 978-3-319-88443-1   ISBN 978-3-319-67516-9   (eBook)
https://doi.org/10.1007/978-3-319-67516-9

Printed on acid-free paper

This Springer imprint is published by Springer Nature
The registered company is Springer International Publishing AG
The registered company address is: Gewerbestrasse 11, 6330 Cham, Switzerland

Preface

The research book is a continuation of our previous books which are focused on the recent advances in computer vision methodologies and technical solutions using conventional and intelligent paradigms.

- Computer Vision in Control Systems-1, Mathematical Theory, ISRL Series, Volume 73, Springer-Verlag, 2015
- Computer Vision in Control Systems-2, Innovations in Practice, ISRL Series, Volume 75, Springer-Verlag, 2015

The research work presented in the book includes multidimensional image models and processing, vision-based change detection, filtering and texture segmentation, extraction of objects in digital images, road traffic monitoring by UAV, video stabilization, image deblurring, structural verification, matrix transformation and numerical system for computer vision.

The book is directed to the Ph.D. students, professors, researchers and software developers working in the areas of digital video processing and computer vision technologies.

We wish to express our gratitude to the authors and reviewers for their contribution. The assistance provided by Springer-Verlag is acknowledged.

Krasnoyarsk, Russian Federation
Canberra, Australia

Margarita N. Favorskaya
Lakhmi C. Jain

Contents

About the Editors

Dr. Margarita N. Favorskaya is a Professor and Head of Department of Informatics and Computer Techniques at Reshetnev Siberian State University of Science and Technology, Russian Federation.

Professor Favorskaya is a member of KES organization since 2010, the IPC member and the Chair of invited sessions of international conferences. She serves as a reviewer in international journals (Neurocomputing, Knowledge Engineering and Soft Data Paradigms, Pattern Recognition Letters, Engineering Applications of Artificial Intelligence), an associate editor of Intelligent Decision Technologies Journal and Computer and Information Science Journal. She is the author or the co-author of 160 publications and 20 educational manuals in computer science. She co-edited three books for Springer recently. She supervised eight Ph.D. candidates and presently supervising five Ph.D. students.

Her main research interests are digital image and videos processing, remote sensing, pattern recognition, fractal image processing, artificial intelligence, information technologies.

Dr. Lakhmi C. Jain is with the Faculty of Education, Science, Technology and Mathematics at the University of Canberra, Australia, and Bournemouth University, UK. He is a Fellow of the Institution of Engineers Australia.

Professor Jain founded the KES International for providing a professional community the opportunities for publications, knowledge exchange, cooperation and teaming. Involving around 5000 researchers drawn from universities and companies worldwide, KES facilitates international cooperation and generates synergy in teaching and research. KES regularly provides networking opportunities for professional community through one of the largest conferences of its kind in the area of KES www.kesinternational.org.

His interests focus on the artificial intelligence paradigms and their applications in complex systems, security, e-education, e-healthcare, unmanned air vehicles and intelligent agents.

Chapter 1
Theoretical and Practical Solutions in Remote Sensing

Margarita N. Favorskaya and Lakhmi C. Jain

Abstract The chapter presents a brief description of chapters that contribute to theoretical and practical solutions for aerial and satellite images processing and the fields close to this scope. One can find the original investigations in the novel tensor and wave models, new scheme of comparative morphology, warping compensation in video stabilization task, image deblurring based on physical processes of blur impacts, fast and robust core structural verification algorithm for feature extraction in images and videos, among others. Each chapter involves practical implementations and explanations.

Keywords Remote sensing · Multidimensional image processing Comparative morphology · Digital halftone images · Object extraction Traffic monitoring · Warping technique · Image deblurring · Feature extraction Double-sided matrix transformation · Counter noise immunity

1.1 Introduction

Nowadays, computer vision techniques play a significant role in many applications. In spite of pioneer investigations in remote sensing since 1970s, the development of theories and algorithms for this purpose remains the crucial issue yet. The recent

M.N. Favorskaya (✉)
Institute of Informatics and Telecommunications, Reshetnev Siberian State University of Science and Technology, 31, Krasnoyarsky Rabochy Avenue, Krasnoyarsk 660037, Russian Federation
e-mail: favorskaya@sibsau.ru

L.C. Jain
Faculty of Education, Science, Technology and Mathematics, University of Canberra, Canberra, ACT 2601, Australia
e-mail: jainlakhmi@gmail.com

L.C. Jain
Bournemouth University, Poole, UK

© Springer International Publishing AG 2018
M.N. Favorskaya and L.C. Jain (eds.), *Computer Vision in Control Systems-3*,
Intelligent Systems Reference Library 135, https://doi.org/10.1007/978-3-319-67516-9_1

achievements in computer vision permit to have another look at the well-known problems in remote sensing, as well as to increase the speed of computations. This field of investigations is very wide and will require great efforts of the researchers in further development and improvement of classical methods for aerial and satellite image processing.

1.2 Chapters Included in the Book

Chapter 2 contains a detailed description of novel tensor and wave models, as well as the interesting modifications of autoregressive models with multiple roots, in order to synthesize the algorithms of multidimensional image processing and their sequences. The consuming time algorithms on the mathematical level were developed for the applications, which process the images with random nature including TV, multispectral, and radar images and imageries. These mathematical models are based on the random field representation and simulation using the Gibbs, autoregressive, tensor and wave models [1–3]. The random fields are considered on rectangular grids of some dimension and also on the surfaces, such as the cylinder, sphere, and paraboloid that extends the application of the proposed models and algorithms to simulate the planets' surfaces. Another question deals with the anomalies' detection in the images. Four equivalent forms of the optimal decision rules are derived to detect the extensive anomalies. Also the unknown parameters of the affine and curvilinear deformations are determined according to coordinates of the obtained fixed point [4].

Chapter 3 presents a new scheme of Comparative Morphology (CM) as a generalization of the Morphological Image Analysis (MIA) scheme originally proposed by Pyt'ev [5] and further developed in [6, 7]. Such comparative morphologies are based on the mathematical shape theories, which solve the tasks of the image similarity estimation, image matching, and change detection by means of some special morphological models and tools. The proposed CM scheme excludes the MIA ideas of shape and projection and, at the same time, preserves the idea of asymmetrical comparative filtering for the robust similarity estimation and change detection. Comparative filter maps two images, reference and test, as input data and forms the filtered version of test image depending on its relation with reference image [8]. In the suggested terms, a comparative filter is called the morphological filter if it implements a smoothing mapping that preserves any constant (flat) image and preserves the test image equal to the reference image. The experiments estimate the qualitative and quantitative change detection in the real images for different scene types and simulated aerial images from the public benchmark. The obtained results of precision and recall demonstrate that the proposed pipeline is useful for change detection in long-range remote sensing data.

Chapter 4 comprises a development of the combined algorithm for detecting texture regions in noisy digital images that includes two main stages. First, the algorithm recovers the digital halftone (color) images with low signal/noise ratios.

Second, it detects the extended regions with the homogeneous statistical charac-teristics. The causal multidimensional multi-valued random Markov processes (i.e. the multi-dimensional Markov chain with several states) was used as an approxi-mation of multispectral images [9]. The remote sensing systems can generate the digital images in different spectral bands containing from three (like RGB color images) to hundreds of components, each of which can be considered as the monochrome (grey-scale) image. The color RGB images can be represented as a special case of multispectral images. Big statistical dependence between image elements may exist in the Digital HalfTone Images (DHTI) belonging to different spectral components [10]. Such statistical relationship between the elements within and between elements of the DHTI color components (RG, GB, BR) can be approximated by 3D multi-valued Markov chain. During processing the DHTI with 2^g brightness levels, the problem of memory storage and working with the transi-tion probability matrices $2^g \times 2^g$ in size is appeared. It was suggest to divide the DHTI represented by g bits binary numbers into g bit binary image or bit planes [11]. This allows to reduce the computational resources owing to working with 2×2 transition probability matrices. The quality of segmentation was improved using the evaluation of the statistical characteristics of each local area within and between adjacent bit binary images for different color component [12]. The pro-posed method allows to segment a noisy image (with signal-to-noise ratio –6 dB) efficiently if the transition probability between elements in the areas does not exceed 0.15. In this case, a segmentation error is less than 8%.

Chapter 5 conducts the investigations for extraction of objects in different types of images like aerial, satellite, and synthetic aperture radar images using the straight edges segments, which are in following grouped in the simple or complex shapes. The problem of feature construction for object description and extraction includes the feature detection and feature description tasks [13]. An advanced edge-based method for automatic detection and localization of straight edge segments was developed that provides an ordered set of straight line segments with their orien-tations and magnitudes. The algorithm automatically adjusts an appropriate mask size for the slope line filter to give the maximum normalized filter output. It pro-duces a set of segments as affine invariant and scale invariant features for object description. The advanced algorithm performs a reliable detection of local edges and accurately obtains the edge intersection points. The new elements are the global seeking for the most valuable edge direction and automatic adjusting of the mask size for the slope line filter to match with the edge length. The comparative analysis of the noisy image shows that the advanced algorithm is inferior to others in feature detection performance. A hierarchical set of features is developed for object description subject to the proposed feature detector. This set contains four levels of line combination, each level relates to the number of combined line segments. The problem of object selection using the straight line segments extraction and grouping includes several steps, when the following elements are determined sequentially: the straight line segments, crossing points of lines, and closed complex structures [14]. The algorithm can extract not only rectangular or triangular objects or parts of them but other types of objects (roads, polygonal structures). The extended experiments with the noisy images were conducted [15]. Applications to the aerial, satellite, and radar images show a good ability to separate and extract rectangular objects like buildings and other line-segment-rich structures. Most of objects are selected somehow or other and the following problem is how to improve grouping process.

Chapter 6 includes the study of the ground traffic monitoring aided by the Unmanned Aerial Vehicles (UAV) based on applying the on-board computer vision systems. At present, the UAV cannot regard as a direct testifier of a specific road traffic situation, for example, collision of cars [16, 17]. The authors consider a possibility of situation classification based on the observed scene after the situation occurrence including the relative positions of cars, their damages, and position and behaviour of people. For the estimation and forecast of development of the present road situations, it is required to classify the traffic situations. For these purposes, the authors solved the following tasks: the selection of classification type and an alphabet forming of situation class, forming descriptions in the observed scene containing attributes of recognized situations (close to the feature dictionary), and choice of the decision-making algorithms that allows ranging the examined situation to corresponding class. Four classes called as the "Observed Scene", "Objects of Traffic Accident (TA)", "External Condition", and "Additional Objects" were introduced. This description has a hierarchic structure and is divided into classes, subclasses, and divisions of various levels depending on a hierarchic level. In general, various types of descriptions, such as spatial, spatiotemporal, temporal, and causal, may be used [18]. The observed scene description is built using the proposed ontology structure of the TA. For decision making, the production model of knowledge representation and corresponding knowledge base were designed. The functional criteria of the losses, flight safety of the UAV, and reliability of class recognition allow to reconstruct the TA using video sequence obtained from the UAV.

Chapter 7 explores the warping techniques in video stabilization task [19]. All warping techniques are classified as the parametric and non-parametric methods. The parametric methods, such as translation, rotation, procrustes, affine, perspective, bilinear, and polynomial, provide the compact representation and fast computation of a warping but cannot perform well the local distortions. Opposite to them, the non-parametric methods like elastic deformations, thin-plate splines, and Bayesian approach demonstrate a heavy computational load and the presence of local optima. Not all of possible warping techniques are suitable for video stabilization task [20]. Both 2D and 3D stabilization algorithms involve the stabilization of video sequence and reconstruction of missing boundary regions. These stages are implemented differently for 2D and 3D stabilization but in each case the stabilization of video sequence requires the global frame warping, while the reconstruction of missing boundary regions in cropped after stabilization frames demands in the local frame warping. Both types of warping are based on the introduced Structure-From-Layered-Motion (SFLM) model based on the compactness theory of visual objects' representation. The SFLMs are associated with all foreground moving objects, partly with background moving objects and render the individual motion types of moving objects [21]. The global warping in 2D stabilization is aimed on the compensation of each frame for removal of the unwanted motion, while 3D stabilization requires the generation of a desired 3D camera path and synthesis of the images along 3D path [22]. The reconstruction stages for 2D/3D stabilization are almost similar and based on the special inpainting methods, including the proposed pseudo-panoramic key frame and the multi-layered motion fields. The experiments confirm the quality of the stabilized test video sequences in terms of the content alignment and the frame size preserving.

Chapter 8 provides the investigate study of image deblurring based on physical processes of blur impacts. One can find the detailed overview of the non-blind deconvolution methods including inverse filtering, Wiener filtering, Kalman recursive filtering, minimum Mean Square Error (MSE) method, and various forced iterative methods, and blind deconvolution methods (various types of regularization methods like Tikhonov regularization algorithm, Lucy-Richardson algorithm, Shepp-Logan intuitive regularization method, and Arsenin method of local regularization) for image restoration. The authors proposed the original approach based on the Distortion Functions (DFs) for the small and large linear blur. The corresponding mathematical models as the systems of the linear algebraic equations are built with multiple illustrating examples. A comparison of the restoration results yielded by MatLAB software package using the known deblurring methods shows that application of the proposed model allows to restore the smaller distorted images with low computing complexity. Also the model of non-linear blur was constructed on the basis of the generalized description for processes of video signal distortion. A non-linear vibrational blur can occur, when the unstable position of the camera, the reciprocating motion of the object, or atmospheric turbulence take place [23]. For this case, five systems of linear equations were obtained. All models are based on the accurate background evaluation and perform a promising base for following investigations.

Chapter 9 involves a fast and robust Core Structural Verification Algorithm (CSVA) for feature extraction in images and videos. The proposed algorithm is based on many-to-one matches' exclusion, the improved Hough clustering of keypoint matches, and cluster verification procedure as the modified RANSAC. The feature detection, feature description, descriptors matching, and structural verification are the main stages of keypoint-based methods. These stages are sequentially discussed for many types of known descriptors with their comparative analysis. The input of the CSVA is a set of the nearest neighbour solution for keypoint descriptors (set M). The CSVA takes three steps, such as the primary outlier elimination (set M_{prime}), Hough clustering (set M_{cl}), and verification of the largest cluster(s) relying on mutual relation of the keypoints of different matches (set M_{final}). The ratio of correct/incorrect matches is a key factor for the successful application of geometrical constraints. Three methods, such as the nearest neighbour ratio test, cross-check, and exclusion of ambiguous n-to-1 matches, are considered in details. Verification stage of the CSVA is based on clustering of feature matches in similarity transform parameter space and consists of the Hough clustering and cluster verification [24]. The experiments were conducted using three datasets. Dataset db1 has 20 challenging image pairs of 3D scenes (mostly under strong viewpoint changes). The pairs were matched by different combinations of the SURF and the SIFT detectors and descriptors. Dataset db2 consists of 120 pairs of aerial and cosmic image pairs under the strong viewpoint, season, day-time, and man-made changes. Several manually matched points were used to calculate a ground truth homography H for every image pair. Dataset db3 has more than 300 pairs of images of static 3D scenes (mostly indoor). Few manually labelled corresponding points were used to calculate etalon fundamental matrices E by the least squares. The proposed algorithm uses some specific information (rigidity of objects in a scene), consume low volume memory and only 3 ms in average on a standard Intel i7 processor for verification of 1000 matches. The CSVA has been successfully applied to practical tasks with minor adaptation, such as the matching of 3D indoor scenes, retrieval of images of 3D scenes based on the concept of Bag of Words, and matching of aerial and cosmic photographs with strong appearance changes caused by season, day-time and viewpoint variation [25, 26].

Chapter 10 covers a double-sided matrix transformation that is efficiently used in telecommunication systems. The issues how to increase the interference immunity, transmission speed as well as to provide for the quality of a transmitted image over a telecommunication channel had been appeared since 1980s. Mironovsky and Slaev contributed significantly in this scope, for example [27–29]. The double-sided matrix transformation in respect of its matrices under some serious restrictions (orthogonality, symmetry) is the subject of interest in many researches. In this chapter, a more general case, for which criteria of invariant image existence are introduced, is analyzed. The scale, rotation, and negative (inverse) strip-invariants are investigated and the methods of finding them are proposed. Moreover, the task of arraying the double-sided transformation on the basis of a given set of invariant images is solved. An effective way of finding invariant images at given matrices is based on the use of eigenvectors of matrices. Such productive approach leaded to the novel results in the strip-method useful in cryptography, steganography, and other applications. The multiple examples are met and fully explained during a whole chapter representation.

Chapter 11 is devoted to the numerical systems that may be used in the counters and decoders of digital tools. Nowadays, the classical binary system is the basis of the counters due to their sufficiently simplicity, reliability, and cheapness. However, the counters based upon the binary system lack a natural informational redundancy, which prevents them from finding easily the errors occurring when they work. That is why various methods of enhancing the counters' noise immunity are often proposed. Fibonacci numeral systems with the Fibonacci numbers serving as the weights in the coding words boast such a redundancy to a higher grade than other systems. In addition to their ability to detect errors arising when functioning, the counters based on Fibonacci numeral systems reveal the high computational speed as well. Based on the Fibonacci numeral system, schemes of various noise-proof digital devices such as processors, counters, summers, and even computers have been developed [30]. However, when implementing those devices and circuits, two distinct representation forms: minimal and maximal, were exploited with possible transitions from the minimal form to the maximal one, and vice versa, to do operations over the Fibonacci numbers [31]. In order to complete such transitions, special digital circuits are necessary that realize the operations of catamorphisms and anamorphisms (or "folds" and "unfolds"). These operations made the Fibonacci devices more complicated and slow. This prohibited the thorough use of all potential capabilities of the Fibonacci codes as implemented in efficient Fibonacci counters [32]. The authors investigated the efficiency of the Fibonacci counters handling the minimal presentation form for the Fibonacci numbers, in the part of their information transmission velocity and noise-immunity. Another task is the description of Fibonacci decoders boasting the minimal apparatus costs.

1.3   Conclusions

The chapter provides a briefly description of ten chapters with original mathematical investigations in computer vision techniques applied in remote sensing and close issues. All included chapters involve the long standing original investigations of the authors in novel tensor and wave models and modifications of autoregressive models in order to synthesize the algorithms of multidimensional image processing, new scheme of comparative morphology, which solve the tasks of the image similarity estimation, image matching, and change detection, development of digital halftone (color) images, extraction of straight edges segments with following grouping in the simple or complex shapes, ground traffic monitoring aided by the unmanned aerial vehicles, warping compensation using the proposed structure-from-layered-motion model, image deblurring based on physical processes of blur impacts, core structural verification algorithm for feature extraction, the novel results in the strip-method useful in cryptography and steganography, and the enhanced approach for the counters' noise immunity in numerical systems based on the Fibonacci numbers. As a rule, each chapter involves large volume of experimental results and justified illustrations.

References

1. Vasil'ev, K.K., Dement'ev, V.E., Andriyanov, N.A.: Application of mixed models for solving the problem on restoring and estimating image parameters. Pattern Recognit. Image Anal. **26** (1), 240–247 (2016)
2. Krasheninnikov, V.R.: Correlation analysis and synthesis of random field wave models. Pattern Recognit. Image Anal. **25**(1), 41–46 (2015)
3. Krasheninnikov, V.R., Kalinov, D.V., Pankratov, Yu.G.: Spiral autoregressive model of a quasi-periodic signal. Pattern Recognit. Image Anal. **8**(1), 211–213 (2001)
4. Krasheninnikov, V.R., Potapov, M.A.: Estimation of parameters of geometric transformation of images by fixed point method. Pattern Recognit. Image Anal. **22**(2), 303–317 (2012)

5. Pyt'ev, Y.P.: Morphological Image Analysis. Pattern Recognit. Image Anal. **3**(1), 19–28 (1993)
6. Vizilter, Y.V., Zheltov, S.Y.: Geometrical correlation and matching of 2D image shapes. ISPRS Ann. Photogramm. Remote Sens. Spat. Inf. Sci. **I-3**, 191–196 (2012)
7. Vizilter, Y.V., Gorbatsevich, V.S., Rubis, A.Y., Zheltov, S.Y.: (2014) Shape-based image matching using heat kernels and diffusion maps. Int. Arch. Photogramm. Remote Sens. Spat. Inf. Sci. **XL**(3), 357–364
8. Vizilter, Y.V., Rubis, A.Y., Zheltov, S.Y., Vygolov, O.V.: Change detection via morphological comparative filters. ISPRS Ann. Photogramm. Remote Sens. Spat. Inf. Sci. **III**(3), 279–286 (2016)
9. Petrov, E.P., Medvedeva, E.V.: Nonlinear filtering of statistically connected video sequences based on hidden Markov chains. J. Commun. Technol. Electr. **55**(3), 307–315 (2010)
10. Petrov, E.P., Trubin, I.S., Medvedeva, E.V., Smolskiy, S.M.: Mathematical models of video-sequences of digital half-tone images. In: Atayero, A.A., Sheluhin, O.I. (eds.) Integrated Models for Information Communication System and Networks: Design and Development, pp. 207–241. IGI Global, Hershey (2013)
11. Petrov, E.P., Trubin, I.S., Medvedeva, E.V., Smolskiy, S.M.: Development of nonlinear filtering algorithms of digital half-tone images. In: Atayero, A.A., Sheluhin, O.I. (eds.) Integrated Models for Information Communication System and Networks: Design and Development, pp. 278–304. IGI Global, Hershey (2013)
12. Medvedeva, E.V., Kurbatova, E.E.: Image segmentation based on two-dimensional Markov chains. In: Favorskaya, M.N., Jain, L.C. (eds.) Computer Vision in Control Systems-2. Innovations in Practice, vol. 75, pp. 277–295. Springer International Publishing, Switzerland (2015)
13. Volkov, V., Germer, R., Oneshko, A., Oralov, D.: Object description and extraction by the use of straight line segments in digital images. In: International Conference on Image Processing, Computer Vision and Pattern Recognition (IPCV'2011), pp. 588–594 (2011)
14. Volkov, V., Germer, R., Oneshko, A., Oralov, D.: Object selection by grouping of straight edge segments in digital images. In: International Conference on Image Processing, Computer Vision and Pattern Recognition (IPCV'2013), pp. 321–327 (2013)
15. Volkov, V., Germer, R.: Straight edge segments localization on noisy images. In: International Conference on Image Processing, Computer Vision and Pattern Recognition (IPCV'2010), vol. II, pp. 512–518 (2010)
16. Kim, N., Chervonenkis, M.: Situational control unmanned aerial vehicles for traffic monitoring. Mod. Appl. Sci. **9**(5), 1852–1913 (2015)
17. Kim, N., Bodunkov, N.: Adaptive surveillance algorithms based on the situation analysis. In: Favorskaya, M., Jain, L.C. (eds.) Computer Vision in Control Systems-2, ISRL, vol. 75, pp. 169–200 (2015)
18. Kim, N.: Automated decision making in road traffic monitoring by on-board unmanned aerial vehicle system. Ind. J. Sci. Technol. **8**(S10), 1–6 (2015)
19. Favorskaya, M., Jain, L.C., Buryachenko, V.: Digital video stabilization in static and dynamic scenes. In: Favorskaya, M.N., Jain, L.C. (eds.) Computer Vision in Control Systems-1, ISRL, vol. 73, pp. 261–309. Springer International Publishing, Switzerland (2015)
20. Favorskaya, M., Buryachenko, V.: Fuzzy-based digital video stabilization in static scenes. In: Tsihrintzis, G.A., Virvou, M., Jain, L.C., Howlett, R.J., Watanabe, T. (eds.) Intelligent Interactive Multimedia Systems and Services in Practice, SIST, vol. 36, pp. 63–83. Springer International Publishing, Switzerland (2015)
21. Favorskaya, M., Buryachenko, V.: Fast salient object detection in non-stationary video sequences based on spatial saliency maps. In: De Pietro, G., Gallo, L., Howlett, R.J., Jain, L. C. (eds.) Intelligent Interactive Multimedia Systems and Services, SIST, vol. 55, pp. 121–132. Springer International Publishing, Switzerland (2016)
22. Favorskaya, M., Buryachenko, V., Tomilina, A.: Global motion estimation using saliency maps in non-stationary videos with static scenes. In: De Pietro, G., Gallo, L., Howlett, R.J.,

Jain, L.C. (eds.) Intelligent Interactive Multimedia Systems and Services, SIST, vol. 55, pp. 133–144. Springer International Publishing, Switzerland (2016)

23. Bogoslovskiy, A.V., Zhigulina, I.V., Bogoslovskiy, E.A., Ponomarev, A.V., Vasilyev, V.V.: Linear Blur. Radiotec, Moscow (in Russian) (2015)

24. Malashin, R.: Correlating images of three-dimensional scenes by clusterizing the correlated local attributes, using the Hough transform. J. Opt. Technol. **81**(6), 327–333 (2014)

25. Malashin, R.: Image retrieval with the use of bag of words and structural analysis. J. Phys: Conf. Ser. **735**(1), 12–16 (2016)

26. Malashin, R.: Matching of aerospace photographs with the use of local features. J. Phys.: Conf. Ser. **536**(1), 12–18 (2014)

27. Mironovsky, L., Slaev, V.: Invariants in metrology and technical diagnostics. Meas. Tech. **39** (6), 577–593 (1996)

28. Mironovsky, L., Slaev, V.: The strip method of transforming signals containing redundancy. Meas. Tech. **49**(7), 631–638 (2006)

29. Mironowsky, L., Slaev, V.: Double-sided noise-immune strip transformation and its root images. Meas. Tech. **55**(10), 1120–1127 (2013); Mironovsky, L., Slaev, V.: Implementation of Hadamard matrices for image processing. In: Favorskaya, M.N., Jain, L.C. (eds.) Computer Vision in Control Systems-1, ISRL, vol. 73, pp. 311–349. Springer International Publishing, Switzerland (2015)

30. Stakhov, A.P.: Fibonacci and Golden Proportion Codes as an Alternative to the Binary Numeral System. Part 1. Academic Publishing, Germany (2012)

31. Borisenko, A.A., Kalashnikov, V.V., Protasova, T.A., Kalashnykova, N.I.: A new approach to the classification of positional numeral systems. In: Neves-Silva, R., Tshirintzis, G.A., Uskov, V., Howlett, R.J., Jain, L.C. (eds.) Frontiers in Artificial Intelligence and Applications (FAIA), vol. 262, pp. 444–450 (2014)

32. Borysenko, O.A., Matsenko, S.M., Polkovnikov, S.I.: A noise-proof Fibonacci counter. Trans. Natl. Technol. Univ. Kharkiv **18**, 77–81 (2013) (in Russian)

Chapter 2
Multidimensional Image Models and Processing

Victor Krasheninnikov and Konstantin Vasil'ev

Abstract The problems of developing mathematical models and statistical algorithms for processing of multidimensional images and their sequences are presented in this chapter. Different types of random fields are taken for the basic mathematical image model. This implies two main problems associated with image modeling, namely, model analysis and synthesis. The main attention is paid to the correlation aspect, i.e. evaluation of the correlation function of a random field generated by a given model and, vice versa, development of a model generating a random field with a predetermined correlation function. For this purpose, new models (tensor and wave) and new versions of autoregressive models (with multiple roots) are suggested. The problems of image simulation on the curved surfaces are considered. The suggested models are used to synthesize the algorithms of multidimensional image processing and their sequences. The tensor filtration of imaging sequences and recursive filtration of multidimensional images, as well as the asymptotic characteristics of efficiency of random field filtration on grids of arbitrary dimension are suggested. The problem of object and anomaly detection on the background of interfering images is considered for the images of any dimension, e.g. for multi-zone data. It is shown that four equivalent forms of the optimal decision rule, which reflect various aspects of detection procedure, exist. Potential efficiency of anomaly detection is analyzed. The problems of alignment and estimation of parameters for interframe geometric image transformations are considered for multidimensional image sequences. A tensor procedure of simultaneous filtration of multidimensional image sequence and their interframe displacements are suggested. A method based on a fixed point of a complex geometric image transformation was investigated in order to evaluate large interframe displacements. Options for adaptive image processing algorithms are also discussed in this chapter. In this context, pseudo-gradient procedures are taken as a basis, as they do not require

V. Krasheninnikov (✉) · K. Vasil'ev
Ulyanovsk State Technical University, 32 Severni Venets Street,
Ulyanovsk 432027, Russian Federation
e-mail: kvrulstu@mail.ru

K. Vasil'ev
e-mail: 1948vasiliev@gmail.com

M.N. Favorskaya and L.C. Jain (eds.), *Computer Vision in Control Systems-3*,
Intelligent Systems Reference Library 135, https://doi.org/10.1007/978-3-319-67516-9_2

preliminary evaluation of any characteristics of the processed data. This allows to develop the high-performance algorithms that can be implemented in real-time systems.

Keywords Multidimensional image model · Autoregressive model
Tensor model · Wave model · Curved surface · Processing · Potential efficiency
Prediction · Filtration · Anomaly detection · Recognition · Adaptive algorithm
Pseudo-gradient algorithm

2.1 Introduction

Nowadays, the problems of automatic analysis of images and their sequences are becoming more and more important. This is due to the rapid development of the aerospace Earth monitoring systems, radio and sonar systems with spatial arrays, medical devices for early disease diagnosis, and computer vision systems. For some applications, images can be represented as data files set on a multidimensional index grid changing in discrete time. This allows to describe certain 2D or multi-dimensional images and their temporal sequences with scalar or vector values, e.g. sequence of TV images, multispectral images, spatial field of wind speeds, and others.

Mathematical models are necessary to formalize image processing problems. Random nature of image values determines the use of probability theory and mathematical statistics methods, i.e. image representation in the form of Random Fields (RFs) [1–5]. This implies the problems of the RF model analysis and synthesis. Composition of information and noise RF is usually used as a model of image observations. This model can also include an additional parameter vector, which allows to consider the peculiarities of image registration, e.g. possible mutual spatial displacement of adjacent frames and various abnormalities (object, signal) [6–9]. The reason for processing the sequences of observations can be development of estimations of an informational RF (prediction, filtration, and interpolation). Another important problem is the estimation of vector parameters, e.g. the image model parameters, spatial displacements, parameters of detected objects, and others.

This chapter presents the results obtained by the authors and their colleagues in modeling and statistical processing of 2D and multidimensional images and their sequences. A number of models [2, 3, 10, 11], including multidimensional ones [7, 12], are known to describe the images and their sequences [13, 14]. Section 2.2 covers the problems of the RF description and simulation. Random field is a set of random variables defined on multidimensional spatial grids with rectangular or more complex cells. For this purpose, the Gibbs, autoregressive, Tensor, and wave models are used [3, 12, 15, 16]. It is difficult to solve synthesis problems using autoregressive models, in particular, even modeling of spatially isotropic RF. However, the use of models with multiple roots of a characteristic equation facil-itates to obtain approximately isotropic RF [16–19]. For linear transformations of image frames, it is desirable to use tensor operations that provide the basis for tensor models, in which every recurrent frame is formed from the previous frame and the frame of random variables [16]. The suggested wave models permit to solve the analysis, synthesis, and simulation problems rather efficiently. In these models, the RF is a result of disturbance interaction, which can occur in random places at random times [16, 20]. These models include many well-known models as special cases. Obtaining of the RF with a given type of correlation is achieved by varying a probability distribution of a large-scale disturbance ratio. Section 2.2 also covers different ways to describe the images on arbitrary surfaces. Autoregressive and wave RF models are used for this purpose [16]. For example, the scan of a cylindrical image can be used to represent the speech and other quasi-periodic signals [21], and images on the sphere can be used to describe a planet relief [22].

There are many papers on image filtering. Many methods of algorithm synthesis are developed taking into consideration a type of informative and noise components of observations [6, 15, 23–25]. At the same time, much attention was paid to reduce the computational complexity of filtration procedures, which, in particular, led to the creation of Kalman filters. However, the use of Kalman filtering in image processing causes significant difficulties due to multidimensionality. These difficulties were partially overcome using line-by-line vector filtration, as well as combination of several one-dimensional filters [12]. Section 2.3 introduces a tensor filter of multidimensional image sequences. This filter was developed on the basis of the RF tensor model described in Sect. 2.2. Asymptotic characteristics of efficiency of random field filtration on arbitrary dimension grids are also introduced, as well as a potential accuracy of multidimensional RF filtration.

Many researches are devoted to the problems of object or signal detection and recognition [6, 8, 9, 11, 26, 27]. The problems of synthesis and analysis of optimal decision rules for detecting point and extensive anomalies in the images, including multi-zone data, and their sequences are discussed in Sect. 2.4. Four equivalent forms of the optimal decision rule based on the likelihood ratio are obtained. These four forms are significantly different in computational complexity for large spatial sizes of images [16]. Characteristics of efficiency for anomaly detection in multidimensional images are also derived.

An important task of image sequence processing is their alignment, when due to various reasons there are spatial deformations of adjacent frames. Numerous approaches how to solve this problem, for example, search for characteristic points, light flow analysis, morphological analysis, and correlation-extreme methods, are proposed in [7, 10, 11, 14, 28]. In Sect. 2.5, the problems of alignment and estimation of parameters of image interframe geometrical transformations are considered for sequences of 2D and multidimensional images. On the basis of the tensor model of image sequences, a tensor filter for simultaneous filtration of images and their interframe displacement estimation are synthesized [16]. The authors also present a new method to estimate the parameters of image deformation using a fixed point of a complex geometric transformation of two images [16, 29, 30]. This complex transformation consists of the actual deformation plus an additional artificially performed geometric transformation. Unknown parameters of actual deformation are determined according to coordinates of the obtained fixed point.

In practice, the image, noise and observation models are usually only partially known, i.e. there is a priory uncertainty. Thus, the synthesis of adaptive processing algorithms is required. For this purpose, many adaptive algorithms were suggested [31, 32]. Section 2.6 outlines the adaptive procedure options, which are included in the algorithms for solving image prediction and alignment [16]. Special attention is paid to the pseudo-gradient adaptation, on the basis of which highly efficient algorithms with comparatively small computational costs are synthesized. These qualities allow their usage in real-time systems, dealing with large images and their sequences. Section 2.7 concluded the chapter.

2.2   Mathematical Models of Images

The mathematical models suitable for multidimensional processing, such as the random fields, tensor models of random fields, autoregressive models of random fields, wave models of random fields, and random fields on surfaces are discussed in Sects. 2.2.1–2.2.5, respectively.

2.2.1   Random Fields

Nowadays, information systems, including spatial sensor devices and digital computing, are widely used. Therefore, the images with discrete spatial and temporal variables will be primarily considered. Without loss of generality, one may assume that the images are the sets of values arranged on multidimensional rectangular grids with a unit step. Two-dimensional and three-dimensional grids are presented in Fig. 2.1a, b, respectively. In general, an image is a set in n-dimensional grid nodes $\Omega = \{\bar{j} = (j_1, \ldots, j_n) : j_k = \overline{1, M_k}, \ k = \overline{1, n}\}$. According to the physical nature, image values may be scalar (e.g. brightness of a monochromatic image), vector (velocity field, color images, displacement field), and more complex (e.g. matrix). If an image value in the node (pixel) $\bar{j}$ is denoted as $x_{\bar{j}}$, then the image is a set of these values on the grid $X = \{x_{\bar{j}} : \bar{j} \in \Omega\}$.

If the data is a time sequence of images, then a sequence can be assumed as a single image, increasing a grid dimension by one unit. For example, the sequence of planar images (Fig. 2.1a) can be regarded as a single three-dimensional image (Fig. 2.1b).

If it is necessary to specify a time variable, let us set it down at the top $X = \{x_{\bar{j}}^i : \bar{j} \in \Omega, \ i \in I\}$. This image is a set on the direct product $\Omega \times I$ of grids Ω and I, where I is a set of time index values. The cross section $x^i = \{x_{\bar{j}}^i : \bar{j} \in \Omega\}$, i.e.

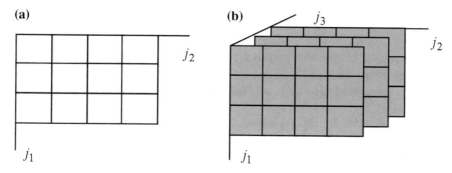

Fig. 2.1 Grids with: **a** two-dimensional image, **b** three-dimensional image

a set of image samples at a fixed value of time index i, is called the ith frame. Each frame is a set on a grid Ω. For example, Fig. 2.1b shows three two-dimensional frames.

Thus, an image can be determined as a function defined on a multi-dimensional grid. The values of the image elements cannot be accurately predicted in advance (otherwise monitoring system would not be necessary). Therefore, it is intrinsic to regard these values as the Random Variables (RVs) using the methods of probability theory and mathematical statistics.

2.2.2 Tensor Models of Random Fields

Consider the RF $X = \left\{ x_{\bar{j}}^i : i \in I, \bar{j} \in \Omega \right\}$ defined on an $(n + 1)$-dimensional grid $\Omega \times I$, where $\Omega = \{ \bar{j} = (j_1, j_2, j_3, \ldots, j_n) \}$ is an n-dimensional $M_1 \times M_2 \times \cdots \times M_n$-grid, $I = \{ i : i = 1, 2, 3, \ldots \}$. Index i may be interpreted as a time index, thus the expression $x^i = \left\{ x_{\bar{j}}^i : \bar{j} \in \Omega \right\}$ is called a cross section of a field X as in ith frame.

Let the sequence of frames be described by a stochastic difference Eq. 2.1, where $\left\{ \xi_{\bar{j}}^i : i \in I, \bar{j} \in \Omega \right\}$ is an updating standard Gaussian field, $\xi^i = \left\{ \xi_{\bar{j}}^i : \bar{j} \in \Omega \right\}$ is the ith field frame, $\varphi^i(x^{i-1}) = \left\{ \varphi_{\bar{j}}^i(x^{i-1}) : \bar{j} \in \Omega \right\}$ is $M_1 \times M_2 \times \cdots \times M_n$-matrix function, $\vartheta^i(x^{i-1}) = \left\{ \vartheta_{\bar{j}\bar{t}}^i(x^{i-1}) : \bar{j}, \bar{t} \in \Omega \right\}$ are tensors of rank $2n$ with group indices forming a perturbation component of the ith frame from ξ^i using the rule of tensor multiplication $\vartheta^i \xi^i = \left\{ \vartheta_{\bar{j}\bar{t}}^i \right\} \{ \xi_{\bar{t}}^i \} = \left\{ \sum \vartheta_{\bar{j}\bar{t}}^i \xi_{\bar{t}}^i \right\}$.

$$x^i = \varphi^i(x^{i-1}) + \vartheta^i(x^{i-1})\xi^i \quad i = 1, 2, \ldots \tag{2.1}$$

Transposing this frame supposes a permutation of its group indices $\vartheta_{ji}^{iT} = \vartheta_{ij}^{i}$. Note that the superscript i indicates a frame number, i.e. time in our case. Thus, index i is considered to be an umbral index and the summation with respect to it is not extended.

Model provided by Eq. 2.1 allows to describe a very broad class of Markov random frame sequences. In particular, a linear model in Eq. 2.1, where tensors P^i and ϑ^i do not depend on x^{i-1}, describes the Gaussian sequence.

$$x^i = P^i x^{i-1} + \vartheta^i \xi^i \tag{2.2}$$

For this field of Covariance Functions (CFs) there exists a multidimensional covariance matrix defined directly

$$V_x(i,j) = M[(x^i - m^i) \times (x^j - m^j)] \quad V_x(i) = V_x(i,i),$$

where $m^i = M[x^i] = \left\{ M[x_{\bar{j}}^i] : \bar{j} \in \Omega \right\}$, and symbol "$\times$" indicates an external matrix multiplication. Thus, $V_x(i)$ and $V_\xi^i = M[\xi^i \times \xi^i]$ are symmetrical $M_1 \times M_2 \times \cdots \times M_n \times M_1 \times M_2 \times \cdots \times M_n$-matrices. For complete definition of a random field with a state Eq. 2.2, it is necessary to set the law of first frame x^0 distribution. It is often a Gaussian distribution with mean m^0 and covariance matrix V_x^0.

Model in Eq. 2.2 with constant tensors $P^i = P$ and $\vartheta^i = \vartheta$ is of particular interest that leads to Eq. 2.3, where (k) determines raising to the kth power.

$$V_x(i, i+k) = P^{(k)} V_x(i,i) = P^{(k)} V_x(i) \tag{2.3}$$

If roots of the characteristic equation $\det(\lambda E - P) = 0$ are less than 1 in modulus, than $P^{(k)} \to 0$ as $k \to \infty$, and from Eq. 2.3 we obtain that $V_x(i, i+k) \to 0$ as $k \to \infty$.

Using z-transformation device, let us write Eq. 2.2 in the form

$$x^i = zPx^i + \vartheta\xi^i$$

or

$$(E - zP)x^i = \vartheta\xi^i.$$

Multiplying the congruence on the left by $(E - zP)^{-1}$, the expression x^i through a perturbing field is obtained.

$$x^i = (E - zP)^{-1} \vartheta \xi^i$$

Hence,

$$x^i(x^i)^T = (E - zP)^{-1} \vartheta \xi^i (\xi^i)^T \vartheta^T (E - z^{-1}P^T)^{-1}$$

that after averaging gives the expression of the tensor spectrum for a stationary field

$$V_x(z) = \sum_{k=-\infty}^{\infty} V_x(0,k)z^k = (E - zP)^{-1} \vartheta V_\xi \vartheta^T (E - z^{-1}P^T)^{-1},$$

which is a Laurent series in powers of z with tensor coefficients. The tensor coefficients can be found from Eq. 2.4, where $C = \{z : |z| = 1\}$ is a unit circumference of a complex plane.

$$V_x(0,k) = \frac{1}{2\pi i} \int_{C_1} (E - zP)^{-1} \vartheta \, V_\xi \vartheta^T (E - z^{-1}P^T)^{-1} z^{k-1} dz \qquad (2.4)$$

It is sufficient to find $V_x = V_x(0,0)$ from Eq. 2.4, other values are obtained from Eq. 2.3 in a view of $V_x(i, i+k) = P^{(k)} V_x$ for a stationary case. To find V_x it is possible to use limit in Eq. 2.3 as $i \to \infty$ instead of integral in Eq. 2.4

$$V_x = PVP^T + \vartheta V_\xi \vartheta^T$$

that is a non-singular system of linear equations relative to tensor components V_x.

Now consider a solution to the model synthesis problem Eq. 2.2, i.e. the problem of finding tensors P^i and ϑ^i, when intraframe $V_x(i,i)$ and interframe $V_x(i-1,i)$ covariance tensors are given. From Eq. 2.3 we obtain

$$V_x(i-1,i) = P^i V_x(i-1,i-1)$$

that is a system of linear equations relative to tensor elements P^i. It is obvious that

$$P^i = V_x(i-1,i)V_x^{-1}(i-1,i-1).$$

Chose a perturbing field with covariance $V_\xi^i = E$. Then the following equation is obtained

$$\vartheta^i \vartheta^{iT} = V_x(i,i) - P^i V_x(i-1,i-1)P^{iT}.$$

For example, for tensor components ϑ^i this equation can be solved on the basis of Gram-Shmidt orthogonalization.

Nonlinear stochastic difference equation may be considered as generalization of the considered tensor model (Eq. 2.2). This equation allows to describe very large class of Markov non-Gaussian RF on n-dimensional grids J_t. Here we have $\left\{\xi_{\bar{l}}^t, \bar{l} \in J_t, t \in T\right\}$, a field of independent, generally speaking, non-Gaussian random variables with known probability density function (PDF) $W(\xi_{\bar{l}}^t)$, $\varphi_{\bar{j}}^t(x_{\bar{l}}^{t-1})$ and $\vartheta_{\bar{j}\bar{l}}^t(x_{\bar{s}}^{t-1})$ are tensors of ranks n and $2n$ correspondingly, which in a general case nonlinearly depend on the values $\{\xi_{\bar{j}}\}$ of the $(t-1)$th frame of a multidimensional RF $\left\{x_{\bar{j}}^{t-1}, t \in T, \bar{j} \in J\right\}$.

2.2.3 Autoregressive Models of Random Fields

Tensor representation assumes that for each moment t of discrete time the RF $\left\{x_{\bar{j}}^t, \bar{j} \in J\right\}$ is formed recurrently on the basis of the previous value $\left\{x_{\bar{j}}^{t-1}, \bar{j} \in J\right\}$ and updating RF $\left\{\xi_{\bar{j}}^t, \bar{j} \in J\right\}$ of independent RV. Despite the fact that the calculations are done recurrently while forming each successive frame $\left\{x_{\bar{j}}^t, \bar{j} \in J\right\}$, it is desirable to conduct linear Eq. 2.1 or nonlinear Eq. 2.2 transformation of all elements $\left\{x_{\bar{j}}^{t-1}, \bar{j} \in J\right\}$, $\left\{\xi_{\bar{j}}^t, \bar{j} \in J\right\}$ determined on an n-dimensional spatial grid.

Thus, tensor models make it possible to describe a big class of non-Gaussian and non-homogeneous RF but lead to an overall computational effort during the RF imitation and processing. In this context, there appear questions about the existence of recurrent not only in time but also in space RF representation and the possibility of constructing optimal recurrent algorithms of statistical analysis for such RF.

In their structure, the random fields are much more complex than stochastic processes. First, implementations of the RF are functions of several variables, the theory of which is more complicated than of one variable. Second, the concept of Markov behavior also becomes much more complicated. A random process can develop in course of time. Model from Eq. 2.1 is a mathematical expression of such development. For Markov sequence, the time interval can be broken at any point i for conditionally independent past $\Gamma^- = \{x^k : k < i\}$ and future $\Gamma^+ = \{x^k : k > i\}$. However, the RF is defined on an n-dimensional domain Ω. For its geometrical partitioning into two parts Γ^- and Γ^+, at least an $(n-1)$-dimensional domain Γ is required. Markov RF suggests that for any set Γ (of a certain class) all RV included in Γ^- are conditionally independent from the RV belonging to Γ^+, when values of Γ are known. It is possible to name Γ^-, Γ, and Γ^+ as past, present, and future only roughly. However, Markov property allows to imagine the RF as one developing in time from Γ^- through Γ to Γ^+, in addition Γ moves along Ω with the course of time. For example, if it is assumed that the lines of a two-dimensional grid Ω represent Γ, then the field is formed line by line.

Further development of this idea allows to generalize AutoRegressive (AR) models of random sequences on the RF. If the procedure for scalar sequence formation $x^0, x^1, x^2, \ldots$ usually corresponds to temporally observed values, then the procedure of field $X = \{x_{\bar{j}} : \bar{j} \in \Omega\}$ formation needs a special attention. For this purpose, it is necessary to order grid nodes Ω. Then, it is possible to say about any two elements of the field that one of the elements precedes the other. If $x_{\bar{i}}$ precedes $x_{\bar{j}}$, let us mark it as $(\bar{i}) < (\bar{j})$. There are many variants of such ordering. In a two-dimensional case a saw-toothed scanning (all lines go in one direction, Fig. 2.2a) and a triangular scanning (while passing to the next line the direction is reversed, Fig. 2.2b) are often used. As a result, the RF scanning is converted into a random sequence.

Linear autoregressive models of random fields. The simplest autoregressive model is a linear stochastic Eq. 2.4 with white Gaussian RF $\{\xi_{\bar{j}}\}$, which corresponds to a well-known autoregressive-moving average equation for random sequences (Eq. 2.5).

$$x_{\bar{j}} = \sum_{\bar{i} \in G_j} \alpha_{\bar{j},\bar{i}} x_{\bar{i}} + \sum_{\bar{i} \in Y_j} \beta_{\bar{j},\bar{i}} \xi_{\bar{i}} \quad \bar{j} \in J \tag{2.5}$$

However, unlike its one-dimensional prototype properties of RF generated by Eq. 2.5 are not thoroughly understood, even for constant coefficient models $\alpha_{\bar{j},\bar{i}} = \alpha_{\bar{i}}$, $\beta_{\bar{j},\bar{i}} = \beta_{\bar{i}}$ and unchanging domains $G_{\bar{j}} = G$ and $Y_{\bar{j}} = Y$ (Eq. 2.6).

$$x_{\bar{j}} = \sum_{\bar{i} \in G} \alpha_{\bar{i}} x_{\bar{j}-\bar{i}} + \sum_{\bar{i} \in Y} \beta_{\bar{i}} \xi_{\bar{j}-\bar{i}} \quad \bar{j} \in J \tag{2.6}$$

An important particular case of Eq. 2.6 is represented by Eq. 2.7 of multidimensional AR.

$$x_{\bar{j}} = \sum_{\bar{i} \in G} \alpha_{\bar{i}} x_{\bar{j}-\bar{i}} + \xi_{\bar{j}} \quad \bar{j} \in J \tag{2.7}$$

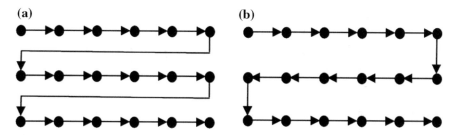

Fig. 2.2 Two-dimensional image scanning: **a** saw-toothed, **b** triangular

The abovementioned equations describe the algorithm of RF formation $\{x_{\bar{j}}\}$ at the point $\bar{j} = (j_1, j_2, \ldots, j_n)$. Besides, it is assumed that at our disposal all values of $x_{\bar{j}-\bar{l}}, \quad \bar{l} \in G, \quad (\bar{j}-\bar{l}) < (\bar{j})$ and the RF calculated before are given as initial conditions.

Model in Eq. 2.7 has a disadvantage, which complicates its analysis. It is probably a large number of summands on the right side of equation. However, it is possible to reduce their number to a minimum in a following way. The simplest AR equation generating n-dimensional field X, which does not fall into independent fields of less dimension, is represented by Eq. 2.8, where $\bar{e}_k = (0, \ldots, 0, 1, 0, \ldots, 0)$ is a unit vector of the kth coordinate axis.

$$x_{\bar{i}} = \sum_{k=1}^{n} \alpha_k x_{\bar{i}-\bar{e}_k} + \beta \xi_{\bar{i}} \tag{2.8}$$

Any model Eq. 2.7 can be reduced to a model of type Eq. 2.8 with a minimal number of summands. For this purpose, let us use vector autoregressive models, which in their linear form are described by Eq. 2.9, where $\bar{x}_{\bar{i}}$ is a value of a vector field in the node $\bar{i}$, $A_{\bar{j}}$, B are the square matrices, $\{\bar{\xi}_{\bar{i}}\}$ is a renewing standard vector field of independent vectors with independent components.

$$\bar{x}_{\bar{i}} = \sum_{\bar{j} \in D} A_{\bar{j}} \bar{x}_{\bar{i}+\bar{j}} + B \bar{\xi}_{\bar{i}} \tag{2.9}$$

Indeed, consider Habibi model [2] as an example

$$x_{i_1 i_2} = \rho_1 x_{i_1-1,i_2} + \rho_2 x_{i_1,i_2-1} - \rho_1 \rho_2 x_{i_1-1,i_2-1} + \sigma \sqrt{(1-\rho_1^2)(1-\rho_2^2)} \, \xi_{i_1 i_2}. \tag{2.10}$$

Its calculation scheme is shown in Fig. 2.3a.

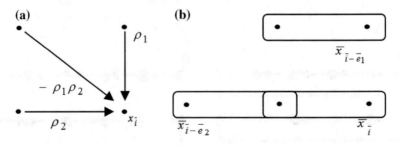

Fig. 2.3 Transition from scalar model to vector model: **a** scalar model, **b** vector model

Let us imagine this model in the form of Eq. 2.9. For this purpose, vectors $\bar{x}_{\bar{i}} = \bar{x}_{ij} = (x_{ij}, x_{i,j-1})^{\mathrm{T}}$ and $\bar{\xi}_{\bar{i}} = (\xi_{ij}, \varphi_{ij})^{\mathrm{T}}$, $\beta = \sigma\sqrt{(1 - \rho_1^2)(1 - \rho_2^2)}$, then $\bar{x}_{\bar{i}-\bar{e}_1} = (x_{i-1,j}, x_{i-1,j-1})^{\mathrm{T}}$ and $\bar{x}_{\bar{i}-\bar{e}_2} = (x_{i,j-1}, x_{i,j-2})^{\mathrm{T}}$ are introduced. Equation 2.10 will be equivalent to the first component of the vector in Eq. 2.11

$$\begin{pmatrix} x_{ij} \\ x_{i,j-1} \end{pmatrix} = \begin{pmatrix} \rho_1 & -\rho_1\rho_2 \\ 0 & 0 \end{pmatrix} \begin{pmatrix} x_{i-1,j} \\ x_{i-1,j-1} \end{pmatrix} + \begin{pmatrix} \rho_2 & 0 \\ 1 & 0 \end{pmatrix} \begin{pmatrix} x_{i,j-1} \\ x_{i,j-2} \end{pmatrix} + \begin{pmatrix} \beta & 0 \\ 0 & 0 \end{pmatrix} \begin{pmatrix} \xi_{ij} \\ \varphi_{ij} \end{pmatrix}$$

$$(2.11)$$

or at obvious notation

$$\bar{x}_{\bar{i}} = A_1\bar{x}_{\bar{i}-\bar{e}_1} + A_2\bar{x}_{\bar{i}-\bar{e}_2} + B\bar{\zeta}_{\bar{i}}$$

that is the minimal vector model of the type Eq. 2.9. Figure 2.3b shows the elements of the field included in the vectors of Eq. 2.11.

At first, let us analyze the main probabilistic characteristics of multidimensional AR models. This class of the RF is generated by linear stochastic difference Eq. 2.12, where $X = \{x_{\bar{i}}, \bar{i} \in \Omega\}$ is a modeled RF defined on an N-dimensional grid $\Omega = \{\bar{i} = (i_1, i_2, \dots i_N) : i_k = 1 \dots M_k, k = 1 \dots N\}$, $\{\alpha_{\bar{j}}, \beta, \bar{j} \in D\}$ are model coefficients, $\Xi = \{\xi_{\bar{i}}, \bar{i} \in \Omega\}$ is a renewing white RF, $D \subset \Omega$ is a casual region.

$$x_{\bar{i}} = \sum_{\bar{j} \in D} \alpha_{\bar{j}} x_{\bar{i}-\bar{j}} + \beta \xi_{\bar{i}} \quad \bar{i} \in \Omega \qquad (2.12)$$

Normal RF distribution with independent components is usually chosen as a renewing field Ξ. In this case RF X also has Gaussian distribution.

In its general form, the task of model Eq. 2.12 analysis was described in [16]. A linear spatial filter with transfer function corresponds to the model Eq. 2.12 is given by Eq. 2.13, where $\bar{z}^{-\bar{j}} = z_1^{-j_1} z_2^{-j_2} \dots z_N^{-j_N}$.

$$H(\bar{z}) = \frac{\beta}{1 - \sum_{\bar{j} \in D} \alpha_{\bar{j}} \bar{z}^{-\bar{j}}} \qquad (2.13)$$

Besides, the RF spectral density of X is written as

$$S_x(\bar{z}) = H(\bar{z}) S_\xi(\bar{z}) H(\bar{z}^{-1}) = \sigma_\xi^2 H(\bar{z}) H(\bar{z}^{-1}),$$

where σ_ξ^2 is the variance of RF Ξ.

The correlation function of X can be calculated by means of backward z-transformation of spectral density

$$R(\bar{r}) = \frac{\beta^2}{(2\pi i)^N} \oint_{C_N} S_x(\bar{z})\, \bar{z}^{\bar{r}-\bar{1}}\, d\bar{z},$$

where $C_N = \{|\bar{z}| = 1\}$ is a unit polycircle in multidimensional complex space.

Analysis of the RF probabilistic properties is simplified if their spectral density can be factorized [12, 16, 33], i.e. $S_x(\bar{z}) = \prod_{k=1}^N S_k(z_k)$. Since in such fields, the transfer function of a multidimensional filter $H(\bar{z}) = \prod_{k=1}^N H_k(z_k)$ and CF $R(\bar{r}) = \prod_{k=1}^N R_k(r_k)$ are also factorized, then it is enough to examine the properties of random sequences generated by one-dimensional AR with the following characteristics $S_k(z_k)$, $H_k(z_k)$ and $R_k(r_k)$, $k = 1 \ldots M_k$.

The drawback of such models is impossibility to describe isotropic RF, e.g. fields with the CF $R(\bar{r}) = R(|\bar{r}|) = R\left(\sqrt{r_1^2 + r_2^2 + \cdots + r_N^2}\right)$. However, the analysis shows that some dimensional models gives the RF close to isotropic ones. In order to obtain close to isotropic RF, the authors [16–18, 33]. Suggested to use one-dimensional filters with multiple roots of characteristic equations $1 - \sum_{j=1}^{n_k} \alpha_{kj} \lambda_k^j = 0$, where n_k, $k = 1 \ldots M$ are the orders of one-dimensional AR.

At first, let us solve these tasks for a one-dimensional model. Consider one-dimensional AR of M length given by Eq. 2.14, where $\{\xi_i\}$ is a Gaussian sequence of independent components with zero mean and variance σ_ξ^2.

$$x_i = \sum_{j=1}^n \alpha_j x_{i-j} + \beta \xi_i \quad i = 1 \ldots M \tag{2.14}$$

To solve the synthesis task, it is necessary to determine the AR coefficients $\{\alpha_j, j = 1 \ldots n;\ \beta\}$ using the given root of a characteristic equation $1/\rho$, its multiplicity n and the desired field variance σ_x^2. In the case of multiple roots, this equation can be written in an operator form Eq. 2.15, where z^{-1} is a shift operator.

$$\left(1 - \rho z^{-1}\right)^n x_i = \beta \xi_i \quad i = 1 \ldots M \tag{2.15}$$

Provided that $z^{-k} x_i = x_{k-i}$, let us rewrite Eq. 2.15 in a view of Eq. 2.16.

$$x_i = \sum_{j=1}^n (-1)^{j+1} C_n^j \rho^j x_{i-j} + \beta \xi_i \quad i = 1 \ldots M \tag{2.16}$$

Comparison of Eqs. 2.15–2.16 permits to write the expression for coefficients $\alpha_j = \alpha_j(\rho, n)$ as Eq. 2.17.

$$\alpha_j(\rho,n) = (-1)^{j+1} C_n^j \rho^j \quad j = 1 \ldots n \tag{2.17}$$

Value of the unknown parameter β, which is an amplification coefficient in transfer function Eq. 2.13, should be selected so as to make the filter stable. Further, it will be shown how to determine β on the basis of the CF model.

One of the tasks of model statistical analysis is to obtain its CF

$$R(k) = \beta^2(\rho,n)\,\rho^k \sum_{\ell=0}^{n-1} g(n,\ell,k)\frac{\rho^{2(n-\ell-1)}}{(1-\rho^2)^{2n-\ell-1}}, \tag{2.18}$$

where

$$g(n,\ell,k) = \frac{(n+k-1)!(2n-\ell-2)!}{\ell!(n-1)!(n-\ell-1)!(n+k-\ell-1)!} \tag{2.19}$$

and coefficient $\beta = \beta(\rho,n)$ can be obtained from equation $R(0) = 1$

$$\beta^2(\rho,n) = \frac{(1-\rho^2)^{2n-1}}{\sum_{\ell=0}^{n-1}\left(C_{n-1}^\ell \rho^\ell\right)^2}. \tag{2.20}$$

Equations 2.18–2.20 give a general view of a Normalized Autocorrelation Function (NAF) of a one-dimensional model Eq. 2.14 for given ρ and n. In order to obtain the CF at variances σ_x^2 and σ_ξ^2, which are not equal to 1, it is necessary to multiply the right hand side of Eq. 2.20 by σ_x^2/σ_ξ^2. Thus, we obtain the expression for coefficient β as well

$$\beta = \frac{\sigma_x}{\sigma_\xi}\sqrt{(1-\rho^2)^{2n-1}/\sum_{\ell=0}^{n-1}\left(C_{n-1}^\ell \rho^\ell\right)^2}. \tag{2.21}$$

Therefore, Eqs. 2.17 and 2.20 completely determine the unknown coefficients of one-dimensional AR model Eq. 2.14 with multiple roots of the characteristic equation.

Now let us consider an N-dimensional case. The RF model for a given variance σ_x^2 is completely determined by a parameter vector $(\rho_1, \rho_2, \ldots, \rho_N)$ and a multiplicity vector $(n_1, n_2, \ldots, n_N)$. Let multidimensional factorable RF be generated by the following AR equations written in an operator form

$$\prod_{k=1}^{N}\left(1-\rho_k z_k^{-1}\right)^{n_k} x_{\bar{i}} = \left(\prod_{k=1}^{N}\sum_{l=1}^{n_k}\alpha_{kl} z_k^{-1}\right)x_{\bar{i}} = \beta\xi_{\bar{i}}, \quad \bar{i}\in\Omega,$$

where N is field dimensionality, ρ_k and n_k are a model parameter and multiplicity root along the kth axis, respectively, Ω is a grid, on which field X is determined.

(a)   **(b)**

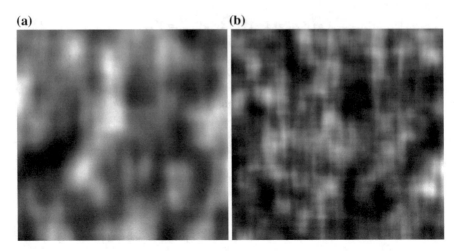

Fig. 2.4 Representation of random fields on the basis of autoregression with multiple roots: **a** multiplicity (2, 2), parameter vector (0.95, 0.95), **b** multiplicity (3, 3), parameter vector (0.95, 0.95)

Coefficients $\alpha_{\bar{j}}$ are the products of corresponding coefficients α_{kj_k} of one-dimensional AR along the kth axis $\alpha_{\bar{j}} = \prod_{k=1}^{N} \alpha_{kj_k}$, where $\bar{j} = (j_1, j_2, \ldots, j_N)$, $j_k = \overline{1 \ldots n_k}$. Coefficient β of a multidimensional model can be found in a similar way $\beta = \frac{\sigma_x}{\sigma_\xi} \prod_{k=1}^{N} \beta_k$, where β_k is a corresponding standardized one-dimensional coefficient.

An example of image frame (600 × 400 elements) obtained on the basis of the model with multiple roots is depicted in Fig. 2.4. Output analysis shows that the changing parameters and the multiplicity might be obtained a wide range of different patterns, which can help to develop complex models of multi-zone images. At the same time, if a root multiplicity grows, then the modeled RF is approximated in its properties to the isotropic RF. This is also confirmed by the cross section shape of equal level CF shown in Fig. 2.5.

Analysis of the CF models with multiple roots shows that the CF cross-sections of the RF obtained by factorable multi-dimensional models become close to hyperellipsoid with the growth of the characteristic equation root multiplicity. To assess the proximity of such RF to isotropic ones, it is desirable to have quantitative estimation of field anisotropy. For this purpose, let us introduce $\tau(\vec{u}) = \int_0^{+\infty} R(t\vec{u}) \, dt$ as the anisotropy of multidimensional RF according to correlation distance τ in $\vec{u}$-direction, where $\vec{u} \in S^{N-1}$ is a point on a hypersphere. In addition, it is possible to suggest the following anisotropy coefficient $A_\tau = \frac{1}{\bar{\tau}} \sqrt{\frac{1}{|S^{N-1}|} \int_{\vec{u} \in S^{N-1}} (\tau(\vec{u}) - \bar{\tau})^2 d\vec{u}}$, $A \geq 0$, where $\bar{\tau} = \frac{1}{|S^{N-1}|} \int_{\vec{u} \in S^{N-1}} \tau(\vec{u}) d\vec{u}$ is a mean angular correlation distance, $|S^{N-1}|$ is an area of hypersphere surface.

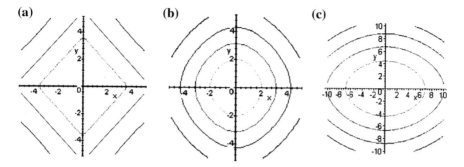

Fig. 2.5 Cross-sections of a correlation function for a two-dimensional model with multiple roots: **a** multiplicity (1, 1), parameters (0.9, 0.9), **b** multiplicity (3, 3), parameters (0.9, 0.9), **c** multiplicity (2, 3), parameters (0.9, 0.9)

The advantage of the proposed anisotropy coefficients is that we should know only field CF because all calculation can be done using standard numerical methods.

2.2.4 Wave Models of Random Fields

Let us consider the RF wave model that generalizes a number of other models and helps to solve the tasks of analysis and synthesis [16, 20, 34] effectively. This model is simple enough and can serve as a basis for simulating images and their sequences with given CF without increasing the number of model parameters.

In a wave model, the RF is determined by Eq. 2.22, where an $(n + 1)$-dimensional domain $\{(\bar{j}, t)\}$ may be discrete or continuous, $\{(\bar{u}_k, \tau_k)\}$ is a discrete Field of Random Points (FRP) in an $(n + 1)$-dimensional continuous space, t and τ_k are interpreted as time, $\bar{\omega}_k$ is a random vector of function f parameters.

$$x_{\bar{j}}^t = \sum_{\{k : \tau_k \leq t\}} f((\bar{j}, t), (\bar{u}_k, \tau_k), \bar{\omega}_k) \qquad (2.22)$$

This field can be represented as the effect of random disturbances or waves $f((\bar{j}, t), (\bar{u}_k, \tau_k), \bar{\omega}_k)$, appearing in random places $\bar{u}_k$ at random time τ_k and changing according to a given law in time and space.

Selection of function f, the FRP parameters and $\bar{\omega}$ allow us to obtain a vast class of fields, which includes the following models:

1. Poisson fields, when $f((\bar{j}, t), (\bar{u}_k, \tau_k), \bar{\omega}_k) = \delta((\bar{j}, t) - (\bar{u}_k, \tau_k))$, δ is a Kronecker delta, and $\{(\bar{u}_k, \tau_k)\}$ is Poisson FRP.
2. Multidimensional filtered Poisson process, when $f((\bar{j}, t), (\bar{u}_k, \tau_k), \bar{\omega}_k) = g((\bar{j} - \bar{u}_k t - \tau_k), \xi_k)$, where $\{\xi_k\}$ is a system of scalar RF. This model can

generate only stationary homogeneous fields and the generating waves may differ from each other only in one parameter ξ_k.

3. Weighted sum model, which is obtained at waves $f((\bar{j}, t), (\bar{u}_k, \tau_k), \bar{\omega}_k) = g((\bar{j}, t), (\bar{u}_k, \tau_k)) \xi_k$, where $\{(\bar{u}_k, \tau_k)\}$ is a set of all grid nodes and g represents corresponding weights of random variables ξ_k. This model corresponds to different RF expansion into a system of basic functions, e.g. [1].

4. Random walk model. The FRP describes a random walk (perhaps with appearance and disappearance) of a set of waves and the choice $\bar{\omega}_k$ determines the dynamics of a wave shape and intensity. Such models can be applied to simulate moving clouds.

Consider a particular case of a wave model, for which correlation tasks of analysis and synthesis can easily be solved (Eq. 2.23, where the FRP is a Poisson one with constant density λ, $\rho_k = |\bar{j} - \bar{u}_k|$ is a distance between $\bar{j}$ and $\bar{u}_k$, $\{R_k\}$ is a system of independent non-negative equally distributed RF with PDF $w(\alpha)$, $\{\xi_k\}$ is a system of independent equally distributed RF).

$$f((\bar{j}, t), (\bar{u}_k, \tau_k), \bar{\omega}_k) = g(\rho_k/R_k) \exp(-\mu/|t - \tau_k|) \xi_k \qquad (2.23)$$

Waves are motionless, independent, have spherical sections in space and exponentially attenuate over time. System $\{\xi_k\}$ determines a wave intensity and $\{R_k\}$ is their spatial scale. Evidently that the generated field X is stationary and homogeneous, has zero mean and isotropic spatial CF. In this case, all summands in Eq. 2.22 are non-correlated and elementary event $\Delta A = \{$in the element $\Delta V = \Delta_{j_1} \Delta_{j_2} \ldots \Delta_{j_n} \Delta \tau$, there appears a point in the FRP, which corresponds to a wave with spatial scale α from the element $\Delta \alpha \}$ with a probability $P(\Delta A) \approx \lambda \Delta V w(\alpha) \Delta \alpha$. Let us express the CF by an integral with respect to variables $\tau, \alpha, j_1, \ldots, j_n$. After integration with respect to τ and taking into account $g(y) = c \exp(-2y^2)$ we obtain Eq. 2.24.

$$V(\rho, t) = \frac{c^2 \pi^{n/2} \lambda}{2^{n+1}} e^{-\mu t} \int_0^\infty \alpha^n \exp\left(-\frac{\rho^2}{\alpha^2}\right) w(\alpha) d\alpha \qquad (2.24)$$

When $\rho = t = 0$, a field variance is found from Eq. 2.24

$$\sigma_n^2 = \frac{c^2 \pi^{n/2} \lambda}{2^{n+1} \mu} M[R^n].$$

It is proportional to the FRP density λ, efficiency interval $1/\mu$ of wave attenuation and mean value of the nth degree of scale R.

Simulation of a discrete field on an n-dimensional grid $\{\bar{j}\}$ with time quantization Δt can be implemented by the following algorithm. At the initial time $t_0 = 0$, the field values in all nodes are equal to zero. At each subsequent moment $t_m = m\Delta t$, a Poisson FRP with density $\lambda \Delta t$ is formed over continuous space or grid, which

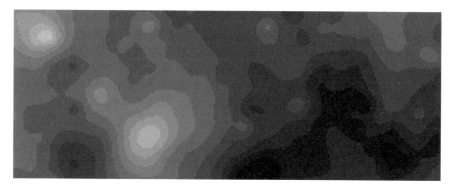

Fig. 2.6 Example of image simulation using a wave model

somehow overlaps $\{\bar{j}\}$. At each generated point $\bar{u}_k$, the RV ξ_k and R_k are formed. After that the following transformation of all field values on grid $\{\bar{j}\}$ is carried out by Eq. 2.25.

$$x_j^{t_m} = x_j^{t_{m-1}} \exp\left(-\mu \cdot \Delta t\right) + \sum_k g(\rho_k/R_k)\,\xi_k \qquad (2.25)$$

In this simulation only large summands (in comparison with a quantization level) can be taken into account. The advantage of this algorithm is its recurrence, as it makes easy to implement a field simulation using computer.

An example of image simulation with the described wave model is represented in Fig. 2.6. It is necessary to underline that Eq. 2.25 actually implements the time-varying images. Therefore, this figure shows only one frame of this process. Each field value is a sums of random numbers of the RV, Thus, generally speaking, the field will not be Gaussian even with Gaussian $\{\xi_k\}$. However, when the model parameter $h = \lambda M[R^n]/\mu$ grows, then the number of summands in Eq. 2.25 with similar distributions increase and the field is normalized.

Now consider the solution of correlation analysis and synthesis task. It follows from Eq. 2.24 that the formed field has an exponential time NAF $e^{-\mu t}$ and space NAF.

$$r(\rho) = \frac{1}{M[R^n]} \int_0^\infty \alpha^n \exp\left(-\rho^2/\alpha^2\right)\omega(\alpha)\,d\alpha \qquad (2.26)$$

Thus, solving analysis tasks, when the PDF $\omega(\alpha)$ is given, the required NAF can be found analytically or by numerical integration. Solving synthesis tasks, when the NAF $r(\rho)$ is given, it is necessary to solve integral Eq. 2.26 with respect to unknown $\omega(\alpha)$. As it is not always possible to find an analytical solution of Eq. 2.26, consider a method of its approximate solution. From Eq. 2.26 it follows that at degenerate distribution ($R = \alpha = \text{const}$) we obtain the CF $\exp\left(-\rho^2/\alpha^2\right)$. Let now an arbitrary non-increasing NAF $r(\rho)$ be given. Let us approximate it with adequate accuracy by a sum of Gaussoids with positive coefficients $r(\rho) \approx h(\rho) = \sum_i q_i \exp(-\rho^2 / \alpha_i^2)$, where $\sum_i q_i = 1$, when $r(0) = 1$. Then, for discrete distribution $P(R = \alpha_i) = k^{-1} q_i / \alpha_i^n$, where $k = \sum_i q_i / \alpha_i^n$, the generated field will have NAF equal to $h(\rho)$. Thus, the generated model allows us approximately to solve a synthesis task by changing only PDF of scale R.

2.2.5 Random Fields on Surfaces

All abovementioned images and the RV were set on rectangular grids of some dimension. In some practical problems, the images can be set on the surface of another type, for example, spherical Earth image or cylindrical image of a rotation shaft. Model representation of such images differs significantly both in the type of a spatial grid and the way of defining correlations. Consider the main peculiarities of presentation of AR images on a cylinder, as well as wave models on almost arbitrary surfaces.

Autoregressive RF model on a cylinder. Consider a cylindrical image, for example, image of a rotation shaft. If it is cut lengthwise and expanded, then it is transformed into a rectangular image. Points along the cross sections are close to each other on the original cylindrical image that is why their values are highly correlated. In the cut image, these points are located at the opposite ends. Thus, the line ends are highly correlated with the new-line characters. However, such images cannot be described by the abovementioned models on the rectangles because the correlations in these models are weakening, when the distances between values increase. Thus, the image points at the line ends do not possess the necessary high correlation. Consider, for example, an image simulated by Habibi model (Fig. 2.7).

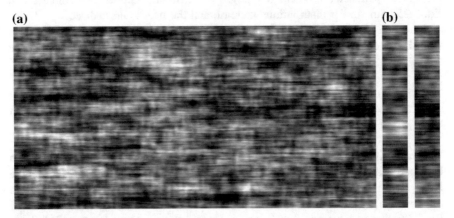

(a)   **(b)**

Fig. 2.7 Simulated images: **a** image simulated by Habibi model, **b** the first and the last five columns of image from (**a**)

Fig. 2.8 Spiral grid of a
cylindrical image

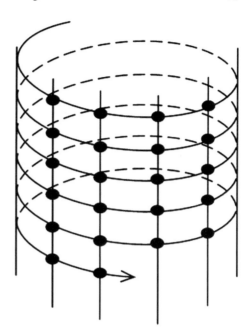

This figure shows that the first and the last columns of the rectangular image are significantly different. This means that, when pasting this image into a cylinder, one will observe great changes in brightness jumps into a cylinder at the junction.

To approximate the model to real cylindrical images, consider a spiral grid on the cylinder shown in Fig. 2.8. Grid lines represent spiral turns. To describe an image set on this grid, let us apply the analog of AR Habibi [2] model Eq. 2.27 where k is a spiral turn number and l is a node number.

$$x_{k,l} = \rho\, x_{k,l-1} + r\, x_{k-1,l} - \rho\, r\, x_{k-1,l-1} + \beta\, \xi_{k,l} \qquad (2.27)$$

Here, $l = 0, \ldots, T$, $x_{k,l} = x_{k+1,l-T}$, when $l \geq T$, T is a period, i.e. the number of points in one turn. It should be pointed out that in model Eq. 2.27 the grid can be also regarded a simple cylindrical grid, i.e. as a sequence of circles.

The resulting model of a cylindrical image can be represented [21] in an equivalent form (Eq. 2.28, where $n = kT + l$) as a model of a random process, which is an image scan along the spiral.

$$x_n = \rho\, x_{n-1} + r\, x_{n-T} - \rho\, r\, x_{n-T-1} + \beta\, \xi_n \qquad (2.28)$$

Obviously, if r value is close to 1, then the neighboring image lines (spiral turns) will be slightly different from each other. Thus, this model can be used to describe and simulate quasi-periodic signals, e.g. speech signals.

It can be shown that the CF of the model Eq. 2.28 is as following

$$V(n) = \beta^2 \left(\frac{1}{(1-r^2)T} \sum_{k=0}^{T-1} \frac{z_k}{(1-\rho z_k)(z_k - \rho)} z_k^n + \frac{s}{(1-\rho^2)(1-rs)(s-r)} \rho^n \right),$$

where $z_k = \sqrt[T]{r} \exp(i2\pi k/T)$ and $s = \rho^T$. In particular, when $n = kT$ we obtain

$$V(kT) = \frac{\beta^2}{(1-\rho^2)(1-r^2)(1-sr)(r-s)} \left((1-s^2)r^{k+1} - (1-r^2)s^{k+1} \right)$$

and find variance, when $k = 0$

$$\sigma^2 = \frac{\beta^2(1+rs)}{(1-\rho^2)(1-r^2)(1-rs)}$$

and another form of the CF

$$V(kT) = \sigma^2 r^k + \frac{s(r^k - s^k)}{(1-\rho^2)(r-s)}.$$

If $0 \leq l \leq T$, we have

$$V(l) = \sigma^2 \rho^l + \frac{r(\rho^{T-l} - \rho^{T+l})}{(1-\rho^2)(r-s)(1-rs)}.$$

The scanning of image implementation obtained by use of model Eq. 2.27 is shown in Fig. 2.9a. It is obvious that the values at the line ends are strongly

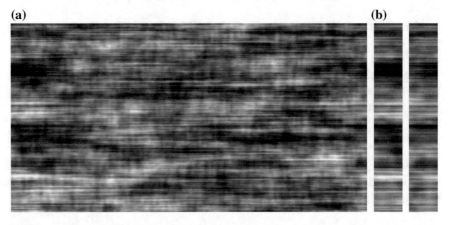

Fig. 2.9 Simulated images: **a** image simulated by model Eq. 2.27, **b** the first and the last five columns of image from (**a**)

correlated (Fig. 2.9b), as it should be for a cylindrical image obtained by image pasting as in Fig. 2.9a.

Wave RF models on an arbitrary surface as cross sections. It was possible to implement the abovementioned autoregressive model on a cylinder and analyze it because there is a simple regular grid on this surface. However, on other surfaces there are no grids with constant configuration and cell size suitable for autoregressive representation. They cannot even be found on a sphere, in spite of its symmetry. Therefore, it is necessary to find other methods of setting the RF on the surfaces. The following approach seems rather promising [16, 20, 34].

Let S be an arbitrary surface. In order to set the RF X on it, one can do a simple thing. Let us take a random field Z in space F containing S and consider X to be values at surface points S of spatial field values Z at these points. In other words, X is a cross-section of field Z with surface S. Thus, it is possible to use any RF model in space. However, one should remember that values must be obtained exactly at the points of surface S. For example, if one uses the AR model set on a rectangular grid, then interpolation into surface points will be necessary. The image on intersection of hyperbolic paraboloid and sphere with three-dimensional autoregressive RF with factorable CF is depicted in Fig. 2.10.

Application of the AR model for the RF construction on surfaces requires calculation of image values at all grid points. Therefore, while simulating images on the surface, a large number of spatial image values will be calculated but not used. Besides, a synthesis of the autoregressive model with predetermined CF is a very complicated process. It prevents imaging on surfaces with a predetermined CF. Wave model is more favorable for imaging on surfaces. This model allows to form values only at desired points (in our case at a set of predetermined points on the surface). The task of correlation synthesis for this model can be solved rather easily.

(a)   **(b)**

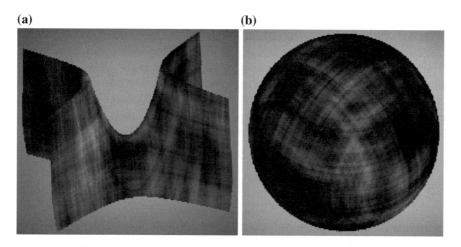

Fig. 2.10 The AR images on: **a** hyperbolic paraboloid, **b** sphere

(a)   (b)

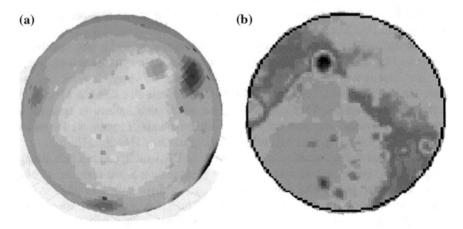

Fig. 2.11 Examples of: **a** relief simulation, **b** real part of lunar surface

Note that the CF of a wave model is a function of the Euclidean distance between spatial points. On the surface, the CF can be determined as a distance function along a geodesic arc. In this case, the CF defined on the surface should be recalculated into the CF according to the Euclidean distance. For example, in the case of a sphere this transformation is easily accomplished.

As an example, consider a relief simulation according to Unified Lunar Control Network (ULCN 2005), consisting of 272,931 points on the lunar surface. These points were obtained from space missions and observations from the Earth. On the basis of these data, an approximating ellipsoid was constructed using the least square method. Deviations from the catalog of the relief heights of this ellipsoid with mean zero samples were taken for image values on an ellipsoid. The CF dependant on the Euclidean distance between two points on an ellipsoid was evaluated according to the data obtained. This CF was approximated by the sum of Gaussoids. The approximation is used to obtain a probability distribution scale R parameter while simulating image on the ellipsoid, i.e. relief deviations from this ellipsoid [22]. Figure 2.11a shows an example of a relief simulation on the ellipsoid segment (greater heights correspond to greater brightness). By comparison, Fig. 2.11b shows some part of the real lunar surface. Visually, these figures are similar. It indicates that the suggested model is rather sufficient.

2.3 Image Filtering

Extraction of useful image component based on their noise observations is an important task, since it gives the opportunity to improve image on the background of noise [4, 12, 15, 23–25]. Furthermore, as it will be shown Sect. 2.4 the anomaly detection problem also makes it necessary to determine the covariance matrices of filtering errors. However, there exist many different approaches to solve the problem of determining the potential accuracy of estimation and construction of multidimensional RF filtering algorithms. In the first case, in order to obtain the relatively simple analytical relations, it is advisable to refer to asymptotic formulae of the Wiener filter. During constructing multidimensional image estimation on the background noise, it is desirable to find the optimal algorithm structures (or structures close to optimal) with low computational cost. It can be done using recursive estimation methods based on the abovementioned image models [16].

The issue of efficiency of optimal image filtering is considered in Sect. 2.3.1, while tensor representation of the Kalman filter is discussed in Sect. 2.3.2.

2.3.1 Efficiency of Optimal Image Filtering

To find an optimal filtering error variance σ_ε^2, consider homogeneous information RF $x_{\bar{j}}$, set on an infinite n-dimensional grid J. Suppose that for use of observations $z_{\bar{j}} = x_{\bar{j}} + \theta_{\bar{j}}$ it is necessary to give the best (with minimal error variance) linear estimation $\hat{x}_{\bar{0}} = \sum_{\bar{j} \in J} h_{\bar{j}} z_{\bar{j}}$ of information RF element $x_{\bar{0}}$. Condition for minimum σ_ε^2 can be written as a system of linear Eq. 2.29 (as an n-dimensional analogue of the Wiener-Hopf equations), where $R(\bar{q}) = M\{x_{\bar{j}}, x_{\bar{j}+\bar{q}}\}$ is a covariation function. In addition, the smallest error variance is $\sigma_\varepsilon^2 = \sigma_\theta^2 h_{\bar{0}}$.

$$h_{\bar{q}} \sigma_\theta^2 + \sum_{\bar{j} \in J} h_{\bar{j}} R(\bar{q} - \bar{j}) = R(\bar{q}), \quad \bar{q} \in J \tag{2.29}$$

Unfortunately, the analytical solution of Eq. 2.29 can be found only for a very small class of "separable" exponential CF $R(\bar{q}) = \sigma_x^2 \prod_{i=1}^{n} \rho_i^{q_i}$ [12]. However, supposing that the cells of spatial grid J are small in comparison with the RF, it is possible to replace the system Eq. 2.29 with one integral equation. Then

$$\sigma_\varepsilon^2 = \sigma_\theta^2 h_{\bar{0}} \cong \sigma_\theta^2 \frac{1}{(2\pi)^n} \int_{-\infty}^{\infty} (f(\bar{\lambda}) \, d\bar{\lambda}) / (\sigma_\theta^2 + f(\bar{\lambda})), \tag{2.30}$$

where

$$f(\bar{\lambda}) = \int_{R^n} R(\bar{u}) \exp(-j(\bar{\lambda}, \bar{u})) \, d\bar{u} \tag{2.31}$$

is the RF spectral density of RF $x(\bar{u})$, $\bar{u}$, $\bar{\lambda} \in R^n$, $(\bar{\lambda}, \bar{u}) = \sum_{k=1}^{n} \lambda_k u_k$, $d\bar{\lambda} = d\lambda_1 d\lambda_2 \ldots d\lambda_n$.

Thus, for analysis of filtering efficiency it is sufficient to find a spectral density in Eq. 2.31 of information RF and to make calculations using Eq. 2.30. The greatest challenge is usually connected with n-multiple integration in Eq. 2.31 and, particularly, in Eq. 2.30. These problems may be significantly lower, when the RF $x(\bar{u})$ is isotropic throughout space R^n or any other subspaces $\Omega_m \subset R^n$. Indeed, after entering spherical coordinates in R^n Eqs. 2.30 and 2.31 for isotropic RF can be written in the forms of Eqs. 2.32 and 2.33, where $J_v(\cdot)$ is Bessel functions of v order, $\Gamma(\cdot)$ is a complete gamma function, $k = |\lambda|$, $\rho = |\bar{u}|$.

$$\frac{\sigma_\varepsilon^2}{\sigma_\theta^2} = \frac{1}{(2\pi)^{n-1}\Gamma(0.5n)} \int_0^\infty \frac{k^{n-1}f(k)}{\sigma_\theta^2 + f(k)} dk \qquad (2.32)$$

$$f(k) = (2\pi)^{n/2} \int_0^\infty R(\rho)\rho^{n-1} \frac{J_{0.5n-1}}{(k\rho)^{0.5n-1}} d\rho \qquad (2.33)$$

For isotropic RF on spaces with an odd number dimensions, the Bessel functions in Eq. 2.32 can be expressed by means of elementary functions. In these cases, the calculations, using Eqs. 2.32 and 2.33, are rather easy. For example, for isotropic RF with

$$R(\rho) = \sigma_x^2 \exp(-\alpha\rho), \quad \rho = \sqrt{\sum_{k=1}^n u_k^2},$$

we obtain Eq. 2.34, where $\gamma = \alpha^n/q$, $q = \sigma_x^2/\sigma_\theta^2$ is information and interference RF variance ratio, $\Phi_1(v) = 2/(1+v^2)$, $\Phi_3(v) = 8\pi/(1+v^2)^2$, $\Phi_5(v) = 64\pi^2/(1+v^2)^3$, $\Phi_7(v) = 96\pi^3/(1+v^2)^4$, etc.

$$\frac{\sigma_\varepsilon^2}{\sigma_\theta^2} = \frac{\gamma q}{(2\pi)^{n-1}\Gamma(0.5n)} \int_0^\infty \frac{v^{n-1}\Phi_n(v)}{\gamma + \Phi_n(v)} dv \qquad (2.34)$$

For isotropic exponentially correlated RF that is set on spaces with even number of dimensions $n = 2N$, there is a simple formula for integral Eq. 2.33

$$f(k) = (2\pi)^M \sigma_x^2 k^{-2(M-1)} (-1)^M \frac{d^M}{d\alpha^M} \left(\frac{(\sqrt{\alpha^2 + k^2} - \alpha)^{M-1}}{\sqrt{\alpha^2 + k^2}} \right).$$

In addition, a minimal variance of filtering error can also be presented as Eq. 2.34, where $\Phi_2(v) = 2\pi/\sqrt{(1+v^2)^3}$, $\Phi_4(v) = 6\pi/\sqrt{(1+v^2)^5}$, etc. Unfortunately, for large dimension n of homogeneous RF rather complicated formulae can be obtained and it is advisable to use numerical methods while calculating Eq. 2.34. At the same time, according to the analysis, if the generalized parameter $\gamma = \alpha^n/q$ increases, then variance of optimal filtering error converges to

$$\sigma_\varepsilon^2/\sigma_x^2 = \beta_n \sqrt[n+1]{\gamma},$$

where $\beta_1 \cong 0.707$, $\beta_2 \cong 0.66$, $\beta_3 \cong 0.63$, $\beta_4 \cong 0.606$, ... $\beta_\infty \cong 0.58$.

Thus, the considered relations give a possibility to obtain rather simple estimations of potential RF filtering accuracy on the background of noise. In addition, the obtained limit values of filtering variance allow to construct the characteristics of optimal algorithms to detect the spatial anomalies on the background of multidimensional interfering images.

2.3.2 Tensor Kalman Filter

Analyzed Wiener filtering procedures can be used in applications with a limited number of k frames and small grids G and J_t. However, there is a significant variety of problems, when observations are carried out continuously and the number of elements in the domain $J_t \otimes T$ is arbitrary large. Nevertheless, a prediction using the weighted summation of all previous to the certain frame observations $\left\{z_{\ni j}^t\right\}$ can be unreasonably time-consuming. In such situation, it is desirable to impose the additional restrictions to the considered image models and use the effective recurrent procedures to construct some optimal prediction.

One of the least overloaded restrictions allowing to find a recurrent solution to the problem is a description of interfering image frame sequence by means of nonlinear tensor stochastic differential Eq. 2.1. Based on observations of frame sequence

$$z_{\ni j}^t = x_{\ni j}^t + \theta_{\ni j}^t \quad \bar{j} \in J_t \quad t = 1, 2, \ldots,$$

representing an additive mixture of information RF $x_{\ni j}^t$ and the RF of white noise $\theta_{\ni j}^t$, it is necessary to find the best estimate $x_{\ni j}^t$ of a recurrent frame of information RF. To find such estimation, let us use the criterion of average maximal gain [16] and an invariant imbedding method. As a result, the recursive estimation rule (Eq. 2.35) can obtained, where $\hat{x}_{\ni j}^t = \varphi_{\bar{j}}^t(\hat{x}_{\bar{l}}^{t-1})$, $\bar{j} \in J_{t-1}$ is an optimal RF prediction, $P_{\ni}^t = (\varphi'(\hat{x}_{\bar{l}}^{t-1}))' \, P^{t-1}(\varphi'(\hat{x}_{\bar{l}}^{t-1}))' + \upsilon'(\hat{x}_{\ni \bar{l}}^t)\upsilon'(\hat{x}_{\ni \bar{l}}^t)$ is a covariance matrix of prediction errors, $x_{\ni j}^1 = 0$; $P_{\ni \bar{jl}}^1 = M\left\{x_{\ni j}^1 x_{\bar{l}}^1\right\}$.

$$\hat{x}_{\bar{j}}^t = \hat{x}_{\ni j}^t + P_{\bar{jl}}^t V_\theta^{-1}(z_{\bar{l}}^t - \hat{x}_{\ni \bar{l}}^t), \quad \bar{l}, \bar{j} \in J_t$$
$$P_{\bar{jl}}^t = P_{\ni \bar{jl}}^t (E + V_\theta^{-1} P_{\ni \bar{jl}}^t)^{-1} \tag{2.35}$$

Equations 2.35 allow to find extrapolated estimates $\hat{x}^t_{\ni j}$ and covariance matrices of extrapolation errors $P^t_{\ni j\bar{l}}$ recursively as new observations $\left\{ z^i_{\bar{j}}, \bar{j} \in J \right\}$ of successive RF frames become available. For Gaussian RF defined by linear stochastic equations, when $\varphi^t_j(x^{t-1}_{\bar{l}}) = P^t_{\bar{j}\bar{l}} x^{t-1}_{\bar{l}}$ and $\upsilon^t_{\bar{j}q}(x^{t-1}_{\bar{l}}) = \upsilon^t_{\bar{j}q}(x^{t-1}_{\bar{l}})$, the procedure of filtration-interpolation together with algorithm Eq. 2.35 gives a strictly optimal solution how to detect the multidimensional signals in image sequence. In this case, domain G coincides with J_t, Eq. 2.35 determines the sequence of interframe observation processing and, while generating the log-likelihood ratio, the weighted summation of observations $z^t_{\bar{j}}, \bar{j} \in J_t$ and predictions $\hat{x}^t_{\ni j}, \bar{j} \in J_t$ of a successive frame of a multidimensional image is used.

Let now X be the field with multiplicative correlation function $M\left\{ x^t_{\bar{j}} \times x^s_{\bar{l}} \right\} = \sigma^2_x \rho^{|t-s|} \prod^n_{k=1} r^{|j_k - t_k|}_k$, where ρ is a time correlation coefficient, r_k is a coefficient along the kth spatial axis, σ^2_x is a field variance. Then, when $t = s$, an intraframe covariance is obtained and, when $t = s - 1$, an interframe covariance $V^{t|t-1}_x = \rho V^{t|t}_x$ is yielded, where $R = R_1 \times R_2 \times \cdots \times R_m$, R_k is a correlation tensor of the kth row.

In this case $\rho^t = \rho E$, $v^{t-1}(v^{t-1})^T = V^t = V_x$, and Eq. 2.35 can be written as

$$\hat{x}^t = \rho \hat{x}^{t-1} + P^t\left(z^t - \rho \hat{x}^{t-1} \right), \quad P^t = P^t_{\ni}\left(E + P^t_{\ni} \right)^{-1},$$
$$P^t_{\ni} = \rho^2 P^t + (1 - \rho^2)\, qR, \quad P^1_{\ni} = qR$$

where $q = \sigma^2_x / V_\theta$ is a signal/noise ratio, tensors $P^t_{\ni}$ and P^t are normalized by noise variance and represent relative error covariance of the extrapolated and current estimates expressed in noise variance.

As an example, consider a field with 3×2 grid J_t, i.e. the case, when frames x^t consist of six points

$$x^t = \begin{pmatrix} x^t_{11} & x^t_{21} & x^t_{31} \\ x^t_{12} & x^t_{22} & x^t_{32} \end{pmatrix}.$$

Thus, correlation matrices of the first and second row are as follows

$$R_1 = \begin{pmatrix} 1 & r_1 & r^2_1 \\ r_1 & 1 & r_1 \\ r^2_1 & r_1 & 1 \end{pmatrix} \quad R_2 = \begin{pmatrix} 1 & r_2 \\ r_2 & 1 \end{pmatrix}.$$

Consequently,

$$V_x = \sigma_x^2 R_1 \times R_2 = \begin{pmatrix} 1 & r_1 & r_1^2 & r_2 & r_2 r_1 & r_2 r_1^2 \\ r_1 & 1 & r_1 & r_2 r_1 & r_2 & r_2 r_1 \\ r_1^2 & r_1 & 1 & r_2 r_1^2 & r_2 r_1 & r_2 \\ r_2 & r_2 r_1 & r_2 r_1^2 & 1 & r_1 & r_1^2 \\ r_2 r_1 & r_2 & r_2 r_1 & r_1 & 1 & r_1 \\ r_2 r_1^2 & r_2 r_1 & r_2 & r_1^2 & r_1 & 1 \end{pmatrix}$$

Note that error covariance matrices $P_{\ni}^t$ and P^t have the same form.

Tensor P^t elements are filtering error covariance x^k, which in this case depend only on four parameters: the correlation coefficients r_1, r_2, and ρ and also signal/noise ratio $q = \sigma_x^2 / V_\theta$. When $q \gg 1$ and $t \to \infty$, the coefficients $P_{j_1 j_2}^t$ converge to limit $\left\{ P_{j_1 j_2}^t \right\} = P_{j_1 j_2}$ rather quickly. Thus, it is often possible to apply limit values at once. It will worsen the filtering results only at the beginning but the amount of computation will be significantly reduced (or the storage space if coefficients $\left\{ P_{j_1 j_2}^t \right\}$ are calculated beforehand).

It is rather important that the deduced tensor filtering Eq. 2.35 can be easily generalized in case of almost arbitrary interaction

$$z_{\bar{j}}^t = S_j^t(x_{\bar{l}}^t, \theta_l^t)\, \bar{j},\ \bar{l} \in J_t$$

of information RF and noise [16]. Besides, using a modified method of invariant imbedding and the abovementioned models it is possible to synthesize recurrent procedures to test multialternative hypotheses such as

$$H_v : \left\{ z^t = S_v^t(x_v^t, \theta^t) \right\}.$$

2.4 Anomalies Detection in Noisy Background

In many applications, it is often desirable to detect anomalies that may appear in the signal, separate image, or recurrent image of frame sequence [6, 8, 14, 26, 35, 36]. For example, these anomalies may be forest fires, pathological changes in medical images, new objects in a security area, etc. Optimal algorithms for signal detection are suggested in Sect. 2.4.1. Efficiency of anomaly detection is represented in Sect. 2.4.2.

Fig. 2.12 Domain G, where and only where a signal becomes apparent, Q is a complimentary domain

2.4.1 Optimal Algorithms for Signal Detection

Let observations of the RF x_j^t with the space-time correlation and additive noise RF θ_j^t, which consists of independent random variables with zero means and variances V_θ, be made by Eq. 2.36, where parameter vector $\bar{\chi}_t$ describes, for example, possible mutual spatial displacements and rotations of nearest-neighbor image frames.

$$z_j^t = x_j^t(\bar{\chi}_t) + \theta_j^t \quad j \in J_t \quad t = 1 \dots k \tag{2.36}$$

Let appearance of a deterministic signal cause changes in the model Eq. 2.36 only in $G \subset J$ (Fig. 2.12). In particular, the domain G can be a part of the last frame observed (Eq. 2.37), where $\{s_{\bar{j}}, \quad \bar{j} \in G\}$ are the values of a useful (detectable) signal.

$$z_{\bar{j}}^k = s_{\bar{j}} + x_{\bar{j}}^k(\bar{x}_t) + \theta_{\bar{j}}^k \quad \bar{j} \in G \tag{2.37}$$

Under considered circumstances, it is necessary to find out a rule to test the hypothesis H_0 on the absence of anomalies in the domain G against alternative assumption H_1 on the validity of the model Eq. 2.37.

At the predetermined probabilistic characteristics of model components Eqs. 2.36 and 2.37, it is possible to estimate corresponding conditional PDF of observations $W(Z|H_0)$ and $W(Z|H_1)$. Therefore, to solve the detection problem one should compare the results with the threshold level Λ_0 of Likelihood Ratio (LR).

$$\Lambda = \frac{W(Z|H_1)}{W(Z|H_0)} \begin{cases} \geq \Lambda_0 - signal, \\ < \Lambda_0 - no\ signal \end{cases} \tag{2.38}$$

To simplify the calculations, let us represent the conditional PDF in a product form $W(Z|H_{0,1}) = W(Z_0|H_{0,1})\,W(Z_G|Z_0, H_{0,1})$, where Z_G is a set of observations in domain G, Z_0 is a set of all observations, which do not belong to the domain of the intended signal. Since $W(Z_0|H_0) = W(Z_0|H_1)$, the LR in Eq. 2.38 can be rewritten in the form

$$\Lambda = \frac{W(Z_G|Z_0, H_1)}{W(Z_G|Z_0, H_0)}. \tag{2.39}$$

Let us approximate the conditional PDF, included in the LR Eq. 2.39, using Gaussian distribution (Eq. 2.40), where $m_0 = \{m_{0\bar{j}}\}$, $V_0 = \{V_{0\bar{i}\bar{j}}\}$, and $m_1 = \{m_{1\bar{j}}\}$, $V_0 = \{V_{0\bar{i}\bar{j}}\}$, $\bar{i}, \bar{j} \in G$ are conditional means and spatial covariance matrices of observations Z_G in the absence and presence of a useful signal, respectively.

$$W(Z_G|Z_0, H_{0,1}) = \frac{1}{(2\pi)^{N/2}\sqrt{\det V_{0,1}}} \exp\left(-0.5\|z^k - m_{0,1}\|^2_{V^{-1}_{0,1}}\right) \tag{2.40}$$

Using the observation models Eqs. 2.36 and 2.37, the following formulae for the conditional mean values is obtained $m_{0\bar{j}} = \overset{k}{\hat{x}}_{\ni\bar{j}}$, $m_{1\bar{j}} = s_{\bar{j}} + \overset{k}{x}_{\ni\bar{j}}$, $\bar{j} \in G$, where $\overset{k}{\hat{x}}_{\ni\bar{j}} = M\left\{x^k_{\bar{j}}|Z_0\right\}$ is the optimal prediction of the RF values $x^k_{\bar{j}}$, $\bar{j} \in G$ obtained on the basis of all available observations Z_0, which do not belong to a signal domain. Matrices V_0 and V_1 are the same $V_0 = V_1 = V = P_{\ni} + V_\theta$, where $P_{\ni}$ is a covariance matrix of optimal prediction errors. Substituting the obtained formulae in Eqs. 2.39 and 2.40 and taking logarithms, it is possible to find the following algorithm for signal detection (Eq. 2.41), where $\lambda = \ln \Lambda_0 + 0,5\|s_{\bar{j}}\|^2_{V^{-1}}$ is a detection threshold.

$$L = \sum_{\bar{i},\bar{j} \in G} s_{\bar{j}} V^{-1}_{\bar{i}\bar{j}} (\overset{k}{z}_{\bar{j}} - \overset{k}{\hat{x}}_{\ni\bar{j}}) \begin{cases} \geq \Lambda_0 - signal, \\ < \Lambda_0 - no\ signal \end{cases} \tag{2.41}$$

Thus, the procedure of anomaly detection includes a compensation of interfering images $\overset{k}{z}_{\bar{j}} - \overset{k}{\hat{x}}_{\ni\bar{i}}$ by subtracting optimal prediction $\overset{k}{\hat{x}}_{\ni\bar{j}}$ from observation $\overset{k}{z}_{\bar{j}}$. The prediction is found on the basis of all observations, which do not belong to domain G. Let us call this type of prediction as "domain prediction" (Fig. 2.13a). After compensation of interference, the RF linear weighted summation of all residuals is carried out.

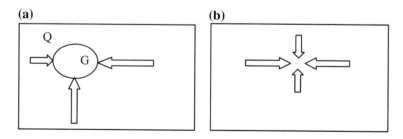

Fig. 2.13 Anomaly detection: **a** domain prediction, **b** point prediction

Another form of optimal detection procedure can be obtained assuming that any useful signal can occupy all image frames, i.e. domain G includes all multi-dimensional grids $J_1, J_2, \ldots, J_t$. Then, the best prediction $x^k_{\ni j} \equiv 0,\, \bar{j} \in G$, and algorithm Eq. 2.41 can be written as Eq. 2.42, where $V_{\bar{l}\bar{j}} = V_{x\bar{l}j} + V_{\theta} E_{\bar{l}j}$, $V_{x\bar{l}j}$, $\bar{l}, \bar{j} \in G$ is a covariance matrix of an interfering image.

$$L = \sum_{\bar{l}\bar{j} \in G} s_{\bar{l}} V_{\bar{l}\bar{j}}^{-1} z^k_{\bar{j}} \begin{cases} \geq \Lambda_0 - signal \\ < \Lambda_0 - no\ signal \end{cases} \tag{2.42}$$

Direct implementation of procedure Eq. 2.42 is difficult because of a large number of computations. However, an expanding the spatial symmetric matrix into the product of two triangle ones $V_{\bar{l}\bar{j}}^{-1} = A_{\bar{l}\bar{v}} A_{\bar{v}\bar{l}}$, $\bar{l}, \bar{v}, \bar{j} \in G$, it is possible to present Eq. 2.42 in the form of Eq. 2.43, which corresponds to the preliminary decorrelation of image sequences $z^t_{\bar{j}}$, $\bar{j} \in G$, and subsequent weighted summation with weights $s_{\bar{l}} A_{\bar{l}\bar{v}}$.

$$L = \sum_{\bar{l}\bar{j},\bar{v} \in G} s_{\bar{l}} A_{\bar{l}\bar{v}} A_{\bar{v}\bar{j}} z^t_{\bar{j}} \begin{cases} \geq \Lambda_0 - signal, \\ < \Lambda_0 - no\ signal \end{cases} \tag{2.43}$$

In many cases, such approach helps to find acceptable in practice quasi-optimal algorithms of decorrelation. At the same time an analysis of adaptive recursive filters, which properties are similar to decorrelation ones, is an important area to look for relatively simple technical or software implementation of system for image sequence processing. The new quality of algorithms Eqs. 2.42 and 2.43 in comparison to Eq. 2.41 is the division of time-consuming operation of optimal prediction or decorrelation, which is not connected with a signal form, and relatively simple weighted summation, which takes into account the type of a useful signal. It allows to solve both problems of anomaly detection with unknown location and more complex problems of multialternative detection (recognition) of several signal types rather easily.

There is one more form (Eq. 2.44) of decision rule, which gives the same result [16]. Let us represent $V_{\bar{l}\bar{j}}^{-1} z^k_{\bar{j}}$ in Eq. 2.42 in the form $V^{-1}\bar{z} = \begin{pmatrix} V_{11} & V_{12} \\ V_{21} & V_{22} \end{pmatrix}^{-1} \begin{pmatrix} \bar{z}_1 \\ \bar{z}_2 \end{pmatrix}$,

where $\bar{z} = \begin{pmatrix} \bar{z}_1 \\ \bar{z}_2 \end{pmatrix}$ is divided into two part like $\bar{z}_1 = (z^k_1, \ldots, z^k_m)^T$ and $\bar{z}_2 = (z^k_{m+1}, \ldots z^k_n)^T$, $V_{ij} = M[\bar{z}_i \bar{z}_j^T]$. Using Frobenious formula we obtain

$$V^{-1}\bar{z} = V^{-1} \begin{pmatrix} \bar{z}_1 \\ \bar{z}_2 \end{pmatrix} = \begin{pmatrix} T^{-1} & -T^{-1}V_{12} \\ -V_{22}^{-1}V_{21}T^{-1} & V_{22} + V_{22}^{-1}V_{21}T^{-1}V_{12}V_{22}^{-1} \end{pmatrix} \begin{pmatrix} \bar{z}_1 \\ \bar{z}_2 \end{pmatrix},$$

where $T = V_{11} - V_{12}V_{22}^{-1}V_{21}$. Then, the first m components of $V^{-1}\bar{z}$ are $T^{-1}(\bar{z}_1 - V_{12}V_{22}^{-1}\bar{z}_2)$, where $V_{12}V_{22}^{-1}\bar{z}_2 = M[\bar{z}_1|\bar{z}_2] = \tilde{\bar{z}}_1$ is the optimal prediction of $\bar{z}_1$ according observation $\bar{z}_2$ and T is covariance matrix of its errors $\Delta = \bar{z}_1 - V_{12}V_{22}^{-1}\bar{z}_2 = \bar{z}_1 - \tilde{\bar{z}}_1$. If $\bar{z}_1$ consists of one element (i.e. $\bar{z}_1 = (z_1)$), then $T^{-1}(\bar{z}_1 - V_{12}V_{22}^{-1}\bar{z}_2) = (z_1 - \tilde{z}_1)/\sigma_1$, where $\tilde{z}_1 = M[z_1|(\bar{z}\backslash z_1)]$ is the optimal prediction of z_1 according to all other elements of $\bar{z}$, σ_1^2 is a variance of error $\Delta_1 = z_1 - \tilde{z}_1$. But any element of $\bar{z}$ can be taken as z_1. Thus,

$$V^{-1}\bar{z} = C^{-1}(\bar{z} - \tilde{\bar{z}}) = C^{-1}\Delta,$$

where $\tilde{\bar{z}}$ is the set of predictions of all elements $z_{\tilde{j}}^k$ and every element is predicted according to all other elements of $\bar{z}$, i.e. $\tilde{z}_{\tilde{j}} = M[z_{\tilde{j}}|(\bar{z}\backslash z_{\tilde{j}})]$, diagonal matrix C consists of variances $\sigma_{\tilde{j}}^2 = M[(\bar{z} - \tilde{\bar{z}})^2]$ of errors $\Delta = \bar{z} - \tilde{\bar{z}}$. Thus $C^{-1}\Delta = \left\{ (z_{\tilde{j}}^k - \tilde{z}_{\tilde{j}}^k)/\sigma_{\tilde{j}}^2 : \tilde{j} \in G \right\}$ and we obtain Eq. 2.44 which is equivalent to Eqs. 2.41–2.43.

$$L = \sum_{\tilde{j} \in G} s_{\tilde{j}} \hat{\delta}_{\tilde{j}} = \sum_{\tilde{j} \in G} s_{\tilde{l}}(z_{\tilde{j}}^k - \tilde{z}_{\tilde{j}}^k)/\sigma_{\tilde{j}}^2 \begin{cases} \geq \Lambda_0 - signal, \\ < \Lambda_0 - no\ signal \end{cases} \quad (2.44)$$

It is based on the "point prediction" $\tilde{z}_{\tilde{j}}^k = M[z_{\tilde{j}}^k|(Z\backslash z_{\tilde{j}}^k)]$ (Fig. 2.13b), which is made taking into account all other points of Z and assuming that there is no signal.

Despite the equality of statistics Eqs. 2.42 and 2.44, there is a fundamental difference between them. In Eq. 2.42, prediction and compensation are made according to observations that do not have a signal. Thus, if there is a signal in G, then it will be distorted only by prediction errors. If these errors are small, the compensation residuals will be close to the values of detectable S_G (it is possible to see the signal with little distortion). In Eq. 2.44 while constructing a point prediction, all other observations are used including those with a signal. Therefore, in the residuals of this compensation every signal value will be distorted not only by prediction errors of an interfering image but by other signal elements as well. Even if prediction errors are small, more distorted signal will be observed.

A significant drawback of all considered algorithms is quite a complex analysis of signal detection efficiency. While searching the ways to simplify this analysis, we managed to obtain one more procedure to detect signals [16]. For this purpose, it is enough to substitute a known connection between the tensor estimations in Eq. 2.41

$$\hat{x}_j^t = \hat{x}_{\ni j}^t + P_{lj}^t V_\theta^{-1}(z_l^t - \hat{x}_{\ni j}^t) \quad (l,j) \in G_0^t \quad t = 1, \ldots, N,$$

where $\hat{x}_j^t$ is an optimal RF estimation in domain $j \in G_0^t$, $t = 1, \ldots, N$, made on the basis of all observations z_l^t, $l \in G_0^t$, $t = 1, \ldots, N$ and P_{lj}^t is a covariance matrix of filtering errors. After elementary transformations and considering that

$P(E + V_\theta^{-1} P_\ni) = P_\ni$, $z_l - \hat{x}_{\ni j} = (E + V_\theta^{-1} P_\ni)(z - \hat{x})$, the decision rule based on the optimal estimates $\hat{x}_j^t$ (Eq. 2.45) can be found.

$$L = s_l^t V_{\theta l j}^{-1}(z_j^t - \hat{x}_j^t) \begin{cases} \geq \Lambda_0 - signal \\ < \Lambda_0 - no\ signal \end{cases} \tag{2.45}$$

Thus, at Gaussian approximation the obtained detection procedure suggests the optimal RF filtration, calculation of covariance matrix of filtering errors, and weighted summation using Eq. 2.45. Since algorithm Eq. 2.45, in contrast to known detectors does not require the time-consuming calculation of covariance matrices of prediction errors, it can be used not only in image processing system but also for probability analysis of detection efficiency. It is important to note that a total amount of computation according to rule Eq. 2.45 is almost the same as in Eq. 2.44 and can be used as another variant to construct a detector at low computational costs and at an unknown spatial position of anomalies. It is especially evident if the relationship between the estimation and point prediction is used

$$x_j^t = \tilde{z}_j^t + P_j^t(1/\sigma_j^2)(z_j^t - \tilde{z}_j^t),$$

where all formula components are scalar and there is no summation over the same lower indices. Substituting this equation in Eq. 2.45, Eq. 2.44 is obtained.

These results allow to specify the conditions, under which a proposed replacement of conditional PDF by normal distribution is valid. First of all, it is a wide class of the Gaussian models Eqs. 2.36 and 2.37. In these cases, the procedures Eqs. 2.41–2.45 are optimal. When model Eqs. 2.36 and 2.37 components are non-Gaussian, the sufficient optimum condition is the possibility to approximate the posterior PDF prediction $x_{\ni j}^k$ using normal distribution. Note that the last condition is fulfilled in many applied tasks of the RF processing with significant space-time correlative relationship and is usually equivalent to the high posteriori prediction accuracy.

2.4.2 Efficiency of Anomaly Detection

On the basis of synthesized optimal algorithms for anomaly detection, the relatively simple quasi-decision rules can be built. These rules use only some part of available observations, different prediction methods or decorrelation. In these cases, as well as while studying the potential possibilities of real detection systems there arises a problem of calculating optimal algorithm characteristics.

In the Gaussian approximation, the conditional statistics distributions Eqs. 2.41–2.44 will also be Gaussian. Therefore, for calculation of the efficiency of anomaly detection it is sufficient to find conditional moments of statistics in Eq. 2.41

$$M\{L/H_0\} = 0 \quad M\{L/H_1\} = S_{\bar{e}}V_{\bar{je}}^{-1}S_{\bar{j}} \quad \sigma_L^2 = S_{\bar{e}}V_{\bar{je}}^{-1}S_{\bar{j}}.$$

Thus, at a given quantile x_F of level P_F of a normal distribution a detection threshold $\lambda = x_F\sigma_L$ and probability of correct detection can be determined by Eq. 2.46, where

$$\Delta_D = \sqrt{S_{\bar{e}}V_{\bar{je}}^{-1}S_{\bar{j}}}, \quad \bar{j}, \bar{e} \in G, \quad \Phi_\circ(x) = (1/\sqrt{2\pi})\int_0^x \exp\left(-z^2/2\right)dz.$$

$$P_D = 0,5 + \Phi_0(\Delta_D - x_F) \qquad (2.46)$$

It can be shown that there is a simple relationship between the spatial covariance matrices $P_{\ni\bar{je}}$ of optimal prediction errors and $P_{\bar{je}} = M\left\{(x_{\bar{j}}^t - \tilde{x}_{\bar{j}}^t)(x_{\bar{e}}^t - \tilde{x}_{\bar{e}}^t)\right\}$ of optimal linear estimation errors $\hat{x}_j^t = \hat{x}_j^t(Z)$. Indeed, analyzing Wiener-Hopf equation for multidimensional discrete RF, it is possible to obtain the equation

$$P_{\ni\bar{je}} = (E_{\bar{j}\bar{q}} - P_{\bar{j}v}V_{\theta\bar{v}\bar{q}}^{-1})^{-1}P_{\bar{q}\bar{e}},$$

where $V_{\theta\bar{v}\bar{q}} = \sigma_\theta^2 E_{\bar{v}\bar{q}}$. Substituting the obtained link between matrices in Eq. 2.46, Eq. 2.47 provides efficacy calculation of signal detection.

$$\Delta_D = \sqrt{S_{\bar{e}}V_\theta^{-1}(E_{\bar{e}j} - V_\theta^{-1}P_{\bar{e}j})S_{\bar{j}}} \qquad (2.47)$$

If there are no interfering images $x_{\bar{j}}$ or if they are estimated precisely $P_{\bar{e}j} = 0$, the deterministic signal detection performance on the background of white noise is determined by a well-known signal/noise ratio $\Delta_D = \sqrt{\sum_{\bar{j}\in G} S_{\bar{j}}^2/V_\theta}$. Errors $x_{\bar{j}}^t$ with time-space correlation decrease the detection accuracy, which is measured by non-zero covariance matrix of optimal RF estimation $x_{\bar{j}}^t$ on the background of white noise $\theta_{\bar{j}}^t$, $\bar{j} \in J$.

Consider two important examples how it is possible to calculate the potential efficiency of point anomaly detection. Suppose that a point signal $s_{\bar{j}} = s_0$ is being detected. It occupies one element J_t of grid G. In this case

$$\Delta_D = \sqrt{Q(1 - \sigma_\varepsilon^2/\sigma_\theta^2)},$$

where $Q = s_0^2/\sigma_\theta^2$ is a signal/noise ratio, σ_ε^2 is an error variance at optimal estimation of one RF element $x_{\bar{j}}^t$, $\bar{j} \in G$, on the basis of all available observations $z_{\bar{j}}^t$, $\bar{j} \in J_t$, $t \in T$. Magnitude of error variance σ_ε^2 can be obtained by known methods of the RF optimal estimation theory, which are observed on the background of noise [16].

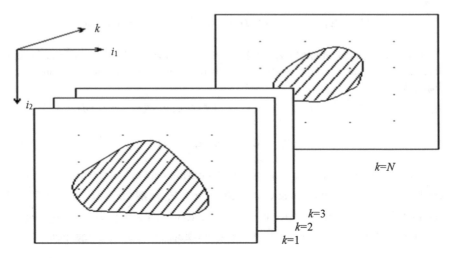

Fig. 2.14 Frames of a multi-zone image

Consider the task of finding a point object based on the results of observations of imagery M_3 of $M_1 \times M_2$ size (Fig. 2.14). Such a problem can arise, for example, during multi-zone observations of one and the same object using different x-systems functioning in different spectral range [7, 11, 12, 37]. In this case, an observation model Eq. 2.36 can be written as following

$$z_{\bar{j}} = x_{\bar{j}} + \theta_{\bar{j}}, \quad \bar{j} = (j_1, j_2, j_3); \quad j_k = 1 \dots M_k.$$

Assume that data sets $z_{\bar{j}}, j_3 = 1 \dots M_3$ are spatially shifted. Therefore, the appearance of the desired signal increase the RF level by s_0 in one and the same element each image frame, of which is numbered (j_1^0, j_2^0) (Fig. 2.14). Covariance functions of interfering images can be written as follows

$$R_x(m_1, m_2, m_3) = M\left\{ x_{j_1, j_2, j_3} x_{(j_1 + m_1), (j_3 + m_3), (j_3 + m_3)} \right\} = R_1(m_1, m_2) \rho^{1(m_3)},$$

where $R_1(m_1, m_2) = R_x(m_1, m_2, m_3 = 0)$ is one and the same covariance function from M_3 image frames, $1(m_3) = 1$ if $m_3 \neq 0$, and $1(m_3 = 0) = 0$. Note that such representation means equality of correlation distances between any pair from M_3 analyzed image frames. After simple calculations it is possible to say that Eq. 2.47 for a given task can be transformed into

$$\Delta_D = \sqrt{M_3} Q \sqrt{1 - F_R(\bar{0}) \frac{(1 + (M_3 - 1)\rho)G_1(\omega_1, \omega_2)}{\sigma_\theta^2 + (1 + (M_3 - 1)\rho)G_1(\omega_1, \omega_2)}},$$

where $\quad Q = s_0 / \sigma_\theta^2; \ G_1(\omega_1, \omega_2) = F\{R_1(\omega_1, \omega_2)\}; \ F(\omega_1, \omega_2)\{\cdot\} \quad$ and
$F_R(m_1, m_2)\{\cdot\}$ are direct and inverse two-dimensional discrete Fourier transformations, $F_R(\bar{0})\{\cdot\} = F_R(m_1 = 0, m_2 = 0)\{\cdot\}$.

The analysis of the obtained equation leads to the following important conclusions about the effectiveness of the point anomaly optimal detection on an arbitrary number of simultaneously processed images. In the absence of interframe correlation ($\rho = 0$), the co-processing of M_3 frames yields to gain a threshold signal in $\sqrt{M_3}$ times in comparison to signal detection on one image frame notwithstanding the covariance function $R_1(m_1, m_2)$ and size $M_1 \times M_2$ of a frame. The interframe correlation ($\rho \neq 0$) leads to the losses in detection efficiency, which correspond to additional increase in variance σ_x^2 of interfering images in $(1 + (M_3 - 1)\rho)$ times.

The abovementioned properties make it easy to recalculate the characteristics of point anomaly detection in a single image frame in the case of co-processing of arbitrary number of mutually-correlated image frames.

2.5 Image Alignment

While synthesizing algorithms for solving different tasks of frame sequence processing $z^1, z^2, \ldots$, it is usually assumed that observations $z_j^1, z_j^2, \ldots$ in the node j of a grid Ω correspond to the same space point. In the real world, the receiver motion, its imperfection, and other factors lead to the fact that $z_{\tilde{j}}^1, z_{\tilde{j}}^2, \ldots$ correspond to the different space points. As a result, the sequence $z^1, z^2, \ldots$ will demonstrate interframe displacement from frame to frame, i.e. Interframe Geometric Transformation (IGT), such as shifts, rotations, etc. If such distortions are not taken into account, the efficiency of processing algorithms can dramatically decrease. In particular, the IGT evaluation is used in video stabilization [28].

However, the IGT can be not only disturbing factors but also contain useful information. For example, analyzing the IGT of frames obtained at different time intervals one can track the aircraft or submarine course under limited visibility.

Further, tensor shift filtering is yielded in Sect. 2.5.1. Random field alignment of images with interframe geometric transformation is described in Sect. 2.5.2. Section 2.5.3 provides a discussion about alignment of two frames of Gaussian random field. Method of fixed point at frame alignment is represented in Sect. 2.5.4.

2.5.1 Tensor Shift Filtering

Tensor Kalman filter described in Sect. 2.3.2, which estimates frame sequence defined by tensor Model, can be extended in order to joint estimation of frames and interframe shift of these frames. Consider this extension. Let a sequence of m-dimensional frames be defined by a linear tensor stochastic equation

$$x^t = \rho^t x^{t-1} + \vartheta^t \xi^t \quad t = 1, 2, \ldots$$

and their observations look like Eq. 2.48, where $y^t = (y_1^t, \ldots, y_m^t)$ is the IGT parameter vector of the tth frame, $x^t(y^t)$ is a proper observation of a frame x^t with parameters y^t, θ^t, is a white Gaussian RF of observation noise.

$$z^t = x^t(y^t) + \theta^t \tag{2.48}$$

Let the sequence of shift vectors be also described by a linear stochastic equation

$$y^t = \Im^t y^{t-1} + \Phi^t \eta^t \quad t = 1, 2, \ldots,$$

with $(m \times n)$-matrices $\Im^t, \Phi^t$ and a white Gaussian generating vector $\eta^t = (\eta_1^t, \ldots, \eta_m^t)$ of a shift model.

Using observations Eq. 2.48, it is necessary to find an estimation of the successive RF frame x^t and estimation y^t, when obtaining a recurrent observation z^t. To find these estimations, let us use the tensor filtering equations as in Eq. 2.35. In these equations, both the frame itself x^t and its parameters y^t are included in the estimated frame x^t. This integration is a combined tensor, so filtering equations become a bit more complicated. Omitting the intermediate calculations, the resulting algorithm (Eq. 2.49) of recurrent estimation of the RF and displacement [16] are given, where $\hat{x}_{\ni}^t = x^t(\hat{y}_{\ni}^t), \hat{y}_{\ni}^t = \Im^t \hat{y}^{t-1}$.

$$\begin{aligned}\hat{x}^t &= \hat{x}_{\ni}^t + P_x^t(V_\theta^t)^{-1}(z^t - \hat{x}_{\ni}^t) + P_B^t(V_\theta^t)^{-1}\frac{d\hat{x}_{\ni}^t}{d\alpha^t}(z^t - \hat{x}_{\ni}^t) \\ \hat{y}^t &= \hat{y}_{\ni}^t + P_y^t(V_\theta^t)^{-1}\frac{d\hat{x}_{\ni}^t}{d\alpha^t}(z^t - \hat{x}_{\ni}^t) + (P_B^t)^T(V_\theta^t)^{-1}(z^t - \hat{x}_{\ni}^t)\end{aligned} \tag{2.49}$$

The recurrent relations between tensor coefficients of Eq. 2.49 are given in Eqs. 2.50–2.51. Note that although filter in Eqs. 2.50–2.51 can solve this task it is rather difficult to use it in real situations. Besides the computational problems, it is connected with model specification Eq. 2.48, namely, defining function $x^t(y^t)$, i.e. how the frame x^t looks like at the IGT parameters y^t. However, if the IGT type is given (for example, shift or rotation), it is much easier to define the function $x^t(y^t)$.

$$\begin{cases} P_x^t A_x^t + P_B^t A_{yx}^t = P_{\ni x}^t, & (P_B^t)^T A_x^t + P_y^t A_{yx}^t = (P_{\ni B}^t)^T \\ P_x^t A_{xy}^t + P_B^t A_y^t = P_{\ni B}^t, & (P_B^t)^T A_{xy}^t + P_y^t A_{yx}^t = P_{\ni B}^t \end{cases} \tag{2.50}$$

$$
\begin{cases}
A_x^t = E + (V_\theta^t)^{-1} P_{\ni x}^t + (V_\theta^t)^{-1} \dfrac{dx^t}{dy^t} (P_{\ni B}^t)^T \\[2ex]
A_y^t = E - (V_\theta^t)^{-1} \left(\dfrac{dx^t}{dy^t}\right)^2 P_{\ni y}^t + (V_\theta^t)^{-1} \dfrac{dx^t}{dy^t} P_{\ni B}^t \\[2ex]
A_{xy}^t = -(V_\theta^t)^{-1} \dfrac{dx^t}{dy^t} P_{\ni y}^t + (V_\theta^t)^{-1} P_{\ni B}^t \\[2ex]
A_{yx}^t = -(V_\theta^t)^{-1} \dfrac{dx^t}{dy^t} P_{\ni x}^t + (V_\theta^t)^{-1} \left(\dfrac{dx^t}{dy^t}\right)^2 (P_{\ni B}^t)^T \\[2ex]
P_{\ni x}^t = \Re^t P_x^{t-1} (\Re^t)^T + V_x^t, \; P_{\ni B}^t = \Re^t P_B^{t-1} (\Im^t)^T \\[2ex]
P_{\ni B}^t = \Im^t P_B^{t-1} (\Im^t)^T + V_y^t, \; V_y^t = \Phi^t (\Phi^t)^T
\end{cases}
\tag{2.51}
$$

2.5.2 Random Field Alignment of Images with Interframe Geometric Transformation

Let us examine the task of the RF alignment, when a parameter distribution for the IGT is not given. For this purpose, consider the IGT model at first.

Let the RF $\dot{X} = \left\{ x_{\bar{u}}^i : \bar{u} \in U, \; i = 1, 2, \dots \right\}$ be set on some continuous domain U at any specific time i. Each frame $x^i = \left\{ x_{\bar{j}}^i : \bar{j} \in \Omega_i \right\}$ of a grid field X is a system of values $x^i = \left\{ x_{\bar{u}}^i : \bar{u} \in U \right\}$ on the grid $\Omega_i = \left\{ \bar{j} : (j_1, \dots j_n) : j_k = \overline{1, M_k} \right\}$. Besides, the position and shape of grids Ω_i can vary with time, while an index size $M_1 \times M_2 \times \dots \times M_n$ remains constant. Some possible positions of two-dimensional grids Ω_{i-1} (continuous line) and Ω_i (dotted line) are depicted in Fig. 2.15.

Each of grids Ω_i may be considered as a system of coordinates. Thus, the task of the RF alignment can be formulated as a task of finding of coordinates' transformation of nodes of grid Ω_i nodes in the system of coordinates Ω_u of domain U. Sometimes, (e.g. to compensate the noise) it is easier to find the transformation of coordinate Ω_i into Ω_{i-1}, i.e. to align each recurrent frame x^i with proceeding frame x^{i-1}.

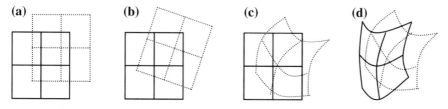

Fig. 2.15 Positions of grids Ω_{i-1} (*continuous line*) and Ω_i (*dotted line*) in two sequenced frames: **a** grid Ω_i is obtained from a rectangular grid Ω_{i-1} by a parallel shift, **b** grid Ω_i is obtained from grid Ω_{i-1} by the shift and rotation, **c** grid Ω_{i-1} is curvilinear, **d** both grids are curvilinear

Consider the task of aligning two frames x^{i-1} and x^i. In general (Fig. 2.15d), it is necessary to estimate a grid form Ω_{i-1} and find transformation of coordinates Ω_i into Ω_{i-1}. Even for stationary fields $\dot{X}$, an estimation of a grid form Ω_{i-1}, i.e. estimation of mutual value ordering x_j^{i-1}, is rather complex and of low accuracy. Let us restrict ourselves to the case, when Ω_{i-1} is a rectangular grid with a unit step (Fig. 2.15a–c). This task simplification is not a rough approximation of a real situation, as value grids are usually close to rectangular.

Let an observation model of field X be Eq. 2.52, where $\theta = \left\{ \theta_j^i \right\}$ is a field of independent RV.

$$z^i = x^i + \theta^i \quad i = 1, 2, \ldots \tag{2.52}$$

If the RF $\dot{X}$ is stationary, then a conditional mutual PDF $w\left(z^{i-1}, z^i | f\right)$ can be obtained, where f is a transformation of coordinates Ω_i in (rectangular) coordinate system Ω_{i-1}. This makes it possible to apply various statistical estimations to combine a pair of frames x^{i-1} and x^i, for example, maximum likelihood estimate by Eq. 2.53.

$$f = \arg \max_f w(z^{i-1}, z^i | f) \tag{2.53}$$

In the general case of transformations f, all coordinates $f(j)$ of all frame values x^i should be estimated in the system Ω_{i-1}. Therefore, function f in Eq. 2.53 contains a large number of parameters and calculation becomes rather complicated.

Alignment task is sufficiently simplified if the type of transformation f is considered. Then, it is desirable only to determine its parameters $\bar{\alpha}$. In such cases, an estimation Eq. 2.53 takes the following form

$$\hat{\bar{\alpha}} = \arg \max_{\bar{\alpha}} w(z^{i-1}, z^i | \bar{\alpha}) \tag{2.54}$$

and contains only a few parameters. Estimation Eq. 2.54 can be used in the case of the RF a priori parametric uncertainty $\dot{X}$ and observation model Eq. 2.52. For this parameter, vector $\bar{\alpha}$ must be supplemented by the RF unknown parameters $\dot{X}$ and an observation model.

2.5.3 Alignment of Two Frames of Gaussian Random Field

Let Gaussian stationary field $\dot{X}$ has a zero mean and the CF as Eq. 2.55, where $\rho(i, |u_1|, \ldots |u_n|)$ is a field $\dot{X}$ correlation coefficient at a distance $|i|$ along the time axis and at a distance $|u_k|$ along the kth spatial axis.

$$V(i, \bar{u}) = M[\dot{x}_v^j \dot{x}_{\bar{v}+\bar{u}}^{j+i}] = \sigma_x^2 \rho(|i|, |u_1|, \ldots |u_n|) \qquad (2.55)$$

In observation model Eq. 2.52, noises θ will also be assumed as Gaussian with zero mean and constant value σ_θ^2.

Let there exist some observations z^{i-1} of a frame x^{i-1} at the nodes of a rectangular grid Ω_{i-1} with a unit step and some observations z^i of a frame x^i at the grid Ω_i nodes. It is necessary to estimate the parameters $\bar{\alpha}$ of IGT x^{i-1} and x^i, i.e. to find the estimation $\hat{\bar{\alpha}}$ of parameters in $f(\bar{j}, \bar{\alpha})$. If we assume that $i = 1$ and chose the grid Ω_1 axis coinciding with the coordinate axis, in which the CF as Eq. 2.55 is set, then the Gaussian joint PDF of a frame x^1 and its observations can easily be calculated from Eqs. 2.52 and 2.55. If parameter vector $\bar{\alpha}$ is given, then position of grid Ω_2 relative to grid Ω_1 becomes definite. Therefore, it is possible to find a joint conditional PDF of observations $Z = (z^1, z^2)$ at given $\bar{\alpha}$ as Eq. 2.56, where $V(\bar{\alpha})$ is a covariance matrix of observations Z, k is a number of elements in Z.

$$w(Z|\bar{\alpha}) = \frac{1}{(2\pi)^{k/2} \det^{1/2}(V(\bar{\alpha}))} \exp\left(-\frac{1}{2} Z V^{-1}(\bar{\alpha})Z\right) \qquad (2.56)$$

It should be noted that calculation of Eq. 2.56 maximum is rather a complicated task which is absolutely unrealizable in real-time systems. In order to simplify the Task, consider the estimation, which can be obtained as a result of only exponent maximization (in Eq. 2.56) provided by Eq. 2.57.

$$\hat{\bar{\alpha}} = \arg \min_{\bar{\alpha}} Z V^{-1}(\bar{\alpha}) Z = \arg \min_{\bar{\alpha}} J(Z, \bar{\alpha}) \qquad (2.57)$$

Functional $J(Z, \bar{\alpha})$ in Eq. 2.57 is the Mahalanobis distance of sample Z from the origin for covariance matrix $V(\bar{\alpha})$. Thus, the estimation in Eq. 2.57 minimizes the Mahalanobis distance of observed Z from the origin. Consider this distance in detail. From Eq. 2.44, it follows that the observation Eq. 2.58 takes place, where $Z^*(\bar{\alpha})$ is an optimal (in our case linear) prediction of observations Z into a point, $\Delta_Z^*(\bar{\alpha})$ are prediction errors, i.e. compensation residues into a point, $C(\bar{\alpha})$ is a diagonal matrix of error variance $\Delta_Z^*(\bar{\alpha})$.

$$J(Z, \bar{\alpha}) = Z C^{-1}(\bar{\alpha})(Z - Z^*(\bar{\alpha})) = Z C^{-1}(\bar{\alpha}) \Delta_Z^*(\bar{\alpha}) \qquad (2.58)$$

If observations z^1 and z^2 are used twice ($z^1 \to z^2$ and $z^2 \to z^1$), then Eq. 2.59 can be obtained, where $\hat{z}^1(\bar{\alpha})$ is an observation prediction of z^1 by z^2, i.e. prediction into domain, $\hat{\Delta}_1(\bar{\alpha})$ are the prediction errors, $N_1(\bar{\alpha})$ is an error covariance $\hat{\Delta}_1(\bar{\alpha})$, $\hat{z}^2(\bar{\alpha})$ is a prediction of z^2 along z^1, $N_2(\bar{\alpha})$ is an error covariance $\hat{\Delta}_2(\bar{\alpha})$ of this prediction.

$$J(z^1, z^2, \bar{\alpha}) = z^1 N_1^{-1}(\bar{\alpha})(z^1 - \hat{z}^1(\bar{\alpha})) + z^2 N_2^{-1}(\bar{\alpha})(z^2 - \hat{z}^2(\bar{\alpha}))$$

$$= z^1 N_1^{-1}(\bar{\alpha})\hat{\Delta}_1(\bar{\alpha}) + z^2 N_2^{-1}(\bar{\alpha})\hat{\Delta}_2(\bar{\alpha}) \qquad (2.59)$$

The evaluation can be somewhat modified by introducing the conditional PDF as the following product

$$w(z^1, z^2 | \bar{\alpha}) = w(z^1 | \bar{\alpha})\, w(z^2 | z^1, \bar{\alpha}).$$

As $w(z^1 | \bar{\alpha}) = w(z^1)$ does not depend on $\bar{\alpha}$, it is enough to maximize conditional PDF $w(z^2 | z^1, \bar{\alpha}) = \frac{1}{(2\pi)^{k/4} \det^{1/2}(N_2(\bar{\alpha}))} \exp\left(-\frac{1}{2}(z^2 - \hat{z}^2(\bar{\alpha}))ZN_2^{-1}(\bar{\alpha})(z^2 - \hat{z}^2(\bar{\alpha}))\right),$
where $\hat{z}^2(\bar{\alpha})$ and $N_2(\bar{\alpha})$ are the same as in Eq. 2.59. From here, it is also possible to obtain a simplified evaluation by the functional in Eq. 2.60, which is the Mahalanobis distance between observations z^2 and their predictions $\hat{z}^2(\bar{\alpha})$ according to z^1.

$$J(z^1, z^2, \bar{\alpha}) = [z^2 - \hat{z}^2(\bar{\alpha})]N_2^{-1}(\bar{\alpha})[z^2 - \hat{z}^2(\bar{\alpha})] \qquad (2.60)$$

Experimental studies show a poor estimation quality obtained using the Mahalanobis distance. It can be explained by the fact that in Eqs. 2.58–2.60 the compensation residues and matrices inversed to $C(\bar{\alpha})$ or to $N_i(\bar{\alpha})$ are multiplied. In addition, the compensation residues can decrease along $\bar{\alpha}$ only to a certain limit. Thus, minimization of these expressions can occur mainly due to increase of matrix elements $C(\bar{\alpha})$ or $N_i(\bar{\alpha})$. This does not happen while the MLM estimates, as the PDF Eq. 2.56, are inversely proportional to the square root of a covariance matrix determinant $V(\bar{\alpha})$.

Much better IGT estimates can be obtained by minimizing the compensation residues $\Delta_Z^*(\bar{\alpha})$, $\hat{\Delta}_1(\bar{\alpha})$, or $\hat{\Delta}_2(\bar{\alpha})$. It is easier to minimize residues

$$\hat{\Delta}_2(\bar{\alpha}) = z^2 - \hat{z}^2(\bar{\alpha}),$$

as the optimal compensation of the recurrent frame z^2 according to the previous frame z^1 is determined. It can often be the main alignment task. Using this approach, it is possible to obtain the estimates in Eq. 2.61 and other estimates, which depend on the used metric. Further, variety of estimates can be obtained using various predictions $\hat{z}_j^2(\bar{\alpha})$, for example, optimal prediction or various interpolations of observation z^1, initially set only on the grid Ω_1.

$$\hat{a} = \arg\ \min_{\bar{\alpha}} M\left[\sum_j \left|z_j^2 - \hat{z}_j^2(\bar{\alpha})\right|\right] \quad \hat{a} = \arg\ \min_{\bar{\alpha}} M\left[\sum_j (z_j^2 - \hat{z}_j^2(\bar{\alpha}))^2\right] \qquad (2.61)$$

Generally speaking, it should be mentioned that the compensation estimates Eq. 2.61 are biased. They do not estimate the IGT parameters $\bar{\alpha}$ but only optimize some compensation in terms of a certain metric. Thus, frames z^1 and z^2 alignment with the use of estimated parameters $\bar{\alpha}$ is a pseudo-alignment in the sense of the best compensation of the selected type. However, in the case of a good choice of prediction function rather effective alignment is often provided.

2.5.4 Method of Fixed Point at Frame Alignment

Many methods of image alignment and the IGT parameter estimation have small operating areas, i.e. they are reasonable only if the estimated parameters are small. Section 2.5.4 describes a method of estimating the IGT parameters with a large operating area. However, the accuracy of this method is low. It can be used to obtain initial parameter approximation, which is then used in a more precise estimation method with a small operating area.

Let mapping $F : W \rightarrow W$ of set W into itself be given. The Fixed Point (FP) of mapping F is any element $w \in W$, for which $F(w) = w$. That is, while mapping F such a point does not change and transforms into itself. Let us apply the concept of the FP to estimate the IGT parameters of images [16, 29, 30].

Consider two images $x(W)$ and $y(J)$ given on integer m-dimensional grids $W = \{w\} = \{(w_1, \ldots, w_m)\}$ and $J = \{j\} = \{(j_1, \ldots, j_m)\}$. Let the IGT type be known and connect positions of these images

$$w = F(j; \bar{\alpha}),$$

where $\bar{\alpha}$ are the unknown IGT parameters, which are subject to estimation. After completing an auxiliary transformation IGT P of image $y(J)$, the image $z(J)$ connected with $x(W)$ by means of a complex transformation is obtained.

$$w = P(F(j; \bar{\alpha})) = H(j; \bar{\alpha})$$

Suppose that this transformation has only one FP v (Eq. 2.62).

$$v = H(v; \bar{\alpha}) \tag{2.62}$$

Then Eq. 2.62 is transformed into a system of m equations relative to parameters $\bar{\alpha}$. If the number of IGT parameters is equal to image dimension, then it is possible to define the parameters from Eq. 2.62. If the IGT has a larger number of parameters, then it is possible to perform K auxiliary transformation P_k, find the FP v_k of each complex transformation H_k, and obtain a system of K equations

$$v_k = H_k(v_k; \text{alpha}), \quad k = 1, \ldots, K,$$

from which the estimates for all transformation F parameters $\bar{\alpha}$ can be obtained. The place of the FP after auxiliary transformation is illustrated by Fig. 2.16. The

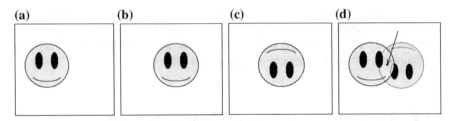

Fig. 2.16 Fixed point of a complex transformation shift-rotation: **a** original image, **b** shift of image **a**, **c** rotation of image (**b**), **d** combination of images (**a**) and (**c**)

original image shows a Smiley (Fig. 2.16a). If this image is shifted right on α_1, the Smiley will shift on α_1 as well (Fig. 2.16b). It is obvious that there is no the FP at a parallel shift. Let us perform an auxiliary rotation of Fig. 2.16b for angle π center-relative. Then, an image in Fig. 2.16c is obtained. Figure 2.16d shows images in Fig. 2.16a, c together. It is evident, that the point marked by an arrow is the same for both Smileys. This is the FP of image transformation of Fig. 2.16a into Fig. 2.16c.

It is essential that the value of the estimated parameters $\bar{\alpha}$ has no effect on the structure of the proposed algorithm and we need to find the FP of a complex transformation that matters. For this purpose, it is necessary that the common part of the images $x(W)$ and $z(J)$ ought not to be too small and contain the FP of the complex IGT, i.e. the restrictions for parameter $\bar{\alpha}$ value are not very strong.

It is obvious that the accuracy of parameter $\bar{\alpha}$ estimation depends on estimation error of the FP position, which has a pixel order that leads to errors of parameter $\bar{\alpha}$ estimation for analyzed IGT. In order to make estimation $\bar{\alpha}$ more accurate, it is possible to use the estimate, which is obtained with the help of a FP in a more precise algorithm as initial approximation. The algorithm can be less time-consuming but with a smaller operating area. Thus, a pair of algorithms allows to improve speed of operation in real-time systems connected with extraction of useful information from image sequences.

Fixed point method for 2D images. Consider the application of the FP method for estimating the IGT parameters of two-dimensional images. The extension of this method on images of larger dimension leads to technical complications only.

Let $x(u, v)$ and $y(i,j)$ be two 2D images with the known IGT (f, g), which relates the positions of these images (Eq. 2.63), where $\bar{\alpha}$ are estimated parameters.

$$u = f(i,j;\bar{\alpha}), \quad v = g(i,j;\bar{\alpha}) \tag{2.63}$$

Conducting auxiliary transformation (p, q) of an image $y(i,j)$, an image $z(i,j)$ connected with $x(u, v)$ by means of a complex transformation is obtained.

$$k = f(p(i,j), q(i,j); \bar{\alpha}) = F(i,j;\bar{\alpha})$$
$$l = g(p(i,j), q(i,j); \bar{\alpha}) = G(i,j;\bar{\alpha})$$

Suppose that this transformation has only one FP (u, v) satisfying Eq. 2.64.

$$u = F(u, v; \bar{\alpha}) \quad v = G(u, v; \bar{\alpha}) \tag{2.64}$$

If one manages to find this FP (at least approximately), then Eq. 2.64 is transformed into a set of equations relative to parameters $\bar{\alpha}$. If the IGT has only two parameters (for example, in the case of a parallel shift), then these parameters can be obtained from two Eq. 2.64. If the IGT has more than two parameters, then it is possible to conduct K auxiliary transformations, find their FP and obtain a set of equations

$$u_k = F_k(u_k, v_k; \bar{\alpha}) \quad v_k = G_m(u_k, v_k; \bar{\alpha}) \quad k = 1, \ldots, K,$$

from which the estimates of all IGT parameters can be obtained.

The most difficult aspect of this method is to find the FP. The necessary criterion for the set of equations Eq. 2.63, i.e. (u, v) immobility, is

$$x(F(u, v; \bar{\alpha}), \ G(u, v; \bar{\alpha})) = z(u, v).$$

However, this condition is not sufficient as there may be other points with values $z(i, j)$ on the image $x(u, v)$ but the FP is only one of them. Therefore, it is necessary to chose an auxiliary transformation for each specific type of the IGT and find the sufficient characteristics of point immobility.

Consider a common type of the IGT: turn on the angle α around the image centre, scale change with coefficient s, and parallel shift on vector (a, b). Both images $x(u, v)$ and $y(i, j)$ are set on an integer grid. Let us place the origin $(0, 0)$ in the central pixel of the grid. Thus, the coordinate transformation (i, j) of an image $y(i, j)$ into coordinates (u, v) of an image $x(u, v)$ has the form of Eq. 2.65.

$$u = a + s(i \cos \alpha - j \sin \alpha) \quad v = b + s(i \sin \alpha + j \cos \alpha) \tag{2.65}$$

Coordinates (u, v) may be fractional, then it may be desirable to calculate values $x(u, v)$ interpolation of the grid image $x(u, v)$. According to the given images $x(u, v)$ and $y(i, j)$, it is necessary to estimate the IGT parameters using Eq. 2.65. As an auxiliary transformation, let us take image rotation $y(i, j)$ around its center pixel (assuming that its coordinates are $(0,0)$) on the angle π, which gives an image $z(i, j) = y(-i, -j)$. In Eq. 2.65, this rotation is equivalent to increase of α by π. Thus, the set Eq. 2.63 has the following form

$$u = a - s(u \cos \alpha - v \sin \alpha), \ v = b - s(u \sin \alpha + v \cos \alpha).$$

It has the only solution (Eq. 2.66).

$$u = [a(1+s\cos\alpha)+bs\sin\alpha]/[(1+s\cos\alpha)^2+(s\sin\alpha)^2]$$
$$v = [b(1+s\cos\alpha)-as\sin\alpha]/[(1+s\cos\alpha)^2+(s\sin\alpha)^2]$$
$$(2.66)$$

If $\alpha \approx 0$, $s \approx 1$, then $u \approx a/2, v \approx b/2$ and an approximate Eq. 2.67 is obtained.

$$a \approx 2u, \quad b \approx 2v \tag{2.67}$$

Thus, having estimated position (u,v) of the FP of image transformation $x(u,v)$ into image $z(i,j)$, the shift parameters (a,b) can be estimated using Eq. 2.67.

Now let us turn to the FP detection. For this purpose, consider an image $\Delta(i,j) = |z(i,j) - x(i,j)|$. Image values $x(u,v)$ and $z(u,v)$ at the FP (u,v) coincide, therefore $\Delta(u,v) = 0$. However, there may be other points, in which $\Delta(u,v) = 0$ because values $x(i,j)$ and $z(i,j)$ are equal at random. At first, let $\alpha = 0$ and $s = 1$ in Eq. 2.65, then it is obvious that image $\Delta(i,j)$ transforms into itself, when rotated at angle π about the FP. That is, this image has central symmetry relative to the FP (u,v). Therefore,

$$\varepsilon(u,v;m,n) = |\Delta(u+m,v+n) - \Delta(u-m,v-n)| = 0$$

for any m and n. However, there may be other Points, at which $\varepsilon(u,v;m,n) = 0$ for some values m and n. It is unlikely, that $\varepsilon(u,v;m,n) = 0$ for many values m and n at once if (i,j) is not the FP. Therefore, the statistics values (Eq. 2.68) are more likely to be small, when a point (i,j) is located near the FP.

$$\varepsilon(i,j) = \sum_{m=0}^{r}\sum_{n=-r}^{r} \varepsilon(i,j;m,n) \tag{2.68}$$

Thus, a position of the minimum point (i,j) of statistics Eq. 2.68 can be considered as a position estimation of the FP (u,v).

If there is rotation and scale changes (subject to $\alpha \neq 0$, $s \neq 1$), then the central symmetry of image $\Delta(i,j)$ is broken. However, if the rotation angle and scale change are small, the symmetry distortion will be small. That is why, the minimum point of statistics in Eq. 2.68 will slightly deviate from the desired point, i.e. there will appear a small error in defining its position. The mutual brightness image distortions $x(i,j)$ and $z(i,j)$, for example, noise, can lead to the secondary errors. Summation in Eq. 2.68 causes smoothing. Thus, a noise impact is slightly reduced.

Additional noise reduction effect can be obtained by considering the approximate central symmetry of an image $\varepsilon(i,j)$ and using statistics similar to Eq. 2.68

$$\delta(i,j) = \sum_{m=0}^{r}\sum_{n=-r}^{r} |\varepsilon(i+m,j+n) - \varepsilon(i-m,j-n).$$

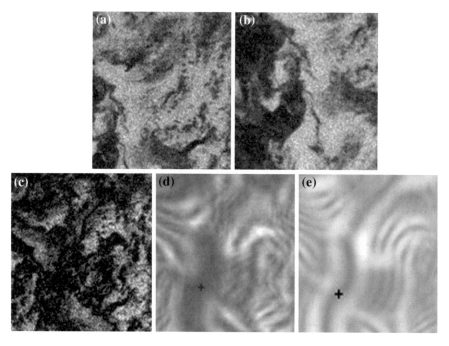

Fig. 2.17 Obtaining of a fixed point at shift, rotation, scale change and noise: **a** and **b** original images, **c** image $\Delta(i,j)$, **d** image $\varepsilon(i,j)$, **e** image $\delta(i,j)$

This method is illustrated by Fig. 2.17. Cloud-cover images in Fig. 2.17a, b have the IGT parameters Eq. 2.65, i.e. shift $a = 43.2$, $b = -38.7$, rotation $\alpha = 0.1$ rad and scale change $s = 0.9$. They are distorted by a strong additive white noise with $\sigma = 30$. Thus, the clouds are slightly visible. Figure 2.17c–e show images $\Delta(i,j)$, $\varepsilon(i,j)$, and $\delta(i,j)$, respectively. Minimum points of images $\varepsilon(i,j)$ and $\delta(i,j)$ are marked with crosses, their coordinates are $(3, -11)$ and $(23, -19)$. Relative shift estimations obtained according to statistics $\varepsilon(i,j)$ and $\delta(i,j)$ are $(6, -22)$ and $(46, -38)$, respectively. The second one is taken into account. It should be considered as a good result for such rigorous estimation conditions.

The tests of the described algorithm showed that if $|\alpha| \leq 0.1$ rad, $|s - 1| \leq 0.1$, and white noise with $\sigma \leq 30$, the estimation error even of large values of shift parameter (tens of pixels) of simulated and real images did not exceed 4–6 pixels.

2.6   Adaptive Algorithms of Image Processing

The image processing algorithms discussed above have been synthesized, mainly, assuming that the image, noise and observation models are thoroughly defined. In practice, such models are usually only partially known, i.e. there exists a priory uncertainty in defining an image model. Thus, the synthesis of adaptive processing algorithms is required. Many adaptive algorithms, in particular, image processing, were developed [31, 32].

Various types of algorithms, such as the pseudo-gradient adaptive, pseudo-gradient adaptive prediction, and pseudo-gradient algorithms of image alignment, are discussed in Sects. 2.6.1–2.6.3, respectively.

2.6.1 Pseudo-Gradient Adaptive Algorithms

According to the purpose of data processing, the adaptive algorithms can be divided into two types: argument and criterion. The starting point for algorithm synthesis is a minimization of average losses, which is formally expressed by quality functional $R(\bar{\alpha}, Z) = J(\bar{\alpha})$, i.e. the criterion, which value should be minimized according to some parameters $\bar{\alpha}$. However, the requirements for this minimization may be different. The objective of argument tasks is the most accurate determination of the minimum point $\bar{\alpha}^*$ (probably, a variable). This type of tasks includes the measurement, filtration, and prediction problems. The objective of criterion tasks is approximation of $J(\bar{\alpha})$ to its minimal value $J^* = J(\bar{\alpha}^*)$, the parameters themselves $\bar{\alpha}$ are irrelevant and may greatly differ from $\bar{\alpha}^*$.

According to the method of optimal parameter estimation, the adaptive algorithms can be divided into identification and non-identification ones. In identification algorithms, the unknown characteristics γ are estimated according to all data available. Then, the estimates $\hat{\gamma}$ obtained are used as the correct ones. As a result, the parameters for the algorithms in the form $\bar{\alpha} = \bar{\alpha}(\hat{\gamma})$ are obtained. This is the core of multiple modified decision rules. The application of these algorithms is complicated by an additional procedure of estimates $\hat{\gamma}$ and instability of further computations to errors of these estimates. In the algorithms without identification, the criterion $J(\bar{\alpha})$ minimization is conducted by variation of $\bar{\alpha}$ without intermediate estimation of any characteristics of the given data. Moreover, $\bar{\alpha}$ may be chosen iteratively during the current processing according to the observations over current values $J(\bar{\alpha})$.

To implement such algorithms, it is necessary to estimate current values $J(\bar{\alpha})$, i.e. the criterion should be observable and this fact limits the application area of the approach. Sometimes it is possible to replace non-observable $J(\bar{\alpha})$ with the other observable criterion $J_1(\bar{\alpha})$. It is required that minimum points $J(\bar{\alpha})$ and $J_1(\bar{\alpha})$ coincide in argument tasks and $J(\bar{\alpha})$ should approximate to $J^* = J(\bar{\alpha}^*)$ as $J_1(\bar{\alpha})$ approximates to $J_1^* = J_1(\bar{\alpha}_1^*)$ in criterion tasks.

Even if the data is thoroughly described, it is not always possible to find the optimal decision rule because of mathematical difficulties. Even if it is possible to find it, it is often unacceptably time consuming. Besides, the initial data model usually describes the reality only approximately. For these reasons, it is often impossible to find and apply the optimal rule under real conditions. Therefore, it is necessary to apply the quasi-optimal implementable rules with a small loss in processing quality. To find such rules, it is possible to use the simplified data models, which describe only their fundamental properties. The obtained rules (algorithms) contain some undetermined parameters $\bar{\alpha}$. These rules should be chosen in such manner that the algorithm gave the best result, when the certain processed data is used.

Let the processing structure be defined and the quality criterion for problem solving be formulated in terms of functional minimization $J(\bar{\alpha})$, which characterizes the average loss, when the processing is performed with parameters $\bar{\alpha}$. In a view of the unexpected uncertainty of data description, it is impossible to determine the optimal parameters $\bar{\alpha}^*$ beforehand. Therefore, the adaptive procedure is desirable, which together with the processing procedure composes the adaptive algorithm. These parameters $\bar{\alpha}$ are determined in terms of a certain implementation (observation) Z of the object processed.

Thus, the adaptation problem is formulated as a problem of function minimization $J(\bar{\alpha}) = J(\bar{\alpha}, Z)$ for a particular data Z. Pseudo-Gradient (PG) algorithms showed good results while solving this problem. There are some numerical methods for extremum search. The most common are various modifications of the gradient algorithm (Eq. 2.69), where $\bar{\alpha}_n$ is an approximation to the minimum point of the functional $J(\bar{\alpha})$ following $\bar{\alpha}_{n-1}$, μ_n is a positive numerical sequence determining the step length, $\nabla J(\bar{\alpha})$ is a gradient of $J(\bar{\alpha})$.

$$\bar{\alpha}_n = \bar{\alpha}_{n-1} - \mu_n \nabla J(\bar{\alpha}_{n-1}) \qquad (2.69)$$

Each step in Eq. 2.69 is made towards the quickest decrease of $J(\bar{\alpha})$, although under certain conditions convergence $\bar{\alpha}_n \to \bar{\alpha}^*$ takes place, it may be very slow.

Multiple computations $\nabla J(\bar{\alpha}_{n-1}, Z)$ prevent the application of these methods in image processing. Each computation usually includes the whole processing procedure of Z with parameters $\bar{\alpha}_{n-1}$. It is possible to significantly decrease the amount of computations, if instead of $\nabla J(\bar{\alpha}_{n-1}, Z)$ a reduction $\nabla Q(\bar{\alpha}_{n-1}) = \nabla J(\bar{\alpha}_{n-1}, Z_n)$ is taken, i.e. to calculate the gradient not across the whole Z but only across some of its part Z_n, for example, in a sliding window. Then, in Eq. 2.69 instead of an exact gradient value its value with a random error $\bar{\delta}_n$ is used and obtained in Eq. 2.70.

$$\bar{\alpha}_n = \bar{\alpha}_{n-1} - \mu_n (\nabla J(\bar{\alpha}_{n-1}, Z) + \bar{\delta}_n) = \bar{\alpha}_{n-1} - \mu_n \nabla Q(\bar{\alpha}_{n-1}) \qquad (2.70)$$

Sequence $\bar{\alpha}_n$ becomes random, thus the very fact of its convergence to $\bar{\alpha}^*$ is random as well. Generally speaking, the random errors $\bar{\delta}_n$ in Eq. 2.70 cannot be a serious obstacle to convergence $\bar{\alpha}_n \to \bar{\alpha}^*$.

The concept of the PG was introduced in [31]. On its basis, a common approach to the analysis and synthesis of algorithms of functional stochastic minimization was developed. The class of the PG algorithms is very wide and it includes all (or nearly all) adaptive and learning algorithms. These algorithms are based on the procedure (Eq. 2.71), where $\bar{\beta}_n$ is a random (particularly, deterministic) direction, generally speaking, depending on previous values $\bar{\alpha}_i$ and n.

$$\bar{\alpha}_n = \bar{\alpha}_{n-1} - \mu_n \bar{\beta}_n \qquad (2.71)$$

Vector $\bar{\beta}_n$ is called the PG of $J(\bar{\alpha})$ at a point $\bar{\alpha}_{n-1}$ if the pseudo-gradient condition

$$[\nabla J(\bar{\alpha}_{n-1})]^T M [\vec{\beta}_n] \geq 0$$

is fulfilled, where the left-hand member is a scalar product, i.e. on the average F is directed at an acute angle to the exact gradient. Algorithm Eq. 2.71 is called the PG if $\bar{\beta}_n$ is the PG at each one-step transition. In this case, the transitions in Eq. 2.71 will be made on the average towards reduction of $J(\bar{\alpha})$ and it is possible to hope for convergence $\bar{\alpha}_n \to \bar{\alpha}^*$ as $n \to \infty$, although some transitions can be made towards increase of $J(\bar{\alpha})$. Indeed, the fulfillment of relatively weak conditions [31] is sufficient for the convergence with probability one for any initial approximation $\bar{\alpha}_0$.

Note that algorithm Eq. 2.71 is much more general than Eq. 2.70, as in Eq. 2.71 the possibility of calculating $J(\bar{\alpha})$ or $\nabla J(\bar{\alpha})$ is not assumed even with a random error, i.e. $J(\bar{\alpha})$ can be also non-observable. Still the availability of the observable PG is required. In particular, even noisy gradient value of another functional $J_1(\bar{\alpha})$ may be chosen as $\bar{\beta}_n$ but it should have the same minimum point as $J(\bar{\alpha})$.

If it is possible for $\bar{\beta}_n$ to depend on previous values $\bar{\alpha}_i$, then it allows to use the PG algorithms to process not only one-dimensional but also multidimensional data in the order of their scanning.

Until now it was assumed that the task is to find the minimum point $\bar{\alpha}^*$ of the functional $J(\bar{\alpha}, Z)$, which is unique for all Z. Such a point exists but the processing will be optimal if data Z is homogeneous. The convergence $\bar{\alpha}_n \to \bar{\alpha}^*$ requires the convergence $\mu_n \to 0$. However, if we bound μ_n from below (for example, take constants $\mu_n = \mu$), then the dispersions of estimate errors $\bar{\alpha}_n$ of parameters $\bar{\alpha}^*$ will stop decreasing and will have the order μ^2, while the parameters $\bar{\alpha}_n$ themselves will oscillate about $\bar{\alpha}^*$. Thus, if a homogeneous data processing is conducted simultaneously with the estimation of $\bar{\alpha}^*$ (when $\mu_n = \mu$), then upon reaching a steady state operating conditions only some quasi-optimal processing, i.e. close to optimal, can be conducted.

If data Z is heterogeneous and there is an abrupt change in its characteristics, then optimal parameter values $\bar{\alpha}^*$ can also change. If a processing proceeds, then it is possible that immediately after such a rapid change the processing quality will get worse and then gradually quasi-optimal results will be achieved once again. When there is only a smooth change in observational characteristics Z (more precisely if optimal parameter values $\bar{\alpha}^*$ change smoothly), which is comparable with the procedure Eq. 2.71 transient rate, it is possible to use the PG algorithms to process the heterogeneous data without their fragmentation into areas of a relatively homogeneous structure.

Thus, the PG algorithms are easy to implement and they can be applied to a wide variety of homogeneous and heterogeneous data. Their adaptation can be performed directly during the processing procedure, thus data delay line is not required. The indicated positive qualities of the PG adaptive algorithms make them preferable in the processing of images and other large data arrays.

Further, the application of the PG algorithms for the prediction and image alignment as examples is considered.

2.6.2 Pseudo-Gradient Adaptive Prediction Algorithms

While solving some tasks of image processing, there often arises an auxiliary problem of their prediction, i.e. the problem of estimation formation x_i^* of an image element x_i according to a certain set (pattern) of observations $Z_{\bar{i}}$, which does not include the predicted element itself. In particular, in the Kalman filters the extrapolated estimate is the prediction. In most real situations, the accuracy of prediction increases with the extension of a pattern but at the same time the computational costs significantly increase.

The prediction is considered to be optimal if the minimum of the mean square prediction error $M[(x_i^* - x_i)^2] = M[\Delta_i^2]$ is achieved. In this case, a conditional mean $M[x_i | Z_{\bar{i}}] = f(Z_{\bar{i}})$, i.e. some function of random variables $Z_{\bar{i}}$, is considered to be the optimal prediction. Type of function f depends on the distribution type of image elements, i.e. on its model. Thus, the optimal prediction may be presented as $x_i^* = f(\bar{\alpha}_{\bar{i}}, Z_{\bar{i}})$, where $\bar{\alpha}_{\bar{i}}$ are the prediction function parameters. For example, in case of the Gaussian images with a zero mean the optimal prediction is linear $x_i^* = \bar{\alpha}_i^T Z_{\bar{i}}$.

If a form of the prediction function is determined, the problem is reduced to its optimization, i.e. it is necessary to find out the optimal values $\bar{\alpha}_i^*$ of parameters $\bar{\alpha}_{\bar{i}}$, whereby a minimum square error is achieved (Eq. 2.72).

$$J(\bar{\alpha}_{\bar{i}}) = M[(f(\bar{\alpha}_{\bar{i}}, Z_{\bar{i}}) - x_i)^2] \qquad (2.72)$$

Let us form the PG algorithm, which minimizes this functional. It needs an observable pseudo-gradient $\bar{\beta}_n$. Non-observable value x_i (the problem under consideration is to estimate x_i) is contained in Eq. 2.72. Therefore, the functional Eq. 2.72 is non-observable, thus it is impossible to obtain the PG as an implementation of its gradient. However, the observations $z_i = x_i + \theta_i$ are observable, their optimal prediction due to independency and centered of noise coincides with the optimal prediction of elements of a true image. Thus, functional Eq. 2.72 can be replaced by an auxiliary functional

$$J_1(\bar{\alpha}_{\bar{i}}) = M\left[(f(\bar{\alpha}_{\bar{i}}, Z_{\bar{i}}) - z_{\bar{i}})^2\right], \qquad (2.73)$$

its observable PG may be as Eq. 2.74.

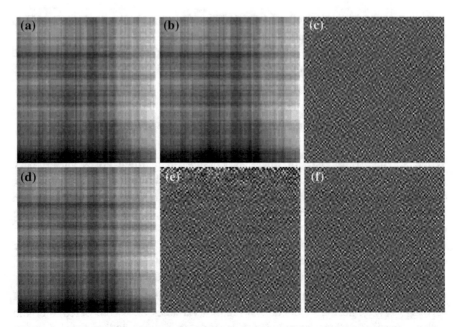

Fig. 2.18 Example of an image prediction: **a** initial image, **b** optimal prediction, **c** optimal prediction errors, **d** the PG prediction, **e** and **f** the PG prediction errors

$$\bar{\beta}_n = \nabla (f(\bar{\alpha}_n, Z_n) - z_n)^2 = 2\Delta_n \frac{\partial f(\bar{\alpha}_n, Z_n)}{\partial \bar{\alpha}_n} \qquad (2.74)$$

Finding a gradient in Eq. 2.74 is rather easy, since the prediction f function is given. For example, if the prediction is linear, then

$$\partial f(\bar{\alpha}_n, Z_n) / \partial \bar{\alpha}_n = \partial f(\bar{\alpha}_n^T Z_n) / \partial \bar{\alpha}_n = Z_n.$$

Note that these algorithms are not associated with any evaluation of image parameters, they are efficient in computation. It enables to implement them in real-time systems.

An example of the PG algorithm application for linear prediction of a homogeneous Gaussian image with a factorable exponential CF is shown in Fig. 2.18. The prediction of an image element is a weighted sum of eight nearest neighbors. The initial image, its optimal prediction, its errors, the adaptive PG prediction, and its errors are represented in Fig. 2.18a–e, respectively. Visually Fig. 2.18b, d are the same, since the adaptive prediction is close to optimal. However, the prediction errors (enlarged for visualization) in Fig. 2.18c, e are different from each other. In the first few (top) lines of the adaptive prediction (Fig. 2.18d) errors are relatively large, as there the process of prediction parameter adjustment takes place. This process takes place rather quickly. Figure 2.18f shows the prediction errors during the second image scanning, which develops from the values of prediction parameters established by the end of the first pass. Analysis of Fig. 2.18c, f allows to make a conclusion that there is no difference between them. It means that the prediction has nearly converged to an optimal one.

Thus, using the adaptive PG algorithms it is possible to obtain a good quality of image prediction at a low computational cost.

2.6.3   Pseudo-Gradient Algorithms of Image Alignment

As it was mentioned in Sect. 2.5, the compensation approach can be a good basis to synthesize algorithms of the IGT parameter estimation, i.e. finding a rather precise estimate of one frame values according to the observations of another frame.

If the IGT type $f(\bar{j}, \bar{\alpha})$ is known, then the problem of frame z^1 and z^2 alignment is sufficiently simplified. It can be formulated as a task of functional minimization (Eq. 2.75), where $\bar{\alpha}$ are parameters of the IGT model, $\hat{z}_{\bar{j}}^2(\bar{\alpha})$ is a prediction into point $z_{\bar{j}}^2$ according to observations z^1.

$$
J(\bar{\alpha}) = M\left[\sum_{\bar{j}} (z_{\bar{j}}^2 - \hat{z}_{\bar{j}}^2(\bar{\alpha}))^2\right] \tag{2.75}
$$

Different interpolations of a grid frame z^1 are usually used as prediction $\hat{z}_{\bar{j}}^2(\bar{\alpha})$ in alignment problems. A set of parameters $\bar{\alpha}^*$, which contributes minimization (Eq. 2.75), is the IGT parameter estimation. The required alignment of frames z^1 and z^2 is obtained by substituting $\bar{\alpha}^*$ into defined by the IGT model transformation $f(\bar{j}, \bar{\alpha})$ of a coordinate system of grid Ω_1, in which frame z^1 is determined, into a coordinate system of frame z^2.

Thus, the problem of the IGT parameter estimation is reduced to the problem of prediction $\hat{z}_{\bar{j}}^2(\bar{\alpha})$ optimization with respect to parameters $\bar{\alpha}$. This problem can be solved using the PG algorithms, as it was done in Sect. 2.6.2.

However, the criterion (Eq. 2.75) can be used only if the brightness distortions are small enough. Suppose that the brightness distortion of the two frames can be approximated by a linear function. In this case, a correlation coefficient between these images should be high. Therefore, the maximum of a local sample correlation coefficient between the observations of the two frames in shifted windows can be taken as a criterion for alignment quality and the PG algorithm can be used to maximize this function.

As it has already been mentioned, the alignment tasks for unknown IGT types are considered to be more complex. Then, it is possible to find some type of transformation $f(\bar{j}, \bar{\alpha})$, which considers parameters $\bar{\alpha}$ as variables. Then the alignment task can be reformulated as a minimization task of functional (Eq. 2.75) with

(a)   **(b)**

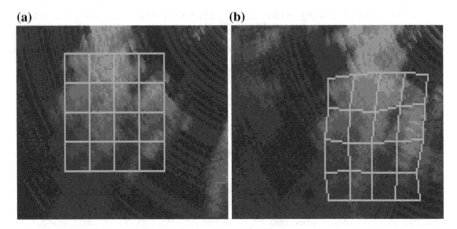

Fig. 2.19 Alignment of radar cloud images: **a** image with rectangle grid, **b** image with distorted grid

variable arguments $\bar{\alpha}$. In such case, the PG algorithms can also be applied, all we need is a parameter μ_n value bounded below. An example of alignment of radar cloud images using the criterion of maximum correlation is depicted in Fig. 2.19.

Figure 2.19b is received by the radar a few minutes after Fig. 2.19a. In this interval, a form of clouds has significantly changed. The clouds did not only move but they changed their shape and density. As a result, their images have the IGT of a general form and significant brightness distortions. To illustrate the alignment result, a conventional rectangular grid is drawn in Fig. 2.19a. Figure 2.19b shows its estimated position (in fact, the estimated position of the elements from Fig. 2.19a in Fig. 2.19b). If the image fragments are compared in the corresponding cells of these both grids, then it is possible to make a conclusion that an alignment is accurate enough. Note that according to the found shifts of image elements it is possible to estimate local speeds of air displacement, build the wind field, which is used, for example, to ensure the flight safety near airports.

Alignment task becomes even more complicated if the brightness distortion is high enough and has of a non-linear character. Then, correlation-extreme criteria are also inapplicable. In this case, the alignment may be performed on the basis of a morphological analysis, i.e. taking into account the image form [10, 11]. In this context, the PG procedures [16] can also be applied.

2.7 Conclusions

New approaches to solve the problems of modeling and statistical analysis of multidimensional data sequences, which can be presented in the form of changing images, are considered in this chapter. A description of images in the form of a random field, which is set on an integer two-dimensional or multidimensional integer grid, is taken as a basis. A number of known model were analyzed and new probabilistic models were developed. It enabled an adequate description of a large class of static and variable images, including ones set on various surfaces. These models are subsequently used to formalize the image processing tasks and synthesize corresponding algorithms. The chapter touches upon the problems of image filtering, detection of point and lengthy anomalies on the background of noise with spatial correlation, image alignment and estimation of parameters of their interframe geometric transformations. Optimal and quasi-optimal algorithms that help to solve these problems were obtained. Their efficiency and computational complexity were analyzed. Special attention is paid to adaptive algorithms of image processing, when there is uncertainty in the description of the processed data.

Acknowledgements The reported study was funded by the Russian Fund for Basic Researches according to the research projects № 16-41-732041 and № 16-41-732027.

References

1. Shalygin, A.S., Palagin, Y.I.: Applied Methods of Statistical Modeling. Mechanical Engineering Leningrad: Mashinostroenie (1986)
2. Habibi, A.: Two-dimensional Bayesian estimate of images. Proc. IEEE **60**(7), 878–883 (1972)
3. Gimel'farb, G.L.: Image Textures and Gibbs Random Fields. Kluwer Academic Publishers, Dordrecht (1999)
4. Woods, J.W.: Two-dimensional Kalman filtering. In: Huang, T.S. (ed.) Two-Dimensional Digital Signal Processing I: Linear Filters. TAP, vol. 42 pp. 155–205 Springer, Berlin, Heidelberg, New York (1981)
5. Yaroslavsky, L.: Digital Picture Processing. An Introduction. Springer, Berlin, Heidelberg (1985)
6. Duda, R.O., Hart, P.E., Stork, D.G.: Pattern Classification, 2nd edn. Wiley-Interscience, New York (2000)
7. Dudgeon, D.E., Mersereau, R.M.: Multidimensional Digital Signal Processing. Signal Processing Series. Prentice-Hall, Englewood Cliffs, New York (1984)
8. Favorskaya, M.N., Levtin, K.: Early smoke detection in outdoor space by spatio-temporal clustering using a single video camera. In: Tweedale, J.W., Jain, L.C. (eds.) Recent Advances in Knowledge-Based Paradigms and Applications. AISC, vol. 234, pp. 43–56. Springer International Publishing, Switzerland (2014)
9. Gonzalez, R.C., Woods, R.E.: Digital Image Processing, 4th edn. Pearson/Prentice-Hall, New York (2017)
10. Serra, J. (ed.): Image Analysis and Mathematical Morphology. Vol 2: Theoretical Advances. Academic Press, London (1988)
11. Vizilter, Y.V., Pyt'ev, Y.P., Chulichkov, A.I., Mestetskiy, L.M.: Morphological image analysis for computer vision applications. In: Favorskaya, M.N., Jain, L.C. (eds.) Computer Vision in Control Systems-1, ISRL, vol. 73, pp. 9–58. Springer International Publishing, Switzerland (2015)
12. Gruzman, I.C., Kirichuk, V.P., Kosikh, G.I., Peretryagin, G.I., Spector, A.A.: Digital Image Processing in Informative Systems. Novosibirsk State Technical University (2000) (in Russian)
13. Huang, T.S. (ed.): Image Sequence Analysis. Springer, Berlin, Heidelberg, New York (1981)

14. Huang, T.S. (ed.): Image Sequence Processing and Dynamic Scene Analysis. Springer, New York (1983)
15. Soifer, V.A. (ed.): Computer Image Processing. Part I: Basic Concepts and Theory. VDM Verlag Dr. Muller E.K. (2009)
16. Vasil'ev, K.K., Krasheninnikov, V.R.: Statistical Analysis of Images. Ulyanovsk State Technical University (2015) (in Russian)
17. Vasil'ev, K.K., Dement'ev, V.E., Andriyanov, N.A.: Doubly stochastic models of images. Pattern Recognit. Image Anal. 25(1), 105–110 (2015)
18. Vasil'ev, K.K., Popov, O.V.: Autoregression models of random fields with multiple roots. Pattern Recognit. Image Anal. 9(2), 327–328 (1999)
19. Vasil'ev, K.K., Dement'ev, V.E., Andriyanov, N.A.: Application of mixed models for solving the problem on restoring and estimating image parameters. Pattern Recognit. Image Anal. 26(1), 240–247 (2016)
20. Krasheninnikov, V.R.: Correlation analysis and synthesis of random field wave models. Pattern Recognit. Image Anal. 25(1), 41–46 (2015)
21. Krasheninnikov, V.R., Kalinov, D.V., Pankratov, YuG: Spiral autoregressive model of a quasi-periodic signal. Pattern Recognit. Image Anal. 8(1), 211–213 (2001)
22. Krasheninnikov, V.R., Mikeev, R.R., Kuzmin, M.V.: The model and algorithm for simulation of planets relief as surfaces image. Radioengineering 175, 192–194 (2012). (in Russian)
23. Dikshit, S.: A recursive Kalman window approach to image restoration. IEEE Trans Acoust. Speech Signal Process. 30(2), 125–140 (1982)
24. Jähne, B.: Digital Image Processing, 6th edn. Springer, Berlin, Heidelberg (2005)
25. Pratt, W.K.: Digital Image Processing. PIKS Inside. 3rd ed. Wiley, New York (2001)
26. Prewitt, J.M.S.: Object enhancement and extraction. In: Lipkin, B.S., Rosenfeld, A. (eds.) Picture Processing and Psychopictorics, pp. 75–149. Academic Press, New York (1970)
27. Zhuravlev, Yu.I.: An algebraic approach to recognition or classifications problems. Pattern Recognit. Image Anal. 8(1), 59–100 (1998)
28. Favorskaya, M., Jain, L.C., Buryachenko, V.: Digital video stabilization in static and dynamic scenes. In: Favorskaya, M.N., Jain, L.C. (eds.) Computer Vision in Control Systems-1, ISRL, vol. 73, pp. 261–309 Springer International Publishing, Switzerland (2015)
29. Krasheninnikov, V.R., Potapov, M.A.: A way to detect the straight line trajectory of an immovable point for estimating parameters of geometrical transformation of 3D images. Pattern Recognit. Image Anal. 21(2), 280–284 (2011)
30. Krasheninnikov, V.R., Potapov, M.A.: Estimation of parameters of geometric transformation of images by fixed point method. Pattern Recognit. Image Anal. 22(2), 303–317 (2012)
31. Polyak, B.T., YaZ, Tsypkin: Optimal pseudogradient adaptation procedure. Autom. Remote Control 8, 74–84 (1980)
32. Widrow, B., Stearns, S.D.: Adaptive Signal Processing. Prentice-Hall Inc., Englewood, Cliffs, NJ (1985)
33. Vasil'ev, K.K.: Statistical analysis of multidimensional images. Pattern Recognit. Image Anal. 9(4), 732–748 (1999)
34. Krasheninnikov, V.R.: Wave image models on the surfaces. In: 8th Open German-Russian Workshop on Pattern Recognition and Image Understanding Nizhny, Novgorod, pp. 154–157 (2011)
35. Krasheninnikov, V.R., Kuznetsov, V.V., Lebedeva, E.Y., Krasheninnikova, N.A.: Optimization of dictionary and model library for recognition of speech commands based on cross-correlation portraits. Pattern Recognit. Image Anal. 23(1), 80–86 (2013)
36. Krasheninnikov, V.R., Kopylova, A.S.: Algorithms for automated processing images of blood serum facies. Pattern Recognit. Image Anal. 22(4), 583–592 (2012)
37. Vasil'ev, K.K., Dement'ev, V.E., Luchkov, N.V.: Analysis of efficiency of detecting extended signals on multidimensional grids. Pattern Recognit. Image Anal. 22(2), 400–408 (2012)

Chapter 3
Vision-Based Change Detection Using Comparative Morphology

Yu. Vizilter, A. Rubis, O. Vygolov and S. Zheltov

Abstract The chapter addresses the theoretical and practical aspects of the scene change detection problem with the use of computer vision techniques. It means detecting new or disappeared objects in images registered at different moments of time and possibly in various lighting, weather, and season conditions. In this chapter, we propose the new scheme of Comparative Morphology (CM) as a generalization of the Morphological Image Analysis (MIA) scheme originally proposed by Pyt'ev. The CMs are the mathematical shape theories, which solve the tasks of the image similarity estimation, image matching, and change detection by means of some special morphological models and tools. The original morphological change detection approach is based on the analysis of difference between the test image and its projection to the shape of reference image. In our generalized approach, the morphological filter-projector is substituted by the comparative morphological filter with weaker properties, which transforms the test image guided by the local shape of reference image. Following theoretical aspects are addressed in this chapter: the comparative morphology, change detection scheme based on morphological comparative filtering, diffusion morphology, and morphological filters based on guided contrasting. Following practical aspects are addressed: the pipeline for change detection in remote sensing data based on comparative morphology and implementation of change detection scheme based on both guided contrasting and diffusion morphology. The chapter also contains the results of qualitative and quantitative experiments on a wide set of real images including the public benchmark.

Yu. Vizilter (✉) · A. Rubis · O. Vygolov · S. Zheltov
State Research Institute of Aviation Systems, 7, Viktorenko Street,
Moscow 125319, Russian Federation
e-mail: viz@gosniias.ru

A. Rubis
e-mail: arubis@gosniias.ru

O. Vygolov
e-mail: o.vygolov@gosniias.ru

S. Zheltov
e-mail: zhl@gosniias.ru

© Springer International Publishing AG 2018
M.N. Favorskaya and L.C. Jain (eds.), *Computer Vision in Control Systems-3*,
Intelligent Systems Reference Library 135, https://doi.org/10.1007/978-3-319-67516-9_3

Keywords Mathematical morphology · Change detection · Diffusion maps
Guided filtering · Shape representation

3.1 Introduction

This chapter addresses the theoretical and practical aspects of the change detection problem. It means detecting new or disappeared objects in images registered at different moments of time and possibly in various lighting, weather, and season conditions. For example, in video surveillance change detection presumes the short-term changes in illumination combined with presence of moving objects. The most popular and challenging change detection problem appears in analysis of bi-temporal or multi-temporal spaceborne or airborne remote sensing data.

A lot of change detection techniques are developed for remote sensing applications [1–4]. There are two main categories of change detection techniques: the pixel-level and object-level. The pixel-based methods usually provide the attractive computational efficiency but relatively low detection characteristics. In contrast, the object-based techniques usually provide the high detection quality but require much more computational efforts. In the chapter, we propose the new scheme of the CM as a generalization of the MIA scheme [5]. Such comparative morphologies are based on the mathematical shape theories, which solve the tasks of the image similarity estimation, image matching, and change detection by means of some special morphological models and tools. Additionally, a new change detection technique based on the CM generalization is presented. The original morphological change detection approach is based on the analysis of difference between the test image and its projection to the shape of reference image. The overview of modern MIA results is given in [6]. In our generalized approach, the morphological filter-projector is substituted by the comparative morphological filter with weaker properties, which transforms the test image guided by the local shape of reference image. Thus, such approach implements some important properties of object-level image comparison immediately in the pixel-level image filtering. Due to this, we can talk about the morphological mid-level change detection. It should provide the desired compromise between the computational efficiency of pixel-based methods and detection quality of object-based techniques.

The theoretical contributions of this chapter are:

- The comparative morphology as a generalized scheme of MIA.
- The new change detection scheme based on morphological comparative filtering.
- The new morphological filters based on guided contrasting.

We need to note that the comparative filtering scheme is close in some sense to the guided filtering scheme [7, 8], but proposed morphological filters additionally satisfy the special mathematical properties.

The practical contributions of this chapter are:

- The new pipeline for change detection in remote sensing data [9].
- The implementation of change detection scheme based on both guided contrasting and diffusion morphology previously proposed in [10].

Qualitative experiments with proposed guided contrasting filters are performed on a wide set of real images. Quantitative experiments with change detection pipeline are performed on the public benchmark containing simulated aerial images.

This chapter is structured as follows. In Sect. 3.2, the related works are briefly described. Section 3.3 contains the theoretical basics, new ideas, schemes, and algorithms. In Sect. 3.4, the experimental results are reported. Section 3.5 concluded the chapter.

3.2 Related Works

There are some well-known reviews of change detection approaches both classical and modern enough [1–4]. In [4], two main categories of methods are pointed: the Pixel-Based Change Detection (PBCD) and Object-Based Change Detection (OBCD) techniques. The PBCD category of change detection methods contains the direct, transform-based, and classification-based comparison of images at the pixel level. Some machine learning techniques are applied at the pixel level too. The OBCD category contains the direct, classified, and composite change detection at the object level. We start our brief overview from the pixel-level techniques and then go to the object-level comparison.

The simplest direct image comparison technique is an image difference calculation from intensity values of original or transformed images [11–13]. Since the relative changes occur in both images, the direction of image comparison should be selected [14]. Image rationing forms regions that are not changed with ratio value approximately equal to 1 [15]. Image regression represents second image as a linear function of first one [16]. A regression analysis, such as least-squares regression, is used for identification of regression parameters [17]. Changes are detected by subtracting regressed image from the original one.

Transform-based imaged comparison presumes the analysis of transformed images. Change Vector Analysis (CVA) was developed for change detection in multiple image bands [18–20]. Change Vectors (CV) are calculated by subtracting pixel vectors of co-registered different-time dates. The direction and magnitude of the CV correspond to the type and power of change. Principal Component Analysis (PCA) is applied for change detection in two main ways. The first one is to apply the PCA to images separately and then compare them using differencing or rationing [21]. The second way is to merge the compared images into one set and then apply the PCA transform. Principal components with negative correlation should correspond to changes in compared images [22]. Tasselled cap

Transformation (KT) is a particular case of spectral transform presented in [23]. It produces the stable spectral components, which allow to develop baseline spectral information for long-term studies of forest disturbances [24] or vegetation change [25]. Different texture-based transforms are developed and used, for example for urban disaster analysis [26] and land use change detection [27].

Classification-based change detection contains the post-classification comparison techniques and composite classification methods. Post-classification comparison presumes that the images are first rectified and classified [28–30]. The supervised [31–33] or unsupervised classification [34] can be used. Then the classified images are compared to measure the changes. Unfortunately, the errors from individual image classification are propagated into the final change map reducing the accuracy of change detection [35–37]. In the composite or direct multidate classification [17, 38], the rectified multispectral images are stacked together and the PCA technique is often applied to reduce a number of spectral components to a fewer principal components [1, 39]. The minor components in the PCA should represent changes [40]. However, due to the fact that temporal and spectral features are fused in the combined dataset, it is difficult to separate the spectral changes from the temporal changes during classification [41].

Machine learning algorithms are extensively utilized in change detection techniques. Artificial Neural Networks (ANN) are usually trained by supervised learning on a large training dataset for generating the complex non-linear regression between input pair of images and output change map [36, 42]. The ANN approach was applied for the land-cover change detection [36, 43], forest change detection [44], and urban change detection [45, 46]. Support Vector Machine (SVM) approach based on well-known SVM technique [47] considers the finding change and no-change regions as a binary classification problem [48]. The algorithm learns from training data and automatically finds the binary classifier parameters in a space of spectral features [49]. The SVM approach is used for land cover change detection [50] and forest cover change analysis [48]. Some other machine learning techniques are applied for change detection via learning to change and non-change separation: the decision tree [29], genetic programming [51], random forest [52–54], and cellular automata [55].

Object-based techniques operate with objects instead of pixels. Direct Object Change Detection (DOCD) is based on the comparison of objects extracted from the compared images. Changes are detected by comparing either the geometrical properties [56, 57], spectral information [58, 59], or extracted features of the image objects [26, 56]. In Classified Objects Change Detection (COCD) approach, the extracted objects are compared based on information about both the geometry and class membership [60–62]. The OBCD framework based on post-classification comparison was proposed in [63]. Different algorithms like the decision-tree and nearest neighbor classifier [29], fuzzy classification [64], and Maximum Likelihood Classification (MLC) are used for extracting objects and independently classifying them. Some applications of the COCD are updating maps or GIS layers [65–68]. The COCD is applied for the forest change detection [69], land cover, and land use change analysis [70]. Multitemporal-object change detection presumes that the joint segmentation is performed once for stacked (composite) images. In [69, 71], the forest change detection is performed via segmentation of stacked multi-date SPOT images. In [71], the change detection is performed based on object segmentation in 12-dimensional red, Near InfraRed (NIR), and Short Wavelength InfraRed (SWIR) data. In [72] and [73], the multi-temporal composite images are used both at segmentation and classification stages for map vegetation change objects. Clustering on multi-date objects for deforestation analysis is proposed in [74].

There are some combined approaches those utilize different combinations of described ideas. In particular, in [75] a change detection is performed via differencing after the PCA. In [76], the pixel-based information is combined with object-based information via pixel labeling based on statistical and semantical models. In [77], the pixel-based and object-based information are fused for suppression of output change map noise. In [78], a change detection approach is proposed based on Multivariate Alteration Detection (MAD) transform and sub-clustering of detected objects via maximum likelihood classification.

This chapter presents a new change detection technique based on generalized ideas of the MIA proposed by Pyt'ev [5, 79] and further developed in [80–83]. It is a developed version of our previous works [9] and [10] extended by some new theory, examples, and illustrations. Note that the terms "morphology", "morphological filter", and "morphological analysis" refer to Mathematical Morphology (MM) proposed by Serra [84] as well as to the MIA. These theories of shape have a common algebraic basis (lattice theory) but different tasks and tools. The overview of the MIA and its relation to the MM is given in [6]. Morphological change detection approach is based on the analysis of morphological difference map formed as a difference between test image and its morphological projection to the shape of sample image. In our generalized approach, the role of morphological projector is played by comparative morphological filter with weaker properties, which transforms the test image guided by the shape of sample image. The shape of sample image is described by mosaic segmentation or by local texture features of objects (regions). Thus, such morphological approach implements some important properties of object-level image comparison immediately in the pixel-level image filtering. Due to this, we can speak about the morphological mid-level change detection procedure. It should provide the desired compromise between the computational efficiency of pixel-based methods and detection quality of object-based techniques.

3.3 Methodology

This section contains the description of proposed mid-level change detection methodology. Thus, the comparative morphology as a generalized scheme of morphological image analysis is represented in Sect. 3.3.1. Section 3.3.2 provides the comparative filtering as a generalization of morphological filtering. The comparative filters based on guided contrasting are given in Sect. 3.3.3. The manifold learning, diffusion maps, and shape matching are discussed in Sect. 3.3.4. The generalized diffusion morphology and comparative filters based on diffusion operators are considered in Sect. 3.3.5. Section 3.3.6 provides the image and shape matching based on diffusion morphology. The change detection pipeline based on comparative filtering is presented in Sect. 3.3.7.

3.3.1 Comparative Morphology as a Generalized Scheme of Morphological Image Analysis

Consider the MIA from the two most generic viewpoints—*problems* to be solved and *tools* provided for solution of these problems. Concerning the problems, the MIA performs the *image comparison by shape*, which presumes the solution of following sub-problems:

- The *similarity estimation* between images and/or shapes.
- The image and/or shape *matching*.
- The *difference/change detection* between images and/or shapes.
- The *comparison by complexity* between images and/or shapes.

The term "shape" means here some image structure, which is stable relative to some type of image transformation. Due to this, the morphological "image comparison by shape" should be invariant or robust relative to such image transformation.

Concerning the mathematical tools, the MIA provides the following specific tricks and approaches:

- The image shape is described by three different ways: as a *structure* (mosaic shape = tessellation of image frame to the set of non-overlapped connected regions), as a *manifold in the image space* (linear subspace), and as an *operator in the image space* (idempotent projector of any image to the subspace containing the images with the same mosaic structure). Mathematically, all these three shape definitions are equivalent because each mosaic shape corresponds to one and only one linear projective operator to the subspace of images of this mosaic shape.
- The algorithm for *shape generation based on image samples* is determined. The simplest MIA version presumes that a mosaic shape is generated based one the unique image sample via some image segmentation algorithm with some regularization. However, we can also apply some other image segmentation techniques based on the joint segmentation of some sample image set.
- The morphological image-to-shape comparison is performed as a *comparison between the image and its projection to the shape*.
- The numerical similarity measure between image and shape is formed as a ratio of image norm and norm of its projection to the shape. It is called the

Morphological Correlation Coefficient (MCC). The image-to-shape matching is performed as a mutual geometrical registration of image fragments based on the maximal value of the MCC.

- The differences between image and shape are extracted by morphological *background normalization procedure*, which presumes the subtraction of projection to the shape from the original image.
- The comparison of images by complexity is based on the fact that the set of shapes can be considered as a complete lattice structure. The relations of shape complexity "more complex by shape" and "less complex by shape" are the partial order relations, such that the most simple and most complex shapes exist, and for each two shapes the supremum and infimum by complexity are determined.
- The relations of shape complexity can be equivalently determined for all three forms of morphological shape description. The rule for mosaic shape ordering by complexity is the following: "The shape **A** is *simpler* than the shape **B** if the regions of **A** can be obtained by the merging of regions of shape **B**. The shape **A** is *more complex* than the shape **B** if the regions of **A** can be obtained by the splitting of regions of shape **B**. If the transformation of **A** to **B** needs both splitting and merging operations, then the shapes **A** and **B** are *non-comparable by shape*". The partial ordering of manifold shapes (shapes as subspaces) by complexity directly corresponds to the set-theoretic inclusion relation: "If $A \subseteq B$, then **A** is *simpler* than **B** and **B** is *more complex* than **A**. If $A \not\subseteq B$ and $B \not\subseteq A$, then the shapes **A** and **B** are *non-comparable by shape*". The partial ordering of morphological projectors is based on the properties of their superposition: "If $p_B\, p_A = p_A$, then **A** is *simpler* than **B** and **B** is *more complex* than **A**".
- The estimation of similarity or dissimilarity of shapes is correspondingly performed by morphological shape correlation coefficients or by some special shape metrics. These coefficients and metrics can be determined both in the space of structural models (for example, metrics on mosaic tessellations) and in the space of images (distances and angles between linear subspaces).

In [6], we described a wide family of different mathematical morphologies with different structural models of shapes and different ways for determination of morphological operators. In this chapter, we propose the new generic scheme of image analysis called the "comparative morphology", which generalizes the listed ideas and tools of the MIA but can be implemented based on any other image models or filters.

Based on the functional analysis of the MIA, we define the CM as a mathematical shape theory, which solves the following tasks:

- The *similarity estimation* between images and/or shapes.
- The image and/or shape *matching*.
- The *difference/change detection* between images and/or shapes.
- The *comparison by complexity* between images and/or shapes via the following models and tools:

- Three equivalent ways for *shape description*, such as the *image structure, set of images* (manifold in the image space), and *morphological filter* (operator in the image space).
- The algorithm for *shape generation* based on the sample image or some set of sample images.
- The morphological *image-to-shape comparison technique* based on comparison between the image and the result of its morphological filtering.
- The MCC, which is the numerical similarity measure between image and the result of its morphological filtering.
- The morphological *image-to-shape matching technique* based on the search of maximal value of the MCC.
- The morphological *background normalization procedure*, which performs the subtraction of the result of morphological filtering from the original image.
- The morphological *technique for comparison by complexity* with relations of shape complexity equivalently determined for all three forms of morphological shape description.
- The morphological *shape correlation coefficients* and *shape metrics* determined both in the space of structural models and in the space of images.

In the next section, we give the formal mathematical description of comparative morphology based on the comparative filter, which is not required to be an idempotent operator and exactly preserving the images of given shape in the MIA sense. Nevertheless, such filters allow implementing the listed functions and tools of comparative morphology, and the Pyt'ev morphological projector can be considered as a partial case of introduced comparative filter.

3.3.2 Comparative Filtering as a Generalization of Morphological Filtering

Consider the mosaic image model utilized in the MIA given by Eqs. 3.1–3.2, where n is a number of connected regions of *tessellation* $\mathbf{F}$ of the image frame $\Omega \subseteq R^2$, $\mathbf{F} = \{F_1, ..., F_n\}$, $\mathbf{f} = (f_1, ..., f_n)$ is the intensity values, $\chi_{Fi}(x, y) \in \{0,1\}$ is support function of ith region.

$$f(x,y) = \sum_{i=1}^{n} f_i \chi_{F_i}(x,y) \tag{3.1}$$

$$\chi_{F_i}(x,y) = \begin{cases} 1 & \text{if } (x,y) \in F_i \\ 0 & \text{otherwise} \end{cases} \tag{3.2}$$

The tessellation should be obtained by some image segmentation procedure. The *mosaic shape* is a set of images with the same frame tessellation (Eq. 3.3).

$$\mathbf{F} = \left\{ f(x,y) = \sum_{i=1}^{n} f_i \chi_{F_i}(x,y),\ \mathbf{f} = \{f_1,\ldots,f_n\},\ \mathbf{f} \in R^n \right\} \qquad (3.3)$$

For any image $g(x,\ y) \in L^2(\Omega)$, the *projection onto the shape F* is defined by Eq. 3.4.

$$g_F(x,y) = P_F g(x,y) = \sum_{i=1}^{n} g_{F_i} \chi_{F_i}(x,y) \quad g_{F_i} = \left(\chi_{F_i}, g \right) / \left\| \chi_{F_i} \right\|^2 \quad i = 1,\ldots,n$$

$$(3.4)$$

The similarity of images $f(x,\ y)$ and $g(x,\ y)$ is estimated by the *normalized* MCC provided by Eq. 3.5, where $P_O f = f_o \equiv \text{mean}(f(x,\ y))$ and $P_O g = g_o \equiv \text{mean}(g(x,\ y))$ are the mean values of projected images.

$$K_M(g,F) = \frac{\|P_F g - P_O g\|}{\|g - P_O g\|} \quad K_M(f,G) = \frac{\|P_G f - P_O f\|}{\|f - P_O f\|} \qquad (3.5)$$

Morphological extraction of differences on image $g(x,\ y)$ relative to shape F is *performed* via comparison of $g(x,\ y)$ with its projection to F, and vice versa (Eq. 3.6).

$$\Delta g_F = |g - P_F g|, \Delta f_G| f - P_G f| \qquad (3.6)$$

Note that this MIA scheme is basically asymmetrical due to different roles of input images: the *test* image is projected to the shape, while the *sample* image determines the shape. We propose the weaker scheme of morphological image analysis that excludes the ideas of *shape* and projection but preserves the idea of asymmetrical comparative filtering for robust similarity estimation and change detection.

The mapping of *sample* image $f(x,\ y)$ and *test* image $g(x,\ y)$ to the filtered version of test $g_f(x,\ y)$ we refer as a *comparative filter* (Eq. 3.7).

$$\psi(f,g) : L^2(\Omega) \times L^2(\Omega) \rightarrow L^2(\Omega) \qquad (3.7)$$

If sample image f is fixed, then comparative filter takes the usual form with one input and one *output* image in the form of Eq. 3.8.

$$\psi_f(g) = \psi(f,g) \qquad (3.8)$$

We call such filter the *morphological comparative filter* if it satisfies the *conditions* provided by Eq. 3.9, where $o(x,\ y) \equiv \text{const}$ is any constant-valued (flat) image.

$$(1) \; \|\psi(f,g)\| \leq \|g\|$$
$$(2) \; \psi(f,f) = f \tag{3.9}$$
$$(3) \; \psi(f,o) = o$$

The first and third conditions describe the *smoothing* properties. The second condition describes the *exact matching* property: filter should preserve the test image if it is equal to sample one.

Due to the properties (Eq. 3.9) the similarity of test $g(x, y)$ and sample $f(x, y)$ can be estimated like in the MIA by the MCC in the form of Eq. 3.10.

$$K_{\psi}(f,g) = \frac{\|\varphi(f,g) - g_o\|}{\|g - g_o\|} \tag{3.10}$$

Morphological Difference Map (MDM) of image $g(x, y)$ relative to image $f(x, y)$ can be *calculated* in analogous way using Eq. 3.11.

$$\Delta_{\psi} g_F = |g - \psi(f,g)| \tag{3.11}$$

It is easy to see that if sample image $f(x, y)$ satisfies the mosaic model (Eq. 3.1), then the mapping $\psi(f, g) = P_F \, g$ satisfies to Eqs. 3.7–3.9. Thus, the morphological projector is a particular case of morphological comparative filter. Consequently, the proposed comparative filtering scheme is a correct generalization of the MIA.

3.3.3 Comparative Filters Based on Guided Contrasting

Consider the comparative filter based on the local linear correlation. Let $w(x, y)$ be a sliding window at position (x, y). The *guided local contrasting filter* is defined aby Eq. 3.12, where $g^{w(x,y)}(u, v)$ is $g(x, y)$ localized in a window $w(x, y)$, $g_o^{w(x,y)}$ is a mean value of $g(x, y)$ in a window $w(x, y)$, $K(f^{w(x,y)}, g^{w(x,y)})$ is a local linear correlation coefficient in a sliding window $w(x, y)$.

$$\varphi^w(f,g)(x,y) = g_o^{w(x,y)}(x,y) + \left| K\left(f^{w(x,y)}, g^{w(x,y)}\right) \right| \left(g(x,y) - g_o^{w(x,y)}(x,y) \right)$$
$$g^{w(x,y)}(u,v) = \begin{cases} g(x,y) & \text{if } (u,v) \in w(x,y) \\ 0 & \text{otherwise} \end{cases}$$
$$g_o^{w(x,y)}(x,y) \equiv mean(g^{w(x,y)}(x,y))$$
$$K\left(f^{w(x,y)}, g^{w(x,y)}\right) = \frac{\left(f^{w(x,y)} - f_o^{w(x,y)}, g^{w(x,y)} - g_o^{w(x,y)}\right)}{\left\|f^{w(x,y)} - f_o^{w(x,y)}\right\| \left\|g^{w(x,y)} - g_o^{w(x,y)}\right\|} \tag{3.12}$$

Such filter (Eq. 3.12) satisfies conditions (Eq. 3.9), preserves the similar details, and smooths the non-similar details in a test image g guided by sample image f. So, the difference map (Eq. 3.11) based on filter (Eq. 3.12) can be applied for change detection. The size of details is determined by the size of window $w(x, y)$. If we need to detect the details of different size, one can use the image pyramid.

In practice, we proceed to *guided contrasting filter with local search* (in a zone $p(x, y)$) that provides the robustness relative to weak geometrical discrepancy of images provided by Eq. 3.13.

$$\varphi^{w \cdot p}(f, g)\,(x, y) = g_o^{w(x,y)}(x, y) + K_{\max}\left(f, g^{w(x,y)}\right)\left(g(x, y) - g_o^{w(x,y)}(x, y)\right)$$
$$K_{\max}\left(f, g^{w(x,y)}\right) = \max_{(u,v) \in p(x,y)}\left|K\left(f^{w(u,v)}, g^{w(x,y)}\right)\right| \tag{3.13}$$

The scheme of comparative filtering based on guided contrasting (Fig. 3.1) demonstrates the main idea of this approach. First, we perform the local smoothing of test image. Second, we estimate the local similarity of extracted details and recover or not recover them depending on this local similarity. In result, similar details should be recovered but non-similar details should be extremely smoothed.

This simple and general idea can be expressed in the form of *generalized guided contrasting filter* using Eq. 3.14, where $a(f, g^{w(x,y)})$ is a *Local Similarity Coefficient* (LSC) of test image fragment $g^{w(x,y)}$ with sample f.

$$\varphi_a^w(f, g)\,(x, y) = g_o^{w(x,y)}(x, y) + a\left(f, g^{w(x,y)}\right)\left(g(x, y) - g_o^{w(x,y)}(x, y)\right)$$
$$a\left(f, g^{w(x,y)}\right) \in [0, 1] \quad a\left(g, g^{w(x,y)}\right) = 1 \quad a\left(o, g^{w(x,y)}\right) = 0 \tag{3.14}$$

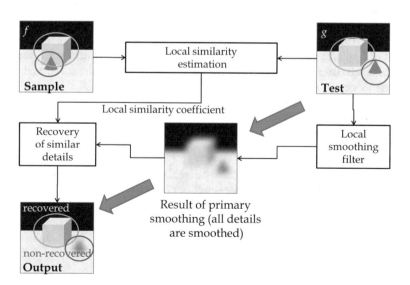

Fig. 3.1 The scheme of comparative filtering based on guided local contrasting

Any LSC with such properties generates the filter satisfying (Eq. 3.9). Different variants of LSC $a(f, g^{w(x,y)})$ can be considered for change detection task:

- The local MCC (Eq. 3.5) [5].
- The LSC based on mutual information [85].
- The local mean square MCC [82].
- The geometrical correlation coefficients [82], etc.

We implement and explore just the guided contrasting filters (Eqs. 3.12–3.13) based on local linear correlation but in the future all of these variants can be implemented and tested both in change detection and in image matching tasks.

The main advantage of proposed guided contrasting filters relative to the MIA projectors is the elimination of sample image segmentation step that allows obtaining the more precise and computationally efficient solutions. Additionally, this approach provides the robustness relative to weak geometrical discrepancy of images.

3.3.4 Manifold Learning, Diffusion Maps and Shape Matching

The application of heat kernels and diffusion maps to shape analysis is initially inspired by the manifold learning technique developed for *NonLinear Dimensionality Reduction* (NLDR). Most interesting NLDR techniques are the following: Isomap [86], Locally Linear Embedding (LLE) [87], kernel PCA [88], Laplacian eigenmaps [89], Hessian LLE [90], manifold sculpting [91] and some other. The terms "heat kernel" and "heat dissipation" were introduced in [89] in the context of Laplacian eigenmap. In this concept, they play the role of some manifold shape characteristics. Based on this, the authors of [92–94] introduced and developed the theory of Diffusion Maps (DMs).

Let the manifold be described by some set of points $\mathbf{X} = \{\mathbf{x}_i\}$ in a high-dimensional space. The solution of the NLDR problem in the DM approach has a following form:

- Generate a neighborhood graph G.
- Form a heat kernel (matrix of pairwise similarity weights) $H = \|h_{ij}\|$ using the rule. If ith and jth points are connected in G, then

$$h_{ij} = \exp\left(-\frac{\|x_i - x_j\|}{2\sigma^2}\right),$$

else $h_{ij} = 0$.

- Normalize the heat kernel, and obtain the *diffusion kernel*

$$\mathbf{P} = \mathbf{M}^{-1}\mathbf{H},$$

where $\mathbf{M} = \|m_{ij}\|$ is a diagonal matrix of column sums for $\mathbf{H}$: $m_{ii} = \sum_i h_{ij}$.
- Select the scale parameter t and form the t-degree diffusion matrix $\mathbf{P}^t$, $t \geq 1$.
- Compute the spectral decomposition of $\mathbf{P}^t$ with eigenvalues

$$1 = \lambda_0 \geq \lambda_1 \geq \lambda_2 \geq \cdots$$

and corresponding eigenfunctions $\{\psi_i\}$.
- Map the data to the low-dimensional vector space via selection of l eigenvalues and forming new coordinates based on corresponding eigenfunctions:

$$x_i = [\lambda_1 \psi_{i1} \quad \lambda_2 \psi_{i2} \quad \cdots \quad \lambda_l \psi_{il}]^T.$$

The normalization of heat kernel provides here both the linear smoothing *diffusion operator P^t* and its interpretation in terms of *Markov chain* with *transition matrix $\mathbf{P}^t$* for t steps of random walking. Thus, the Euclidean distance in this new space can be interpreted as a probability of point-to-point transition in t steps of this random walking. Such distance is called a *diffusion distance*, and it is very popular now in the area of high dimensional data analysis and machine learning.

Being inspired by the NLDR task, the DM approach was later successfully applied for other types of data analysis problems. In particular, the following image restoration technique was outlined in [94]. Let $I(p)$ be a 2D image with $p = (x_p, y_p)$ and let pixels be described by some feature vector $v(p)$. Then, for given $\varepsilon > 0$ the *diffusion kernel* is defined as:

$$A_{p,q} = \frac{\exp - \frac{\|v(p)-v(q)\|^2}{\varepsilon}}{\sum_q \exp \frac{-\|v(p)-v(q)\|^2}{\varepsilon}}.$$

The filtering of image I by this *diffusion filter* has a usual form:

$$I_A(p) = \sum_q A_{p,q} I(q).$$

Figure 3.2 demonstrates the examples of the TV and IR images diffusion filtering. In this case, like in paper [94], $v(p)$ is just a 5×5 vector of grayscale image values in a 5×5 pixel neighborhood. Both for the TV and the IR images, the shape was preserved and the noise was essentially removed. The reason of such success in noise suppression is an adaptive smoothing with high kernel weights for similar neighbors and low or zero weights for dissimilar.

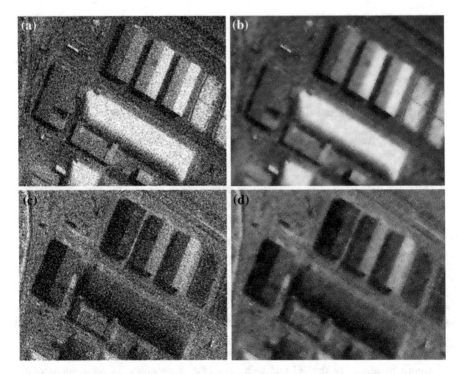

Fig. 3.2 Examples of diffusion filtering for denoising images: **a** TV noisy image, **b** TV filtered image, **c** IR noisy image, **d** IR filtered image

Note that idea of linear smoothing with adaptive kernel is utilized in different image restoration techniques, for example in [95, 96]. At the same time, the DM approach provides a unified way for the shape description, shape comparison, and shape-based image restoration.

Let us look at the data shape matching techniques based on the heat kernels, diffusion maps, and their spectral features. A number of such approaches including Heat Kernel Signature (HKS), heat kernel spectrum (set of eigenvalues), Heat Kernel Signature Distribution (HKSD), and Heat Trace (HT) are proposed and developed in [97–100] and other papers. In the brilliant the overview [101], the classification and unified mathematical description of these techniques are given. Moreover, the special Gromov-Wasserstein distances are proposed for shape matching based on such spectral characteristics, and the stability (robustness) of such distances and matching procedures are theoretically proved [101]. These theoretical results are supported by impressive experiments with 3D models and real data collections. These results allow supposing that the combination of the DM and the MIA approaches could create some effective 2D image and shape matching tools.

3.3.5 Generalized Diffusion Morphology and Comparative Filters Based on Diffusion Operators

Let the image be a 2D function

$$f(x, y) : \Omega \to R, \quad \Omega \subset R^2,$$

where R is a set of real numbers, R^2 is an image plane, Ω is a rectangular frame region of image plane. Images are elements of Hilbert space $L^2(\Omega)$ with *scalar product* (f, g) and *norm* $\|f\| = (f, f)^{1/2}$.

Let us introduce the generalized definitions of basic MIA notions by substitution of "mosaic shape" to "diffusion shape" in order to obtain the relaxed version of the MIA. The *relational model* of *diffusion shape* F for image f is a heat kernel $h_F(x, y, u, v): \Omega \times \Omega \to [0,1]$, such that

$$h_F(x, y, x, y) \geq h_F(x, y, u, v) \, h_F(x, y, u, v) \geq h_F(u, v, x, y), \tag{3.15}$$

and the unique *basic similarity measurement function* η exists providing

$$\eta(f(x, y), f(u, v)) = h_F(x, y, u, v).$$

The *operator model* of *diffusion shape* F is a diffusion operator P_F with normalized diffusion kernel $p_F(x, y, u, v)$ given by Eq. 3.16.

$$\begin{aligned}
&P_F g(x, y) = \iint_{\Omega} p_F(x, y, u, v) \, g(u, v) du \, dv \\
&p_F(x, y, u, v) \geq 0 \quad p_F(x, y, x, y) \geq p_F(x, y, u, v) \\
&\iint_{\Omega} p_F(x, y, u, v) du \, dv = 1 \\
&\iiiint_{\Omega \times \Omega} \int p_F^2(x, y, u, v) \, dx \, dy \, du \, dv < \infty
\end{aligned} \tag{3.16}$$

Thus, any relational model h_F corresponds to the unique operator model with normalized kernel p_F:

$$p_F(x, y, u, v) = h_F(x, y, u, v) / \iint_{\Omega} h_F(x, y, a, b) da \, db \,.$$

The *spatial model* of *diffusion shape* F for image $f(x, y)$ with precision n is an eigenspace of diffusion operator P_F

$$F = span\{\lambda_1 \varphi_1(x, y), \ldots, \lambda_n \varphi_n(x, y)\},$$

where $\{\lambda_1, \ldots, \lambda_n\}$ are n first eigenvalues, $\{\varphi_1(x, y), \ldots, \varphi_n(x, y)\}$ are n first eigenfunctions of *morphological diffusion operator* P_F:

$$P_F\varphi_i(x,y) = \lambda_i\varphi_i(x,y) \quad i = 1,\ldots,n$$

It is easy to see that, in particular case, a heat kernel is represented in the form of Eq. 3.17.

$$h_F^*(x,y,u,v) = \begin{cases} 1 & \text{if } f(x,y) = f(u,v) \\ 2 & \text{otherwise} \end{cases} \tag{3.17}$$

It expresses the binary relation "points of equal values", and the diffusion-based morphological definitions stated above degrade to the MIA definitions given in Sect. 3.2. Really, in this case the similarity relation η becomes an equivalent relation in the image frame points splitting them to a set of non-overlapping regions $\mathbf{F} = \{F_1, \ldots, F_n\}$, where n is a number of regions of frame tessellation $\mathbf{F}$. Hence, the morphological diffusion operator becomes a morphological projector $P_F = P_F\,P_F$ with n 1-valued eigenvalues and characteristic (support) functions $\varphi_i(x,y) = \chi_{Fi}(x,y)$ (Eq. 3.2) of n regions as eigenfunctions. Thus, the diffusion shape in this particular case becomes a mosaic shape F in the form of Eq. 3.3, and for any $g(x,y) \in L^2(\Omega)$ the diffusion filtering P_F becomes a morphological projection onto the mosaic shape F of the classic form (Eq. 3.4). Thus, the Pyt'ev MIA approach is a particular case of generalized *diffusion morphology* described in this section. Therefore, all shape analysis schemes and tools of the MIA can be recovered on this wider basis just using the diffusion operator instead of Pyt'ev morphological projector.

Let us note that the formal definition of diffusion morphology can start directly from a heat kernel of the classic form

$$h_{ij} = \exp(-d_{ij}/\varepsilon),$$

where d_{ij} is a *basic distance* between ith and jth points of discrete digital image.

In this formulation of diffusion Morphology, the MIA case corresponds to a special selection of basic distance as a *discrete distance by pixel value*:

$$d_{ij} = \begin{cases} 1 & \text{if } f(x_i,y_i) = f(x_j,y_j) \\ \infty & \text{otherwise} \end{cases}.$$

Then the heat kernel takes a form $h_{ij} = h_F*(x_i, y_i, u_j, v_j)$ of binary relation "points of equal values" that forces the transformation of diffusion operator to the mosaic projector.

The final note here is that different choice of descriptors and metrics for pixel or/and region comparison provides the design of different diffusion morphologies with different semantic properties. Additionally, one can use the diffusion filter P^t with scale parameter t for morphological scale space analysis.

Thus, this generalized morphological approach provides more information about the image shape than just the information about shape of frame tessellation exploited by the MI or original MIA. And if this new information is robust relative to noise, it will support the higher quality of matching.

In general case, a diffusion model (Eq. 3.15) is formed as a heat kernel proposed in [94]:

$$h_f(x, y, u, v) = \exp\left(-\frac{\left\| \mathbf{v}_f(x, y) - \mathbf{v}_f(u, v) \right\|^2}{\varepsilon}\right), \qquad (3.18)$$

where $\mathbf{v}_f(x, y)$ is a feature vector describing the sample image $f(x, y)$ in some neighborhood (local window) $w(x, y)$ around the point (x, y), $\varepsilon > 0$ is a tuning parameter controlling the sensibility to feature vectors similarity. However, the diffusion filtering with such heat kernel (Eq. 3.18) is a time-consuming procedure. Thus, we prefer the diffusion filters proposed in [9] and based on point feature descriptor iLBP (intensity + LBP):

$$iLBP(x, y) = (m(x, y), LBP(x, y)), \qquad (3.19)$$

where $m(x, y)$ is a mean value of image in a window $w(x, y)$, $LBP(x, y)$ is a threshold Local Binary Pattern (LBP) [102] calculated as a binary vector for central pixel (x, y) based on a comparison of its value and values of its neighbors in a window $w(x, y)$. If the value of neighbor pixel is less than the value of central pixel and the difference between them is greater than threshold, then the corresponding bit is set to 1, otherwise it is set to 0. Correspondingly, the heat kernel is represented by Eq. 3.20, where d_{ham} is Hamming distance, β is a tuning parameter balancing the importance of intensity and LBP parts in iLBP.

$$h_f(x, y, u, v) = \exp(-\beta |m_f(x, y) - m_f(u, v)|^2 \\ - d_{ham}(LBP_f(x, y), LBP_f(u, v)) \qquad (3.20)$$

The use of the iLBP allows both increasing the computational speed and obtaining heat kernels very similar to Eq. 3.18.

As stated above, the projective mapping $\psi(f, g) = P_F\, g$ (Eq. 3.4) satisfies to definition of morphological comparative filter and properties (Eq. 3.9). From the functional point of view, the generalized diffusion mapping $\psi(f, g) = P_f\, g$ (Eq. 3.16) should be a kind of comparative filtering too because it is a smoothing filter and it transforms the test image in accordance with diffusion shape of sample image. Unfortunately, it does not formally match the exact matching property (Eq. 3.9). However, in this case we can talk about the *soft matching* property: the diffusion filters preserve images of sample shape essentially better than the images of other shapes. So, we may refer the diffusion filters as *soft comparative filters* and use the corresponding *diffusion difference map* in the form of Eq. 3.11 for solution of change detection task.

3.3.6 Image and Shape Matching Based on Diffusion Morphology

In Pyt'ev morphology, the comparison of image $g(x, y)$ and shape of image $f(x, y)$ is performed using the normalized morphological correlation coefficient of the following form

$$K_M(g, F) = \frac{\|P_F g\|}{\|g\|},$$

and this coefficient satisfies the property $K_M(f, F) = 1$ due to the fact that $f = P_F f$. The diffusion morphological operator is not a projector, but it is a smoothing filter with $\|P_F f\| \leq \|f\|$. Nevertheless, it is natural to suppose that the smoothing power of P_F will be essentially less for images with similar shapes than for images with different shapes. So, the *Morphological Diffusion Correlation Coefficient* (MDCC) is defined as a ratio of Pyt'ev morphological coefficients

$$K_{MD}(g, F) = \frac{K_M(g, F)}{K_M(f, F)} = \frac{\|P_F g\| \, \|f\|}{\|P_F f\| \, \|g\|},$$

where $K_M(f, F)$ describes the power of self-smoothing of f by F. Note that the MDCC is a correct generalization of the MCC because in case of projective morphology $\|f\| = \|P_F f\|$ and $K_{MD}(g, F) = K_M(g, F)$.

As in the MCC, for elimination of non-informative part of image brightness images should be normalized before comparison:

$$K_{MD}(g, F) = \frac{\|P_F g - P_O g\| \, \|f - P_O f\|}{\|P_F f - P_O f\| \, \|g - P_O g\|},$$

where $P_O f$ is a morphological filtering of image f by the "empty" diffusion shape O. However, in diffusion morphology such normalization is not the trivial subtraction of global mean value (like in Pyt'ev MIA). It is a subtraction of mean value in a sliding window determined by support neighborhood (effective size) of heat kernel. This subtraction preserves the local informative features only (in the corresponding scale of analysis). These informative elements of image g will be passed (if the shape G is similar to shape F) or extremely smoothed (if the shape G is essentially different) by the diffusion filter P_F. This trick is called *morphological image normalization* (Fig. 3.3).

If the effective size of heat kernel is small, such diffusion image-to-shape matching technique uses the local features only like the points-based and contour-based matching techniques. Fortunately, as we stated above, the theory of diffusion maps has a natural instrument for multi-scale data analysis—parameter t (number of Markov random walking steps). The description of the image shape by

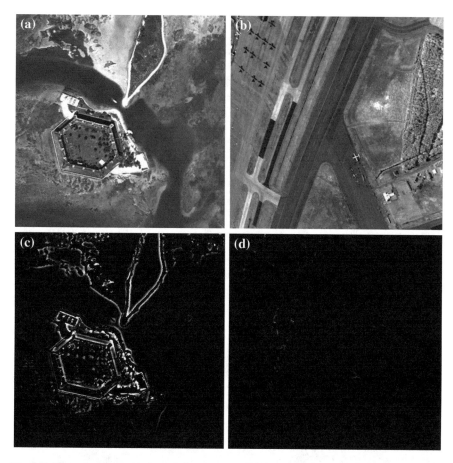

Fig. 3.3 Example of morphological image normalization: **a** image f, **b** image g, **c** image f normalized by self-shape $F(P_F f - P_O f)$, **d** image g normalized by shape $F(P_F g - P_O g)$

the set of different scale diffusion operators $\{P^t\}$ allows performing the *morphological scale-space analysis*.

Finally, as well as for Pyt'ev correlation coefficient, we prefer to use the square of the MDCC instead of the MDCC because both $K_M^2(g, F)$ and $K_{MD}^2(g, F)$ can be interpreted as statistical *coefficient of determination* between the model (shape F) and observed data (image g). Thus, the generalized morphological technique for image-to-shape matching is formed.

3.3.7 Change Detection Pipeline Based on Comparative Filtering

For the task of change detection in long-range (spaceborne or airborne) remote sensing data, we propose the new change detection pipeline based on comparative filtering. It contains the following steps (Fig. 3.4):

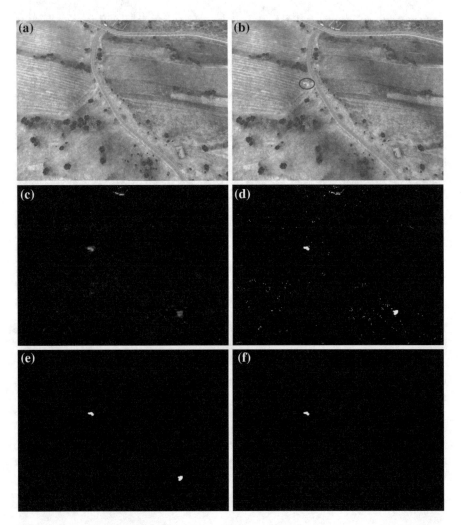

Fig. 3.4 Stages of change detection pipeline: **a** reference image, **b** test image, **c** morphological difference map (MDM), **d** binarized MDM, **e** change proposals (regions of filtered binarized MDM), **f** accepted change proposals

Step 1. Comparative filtering using the image pyramid (Eq. 3.7).
Step 2. Calculation of corresponding morphological difference map (Eq. 3.11).
Step 3. Binarization and filtering of morphological difference map.
Step 4. Forming change proposals.
Step 5. Testing change proposals using local morphological correlation coefficient (Eq. 3.5).
Step 6. Forming the output binary map of changes.

At the first step of pipeline, we apply the morphological comparative filters described above. The pipeline is the same both for guided contrasting (Eqs. 3.12–3.13) and for diffusion filtering (Eqs. 3.16–3.20). The use of image pyramid allows detecting details of different size. Morphological difference map obtained at second step is binarized using a graph cut technique [103]. Then, this binary image is filtered sequentially by morphological closing and opening filters with small disk-shaped structured element [84]. Such filtering allows to delete noisy regions of binarized difference map.

The list of change proposals is formed via calculation of minimal bounding rectangles for all connected regions of filtered binarized difference map. Then the each change proposal is checked by the value of local MCC (Eq. 3.5) compared with a pre-learned threshold. Local MCC is calculated in a weakly expanded proposal rectangle. The expansion of rectangle is performed in order to add some small neighbourhood of the proposed object. If the value of the MCC is greater than threshold, then corresponding connected region of filtered binarized difference map is painted in the output binary map of relative changes.

Note that this scheme is asymmetrical such that it provides the detection of changes in (new) test image g relative to (old) reference image f. If we need to find all differences in both images, it is required to repeat the procedure two times: first, using g as test and f as reference, and, second, using g as reference and f as a test.

We also need to comment the presence of traditional MCC in our pipeline. We have stated above that the MCC is not stable and/or fast due to problems with image segmentation. But this problem exists at the global image scale. Coarse or unstable segmentation leads to false object proposals or lost objects. However, the mosaic shape of small local areas containing the change proposals is simple enough. The simplest histogram-based segmentation with small number of levels provides local segmentation that looks fine for proposals testing. The properties of morphological projector are essentially different from the properties of comparative filters. Thus, testing proposals created by comparative filters via the MCC means the combination of evidences from independent information sources. Such combination is preferable from the statistical point of view. It provides the more reliable detection results.

3.4   Experiments

The results of experimental exploration of both comparative filtering and proposed change detection pipeline are reported in this section. In Sect. 3.4.1, some examples of guided contrasting and corresponding morphological difference map forming are demonstrated applying to real images for different scene types and change detection cases. In Sect. 3.4.2, the results of change detection experiments on the public benchmark containing simulated aerial images are described.

3.4.1   Qualitative Change Detection Experiments

A lot of qualitative experiments with comparative filters based on guided contrasting are performed on a wide set of real images. Different types of scenes and image acquisition conditions are considered. Figures 3.5, 3.6 and 3.7 demonstrate some examples of morphological difference map forming based on comparative guided contrasting filtering. Different columns illustrate scenes and changes of different types. In the first column of Fig. 3.5, the example demonstrates the building construction case that requires comparison of buildings at different stages of construction based on images captured in different weather and season conditions from the close but not exactly the same viewpoint. The second column of Fig. 3.5 shows the example of changes in indoor (in-office) close-range scene. The third column of Fig. 3.5 shows the morphological difference maps. As a result, some new objects appear on a table in a test image.

In the Fig. 3.6, the examples of outdoor video surveillance change detection case are shown. Such case presumes the short-term changes in illumination conditions combined with presence of moving or appeared/disappeared objects. Figure 3.7 demonstrates the most popular case of change detection task in remote sensing imagery (spaceborne or airborne). Such qualitative experiments allow concluding that in all considered cases the proposed approach for difference map forming based on guided contrasting filtering provides reasonable scene change proposals and demonstrates the enough robustness relative to changes in lighting and other image capturing conditions. At the same time, some true image shape changes are extracted, which are not required to be detected as the scene changes from the semantical point of view. For example, clouds and flowers in a building construction case really appear in a scene but they should not be of interest regarding the building construction stage comparison. Thus, some additional analysis of formed morphological difference map is needed for final testing of the formed change proposals based on other type of task-specific information. It means that guided contrasting filtering and corresponding morphological difference maps can be useful as the parts of different task-oriented change detection pipelines. Such pipeline for change detection in remote sensing data is described above. The results of its testing are reported in next Sect. 3.4.2.

Fig. 3.5 Examples of morphological difference maps based on guided contrasting: **a** and **b** reference images, **c** and **d** test images, **e** and **f** morphological difference maps

3.4.2 Quantitative Change Detection Experiments

In our experiments with proposed change detection pipeline for long-range remote sensing, we use the public Change Detection dataset introduced in [103] (Fig. 3.8). This dataset contains 1000 pairs of 800 × 600 simulated aerial images and 1000 corresponding 800 × 600 ground truth masks. Each pair consists of one reference image and one test image. Some of image pairs contain the scene changes and illumination differences. The dataset consists of 100 different scenes with moderate surface relief and several objects (trees, Buildings, etc.). Each scene is rendered

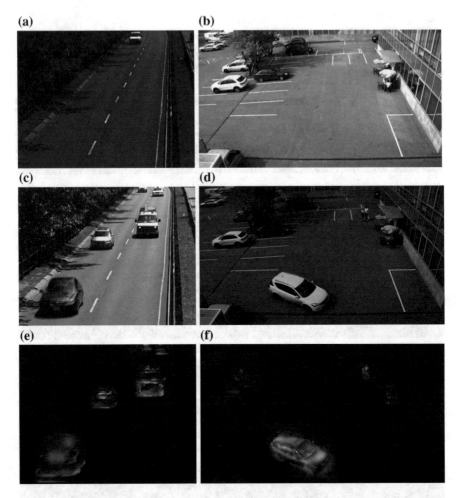

(a)   **(b)**

(c)   **(d)**

(e)   **(f)**

Fig. 3.6 Examples of morphological difference maps based on guided contrasting: **a** and **b** reference images, **c** and **d** test images, **e** and **f** morphological difference maps

with various viewpoints. The cameras are distributed at steps of 10° on a circle of radius 100 m at approximately 250 m high, and with a fixed tilt of about 70°. All images are modelled with a ground resolution of about 50 cm per pixel.

The methodology of our experiments is the following. We select a subset of 100 reference and test image pairs for 50 different scenes with 0° relative camera angle. As proposed in [103], we compare the detection results with respect to the ground truth at pixel level but calculate the precision and recall values at the object (region) level. In order to do this, we form the list of ground truth objects and list of detected objects (accepted regions of filtered binarized morphological difference map). Then we perform the object-to-object comparison via computing of object intersection area. If the intersection area is more than 50%, then we decide that objects match

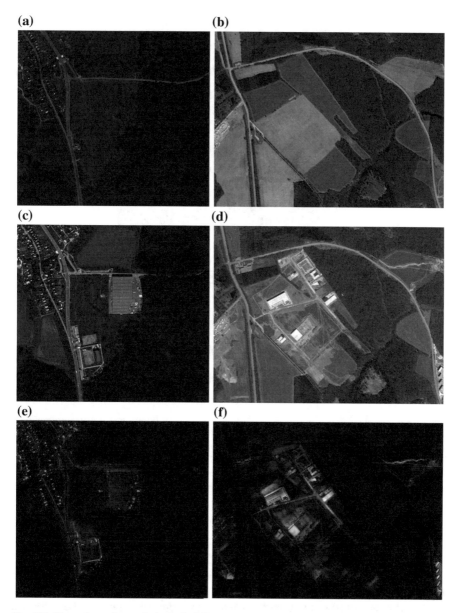

Fig. 3.7 Examples of morphological difference maps based on guided contrasting: **a** and **b** reference images, **c** and **d** test images, **e** and **f** morphological difference maps

each other. The numbers of true and false object detections determine the corresponding precision and recall values.

We implement and test our pipeline with following parameters: the guided contrasting window size is 7 × 7 pixels, a number of pyramid levels is 3, the size of

(a)   (b)

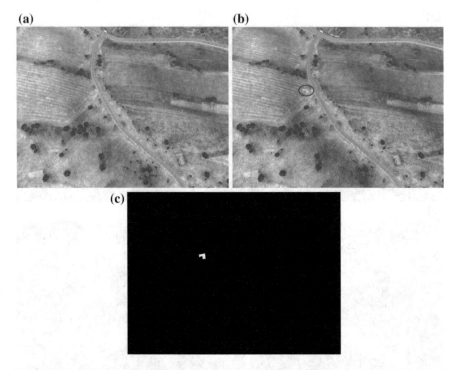

(c)

Fig. 3.8 Example of simulated data from benchmark: **a** reference image, **b** test image, **c** ground truth mask

disk structuring element in the MM opening and closing is 5 pixels, the threshold value for morphological correlation coefficient at the final testing step is 0.5.

The obtained results are:

- For change detection based on diffusion morphological filtering, Precision = 0.61 and Recall = 0.6.
- For change detection based on guided contrasting, Precision = 0.6 and Recall = 0.64.

The reported result of approach [103] on this database is about Precision = 0.51 and Recall = 0.52 (by the graph). It is not the totally correct comparison because this chapter utilizes a little bit different testing methodology but it seems reasonable to state that our results are at least not worse than competing ones. Therefore, we can conclude that the proposed pipeline is useful for change detection in long-range remote sensing data.

3.5 Conclusions

This chapter presents the new scheme of morphological comparison filtering that generalizes the MIA scheme proposed by Pyt'ev. Such comparative morphologies are the mathematical shape theories, which solve the tasks of the image similarity estimation, image matching, and change detection by means of some special morphological models and tools. Our CM scheme excludes the MIA ideas of shape and projection but preserves the idea of asymmetrical comparative filtering for robust similarity estimation and change detection. For implementation of this scheme, we propose a new class of morphological filters based on guided contrasting. The idea of guided contrasting is that the local contrast (energy) of filtered test image is controlled by its local similarity with reference image. If we use the similarity estimation with local search, then such filters demonstrate the robustness relative to weak geometrical discrepancy of compared images. Another essential advantage of guided contrasting filters relative to traditional morphological projectors is the elimination of unstable image segmentation step.

Also the new change detection pipeline based on comparative filtering is proposed for analysis of bi-temporal long-range (spaceborne or airborne) remote sensing data. This pipeline contains a comparative filtering on the image pyramid, calculation of morphological difference map, binarization, extraction of change proposals, and testing change proposals using local morphological correlation coefficient. We implement this pipeline based on both guided contrasting filters and morphological diffusion filters for shape-based matching.

Qualitative experiments with guided contrasting filtering in different change detection tasks demonstrate that they provide the reasonable scene change proposals and demonstrate the enough robustness relative to changes in lighting and other image acquisition conditions. Quantitative experiments on the public benchmark, containing simulated aerial images, demonstrate that the proposed pipeline is useful for change detection in long-range remote sensing data.

The future work on proposed morphological schemes will be connected with development and implementation of new comparative filters and new change detection pipelines for different types of tasks. In particular, new guided contrasting filters can be obtained via different combinations of several smoothing procedures and local similarity estimators. The main experimental work will consist in massive testing of implemented comparative filters and corresponding change detection pipelines on the large datasets containing both simulated and real images.

Acknowledgements This work was supported by Russian Science Foundation (RSF), Grant 16-11-00082.

References

1. Singh, A.: Review article digital change detection techniques using remotely-sensed data. Int. J. Remote Sens. **10**(6), 989–1003 (1989)
2. Lu, D., Mausel, P., Brondízioc, E., Moran, E.: Change detection techniques. Int. J. Remote Sens. **25**(12), 2365–2401 (2004)
3. Chen, J., Lu, M., Chen, X., Chen, J., Chen, L.: A spectral gradient difference based approach for land cover change detection. ISPRS J. Photogramm. Remote Sens. **85**, 1–12 (2013)
4. Hussain, M., Chen, D., Cheng, A., Wei, H., Stanley, D.: Change detection from remotely sensed images: from pixel-based to object-based approaches. ISPRS J. Photogramm. Remote Sens. **80**, 91–106 (2013)
5. Pyt'ev, Y.P.: Morphological image analysis. Pattern Recognit. Image Anal. **3**(1), 19–28 (1993)
6. Vizilter, Y., Pyt'ev, Y., Chulichkov, A., Mestetskiy, L.: Morphological image analysis for computer vision applications. In: Favorskaya, M.N., Jain, L.C. (eds.) Computer Vision in Control Systems-1. Mathematical Theory. Intelligent Systems Reference Library. vol. 73, pp. 9–58. Springer International Publishing, Switzerland (2015)
7. He, K., Sun, J., Tang, X.: Guided image filtering. In: Daniilidis, K., Maragos, P., Nikos Paragios, N. (eds.) Computer Vision—ECCV 2010, LNCS, vol. 6311, pp. 1–14. Springer, Heidelberg (2010)
8. Zhang, Q., Shen, X., Xu, L., Jia, J.: Rolling guidance filter. In: Fleet, D., Pajdla, T., Schiele, B., Tuytelaars, T. (eds.) Computer Vision—ECCV 2014, LNCS, vol. 8691, pp. 815–830. Springer, Heidelberg (2014)
9. Vizilter, Y.V., Gorbatsevich, V.S., Rubis, A.Y., Zheltov, S.Y.: Shape-based image matching using heat kernels and diffusion maps. Int. Arch. Photogramm. Remote Sens. Spat. Inf. Sci. **XL**(3), 357–364 (2014)
10. Vizilter, Y.V., Rubis, A.Y., Zheltov, S.Y., Vygolov, O.V.: Change detection via morphological comparative filters. ISPRS Ann. Photogramm. Remote Sens. Spat. Inf. Sci. **III**(3), 279–286 (2016)
11. Quarmby, N.A., Cushnie, J.L.: Monitoring urban land cover changes at the urban fringe from SPOT HRV imagery in south-east England. Int. J. Remote Sens. **10**(6), 953–963 (1989)
12. Coppin, P.R., Bauer, M.E.: Digital change detection in forest ecosystems with remote sensing imagery. Remote Sens. Rev. **13**, 207–234 (1996)
13. Lu, D., Mausel, P., Batistella, M., Moran, E.: Land-cover binary change detection methods for use in the moist tropical region of the Amazon: a comparative study. Int. J. Remote Sens. **26**(1), 101–114 (2005)
14. Gao, J.: Digital Analysis of Remotely Sensed Imagery. McGraw-Hill, New York (2009)
15. Howarth, P.J., Wickware, G.M.: Procedures for change detection using Landsat digital data. Int. J. Remote Sens. **2**(3), 277–291 (1981)
16. Ludeke, A.K., Maggio, R.C., Reid, L.M.: An analysis of anthropogenic deforestation using logistic regression and GIS. J. Environ. Manag. **31**(3), 247–259 (1990)
17. Lunetta, R.S.: Applications, project formulation, and analytical approach. In: Lunetta, R.S., Elvidge, C.D. (eds.) Remote Sensing Change Detection: Environmental Monitoring Methods and Applications, pp. 1–19. Taylor & Francis, London (1999)
18. Chen, J., Gong, P., He, C., Pu, R., Shi, P.: Land-use/land-cover change detection using improved change-vector analysis. Photogramm. Eng. Remote Sens. **69**(4), 369–379 (2003)
19. Nackaerts, K., Vaesen, K., Muys, B., Coppin, P.: Comparative performance of a modified change vector analysis in forest change detection. Int. J. Remote Sens. **26**(5), 839–852 (2005)
20. Bayarjargal, Y., Karnieli, A., Bayasgalan, M., Khudulmur, S., Gandush, C., Tucker, C.J.: A comparative study of NOAA–AVHRR derived drought indices using change vector analysis. Remote Sens. Environ. **105**(1), 9–22 (2006)

21. Richards, J.A.: Thematic mapping from multitemporal image data using the principal components transformation. Remote Sens. Environ. **16**(1), 35–46 (1984)
22. Deng, J.S., Wang, K., Deng, Y.H., Qi, G.J.: PCA-based land-use change detection and analysis using multitemporal and multisensor satellite data. Int. J. Remote Sens. **29**(16), 4823–4838 (2008)
23. Kauth, R.J., Thomas, G.S.: The tasselled cap—a graphic description of the spectral-temporal development of agricultural crops as seen by Landsat. In: Symposium on Machine Processing of Remotely Sensed Data, vol. 4(B), pp. 41–51 (1976)
24. Jin, S., Sader, S.A.: Comparison of time series tasseled cap wetness and the normalized difference moisture index in detecting forest disturbances. Remote Sens. Environ. **94**(3), 364–372 (2005)
25. Rogan, J., Franklin, J., Roberts, D.A.: A comparison of methods for monitoring multitemporal vegetation change using thematic mapper imagery. Remote Sens. Environ. **80**(1), 143–156 (2002)
26. Tomowski, D., Ehlers, M., Klonus, S.: Colour and texture based change detection for urban disaster analysis. In: Joint Urban Remote Sensing Event (JURSE'2011), pp. 329–332 (2011)
27. Erbek, F.S., Özkan, C., Taberner, M.: Comparison of maximum likelihood classification method with supervised artificial neural network algorithms for land use activities. Int. J. Remote Sens. **25**(9), 1733–1748 (2004)
28. Bouziani, M., Goïta, K., He, D.-C.: Automatic change detection of buildings in urban environment from very high spatial resolution images using existing geodatabase and prior knowledge. ISPRS J. Photogramm. Remote Sens. **6591**, 143–153 (2010)
29. Im, J., Jensen, J.R.: A change detection model based on neighborhood correlation image analysis and decision tree classification. Remote Sens. Environ. **99**(3), 326–340 (2005)
30. Jensen, J.R.: Introductory Digital Image Processing: A Remote Sensing Perspective. Prentice Hall, Toronto (2005)
31. Miller, A.B., Bryant, E.S., Birnie, R.W.: An analysis of land cover changes in the Northern Forest of New England using multitemporal Landsat MSS data. Int. J. Remote Sens. **19**(2), 245–265 (1998)
32. Yuan, F., Sawaya, K.E., Loeffelholz, B.C., Bauer, M.E.: Land cover classification and change analysis of the Twin Cities (Minnesota) Metropolitan Area by multitemporal landsat remote sensing. Remote Sens. Environ. **98**(2–3), 317–328 (2005)
33. Ji, W., Ma, J., Twibell, R.W., Underhill, K.: Characterizing urban sprawl using multi-stage remote sensing images and landscape metrics. Comput. Environ. Urban Syst. **30**(6), 861–879 (2006)
34. Ghosh, A., Mishra, N.S., Ghosh, S.: Fuzzy clustering algorithms for unsupervised change detection in remote sensing images. Inf. Sci. **181**(4), 699–715 (2011)
35. Chan, J.C.-W., Chan, K.-P., Yeh, A.G.-O.: Detecting the nature of change in an urban environment: a comparison of machine learning algorithms. Photogramm. Eng. Remote Sens. **67**(2), 213–225 (2001)
36. Dai, X.L., Khorram, S.: Remotely sensed change detection based on artificial neural networks. Photogramm. Eng. Remote Sens. **65**(10), 1187–1194 (1999)
37. Lillesand, T.M., Kiefer, R.W., Chipman, J.W.: Remote Sensing and Image Interpretation, 6th edn. Wiley, Hoboken, NJ (2008)
38. Lunetta, R.S., Knight, J.F., Ediriwickrema, J., Lyon, J.G., Worthy, L.D.: Land-cover change detection using multi-temporal MODIS NDVI data. Remote Sens. Environ. **105**(2), 142–154 (2006)
39. Mas, J.F.: Monitoring land-cover changes: a comparison of change detection techniques. Int. J. Remote Sens. **20**(1), 139–152 (1999)
40. Collins, J.B., Woodcock, C.E.: An assessment of several linear change detection techniques for mapping forest mortality using multitemporal landsat TM data. Remote Sens. Environ. **56**(1), 66–77 (1996)
41. Schowengerdt, R.A.: Techniques for Image Processing and Classification in Remote Sensing. Academic Press, New York (1983)

42. Gopal, S., Woodcock, C.: Remote sensing of forest change using artificial neural networks. IEEE Trans. Geosci. Remote Sens. **34**(2), 398–404 (1996)
43. Abuelgasim, A.A., Ross, W.D., Gopal, S., Woodcock, C.E.: Change detection using adaptive fuzzy neural networks: environmental damage assessment after the Gulf war. Remote Sens. Environ. **70**(2), 208–223 (1999)
44. Woodcock, C.E., Macomber, S.A., Pax-Lenney, M., Cohen, W.B.: Monitoring large areas for forest change using Landsat: generalization across space, time and Landsat sensors. Remote Sens. Environ. **78**(1–2), 194–203 (2001)
45. Liu, X., Lathrop, R.G.: Urban change detection based on an artificial neural network. Int. J. Remote Sens. **23**(12), 2513–2518 (2002)
46. Pijanowski, B.C., Pithadia, S., Shellito, B.A., Alexandridis, K.: Calibrating a neural network-based urban change model for two metropolitan areas of the Upper Midwest of the United States. Int. J. Geograph. Inf. Sci. **19**(2), 197–215 (2005)
47. Vapnik, V.N.: The Nature of Statistical Learning Theory, 2nd edn. Springer, New York (2000)
48. Huang, C., Song, K., Kim, S., Townshend, J.R.G., Davis, P., Masek, J.G., Goward, S.N.: Use of a dark object concept and support vector machines to automate forest cover change analysis. Remote Sens. Environ. **112**(3), 970–985 (2008)
49. Bovolo, F., Bruzzone, L., Marconcini, M.: A novel approach to unsupervised change detection based on a semisupervised SVM and a similarity measure. IEEE Trans. Geosci. Remote Sens. **46**(7), 2070–2082 (2008)
50. Nemmour, H., Chibani, Y.: Multiple support vector machines for land cover change detection: an application for mapping urban extensions. ISPRS J. Photogramm. Remote Sens. **61**(2), 125–133 (2006)
51. Makkeasorn, A., Chang, N.-B., Li, J.: Seasonal change detection of riparian zones with remote sensing images and genetic programming in a semi-arid watershed. J. Environ. Manag. **90**(2), 1069–1080 (2009)
52. Pa, M.: Random forest classifier for remote sensing classification. Int. J. Remote Sens. **26**(1), 217–222 (2005)
53. Sesnie, S.E., Gessler, P.E., Finegan, B., Thessler, S.: Integrating Landsat TM and SRTM-DEM derived variables with decision trees for habitat classification and change detection in complex neotropical environments. Remote Sens. Environ. **112**(5), 2145–2159 (2008)
54. Smith, G.M.: The development of integrated object-based analysis of EO data within UK national land cover products object-based image analysis. In: Blaschke, T., Lang, S., Hay, G. J. (eds.) Object-Based Image Analysis, LNGC, pp. 513–528. Springer, Berlin, Heidelberg (2008)
55. Yang, Q., Li, X., Shi, X.: Cellular automata for simulating land use changes based on support vector machines. Comput. Geosci. **34**(6), 592–602 (2008)
56. Lefebvre, A., Corpetti, T., Hubert-Moy, L.: Object-oriented approach and texture analysis for change detection in very high resolution images. In: IEEE International Geoscience and Remote Sensing Symposium (IGARSS'2008), pp. IV.663–IV.666 (2008)
57. Zhou, W., Troy, A., Grove, M.: Object-based land cover classification and change analysis in the baltimore metropolitan area using multitemporal high resolution remote sensing data. Sensors **8**(3), 1613–1636 (2008)
58. Miller, O., Pikaz, A., Averbuch, A.: Objects based change detection in a pair of gray-level images. Pattern Recognit. **38**(11), 1976–1992 (2005)
59. Hall, O., Hay, G.J.: A multiscale object-specific approach to digital change detection. Int. J. Appl. Earth Observ. Geoinf. **4**(4), 311–327 (2003)
60. Chant, T.D., Kelly, M.: Individual object change detection for monitoring the impact of a forest pathogen on a hard wood forest. Photogramm. Eng. Remote Sens. **75**(8), 1005–1013 (2009)
61. Hazel, G.G.: Object-level change detection in spectral imagery. IEEE Trans. Geosci. Remote Sens. **39**(3), 553–561 (2001)

62. Li, J., Narayanan, R.M.: A shape-based approach to change detection of lakes using time series remote sensing images. IEEE Trans. Geosci. Remote Sens. **41**(11), 2466–2477 (2003)
63. Blaschke, T.: Towards a framework for change detection based on image objects. Göttinger Geographische Abhandlungen **113**, 1–9 (2005)
64. Durieux, L., Lagabrielle, E., Nelson, A.: A method for monitoring building construction in urban sprawl areas using object-based analysis of Spot 5 images and existing GIS data. ISPRS J. Photogramm. Remote Sens. **63**(4), 399–408 (2008)
65. Hansen, M.C., Loveland, T.R.: A review of large area monitoring of land cover change using Landsat data. Remote Sens. Environ. **122**, 66–74 (2012)
66. Holland, D.A., Sanchez-Hernandez, C., Gladstone, C.: Detecting changes to topographic features using high resolution imagery. In: XXXVII ISPRS Congress Proceedings, pp. 1153–1158 (2008)
67. Xian, G., Homer, C.: Updating the 2001 national land cover database impervious surface products to 2006 using Landsat imagery change detection methods. Remote Sens. Environ. **114**(8), 1676–1686 (2010)
68. Xian, G., Homer, C., Fry, J.: Updating the 2001 national land cover database land cover classification to 2006 by using Landsat imagery change detection methods. Remote Sens. Environ. **113**(6), 1133–1147 (2009)
69. Desclée, B., Bogaert, P., Defourny, P.: Forest change detection by statistical object-based method. Remote Sens. Environ. **102**(1–2), 1–11 (2006)
70. Gamanya, R., De Maeyer, P., De Dapper, M.: Object-oriented change detection for the city of Harare, Zimbabwe. Expert Syst. Appl. **36**(1), 571–588 (2009)
71. Bontemps, S., Bogaert, P., Titeux, N., Defourny, P.: An object-based change detection method accounting for temporal dependences in time series with medium to coarse spatial resolution. Remote Sens. Environ. **112**(6), 3181–3191 (2008)
72. Conchedda, G., Durieux, L., Mayaux, P.: An object-based method for mapping and change analysis in mangrove ecosystems. ISPRS J. Photogramm. Remote Sens. **63**(5), 578–589 (2008)
73. Stow, D., Hamada, Y., Coulter, L., Anguelova, Z.: Monitoring shrubland habitat changes through object-based change identification with airborne multispectral imagery. Remote Sens. Environ. **112**(3), 1051–1061 (2008)
74. Duveiller, G., Defourny, P., Desclée, B., Mayaux, P.: Deforestation in Central Africa: Estimates at regional, national and landscape levels by advanced processing of systematically-distributed Landsat extracts. Remote Sens. Environ. **112**(5), 1969–1981 (2008)
75. Al-Khudhairy, D.H.A., Caravaggi, I., Giad, S.: Structural damage assessments from Ikonos data using change detection, object-oriented segmentation, and classification techniques. Photogramm. Eng. Remote Sens. **71**(5), 825–837 (2005)
76. Niemeyer, I., Nussbaum, S.: Change detection: The potential for nuclear safeguards verifying treaty compliance. In: Avenhaus, R., Kyriakopoulos, N., Richard, M., Stein, G. (eds.) Verifying Treaty Compliance, pp. 335–348. Springer, Berlin Heidelberg (2006)
77. McDermid, G.J., Linke, J., Pape, A.D., Laskin, D.N., McLane, A.J., Franklin, S.E.: Object-based approaches to change analysis and thematic map update: challenges and limitations. Can. J. Remote Sens. **34**(5), 462–466 (2008)
78. Niemeyer, I., Marpu, P.R., Nussbaum, S.: Change detection using object features. In: Blaschke, T., Lang, S., Hay, G.J. (eds.) Object-Based Image Analysis: Spatial Concepts for Knowledge-Driven Remote Sensing Applications, LNGC, pp. 185–201. Springer, Berlin, Heidelberg (2008)
79. Pyt'ev, Y.P.: The morphology of color (multispectral) images. Pattern Recognit. Image Anal. **7**(4), 467–473 (1997)
80. Pyt'ev, Y.P., Falomkin, I.I., Chulichkov, A.I.: Morphological compression of grayscale images of text. Pattern Recognit. Image Anal. **16**(3), 523–528 (2006)
81. Evsegneev, S.O., Pyt'ev, Y.P.: Analysis and recognition of piecewise constant texture images. Pattern Recognit. Image Anal. **16**(3), 398–405 (2006)

82. Vizilter, Y.V., Zheltov, S.Y.: Geometrical correlation and matching of 2D image shapes. ISPRS Ann. Photogramm. Remote Sens. Spat. Inf. Sci. **I-3**, 191–196 (2012)
83. Pyt'ev, Y.P.: Oblique projectors and relative forms in image morphology. J. Comput. Math. Math. Phys. **53**(1), 1916–1937 (2013)
84. Serra, J.: Image Analysis and Mathematical Morphology. Academic Press, Inc., Orlando (1982)
85. Maes, F., Collignon, A., Vandermeulen, D., Marchal, G., Suetens, P.: Multimodality image registration by maximization of mutual information. IEEE Trans. Med. Imaging **16**(2), 187–198 (1997)
86. Tenenbaum, J.B., de Silva, V., Langford, J.C.: A global geometric framework for nonlinear dimensionality reduction. Science **290**(5500), 2319–2323 (2000)
87. Roweis, S.T., Saul, L.K.: Nonlinear dimensionality reduction by locally linear embedding. Science **290**(5500), 2323–2326 (2000)
88. Scholkopf, B., Smola, A.J., Muller, K.-R.: Kernel principal component analysis. In: Schölkopf, B., Burges, C.J.C., Smola, A.J. (eds.) Advances in Kernel Methods: Support Vector Learning, pp. 327–352. MIT Press (1999)
89. Belkin, M., Niyogi, P.: Laplacian eigenmaps and spectral techniques for embedding and clustering. Adv. Neural. Inf. Process. Syst. **14**, 585–591 (2001)
90. Donoho, D., Grimes, C.: Hessian eigenmaps: locally linear embedding techniques for high dimensional data. Proc. Natl. Acad. Sci. **100**(10), 5591–5596 (2003)
91. Gashler, M., Ventura, D., Martinez, T.: Iterative non-linear dimensionality reduction by manifold sculpting. Adv. Neural Inf. Process. Syst. **20**, 513–520 (2008)
92. Lafon, S.: Diffusion Maps and Geometric Harmonics. Ph.D. thesis, Yale University, Department of Mathematics & Applied Mathematics (2004)
93. Coifman, R., Lafon, S.: Diffusion maps. Appl. Comput. Harmon. Anal. **21**(1), 5–30 (2006)
94. Coifman, R., Lafon, S., Maggioni, M., Keller, Y., Szlam, A., Warner, F., Zucker, S.: Geometries of sensor outputs, inference and information processing. Proc. SPIE Intell. Integr. Microsyst. **6232** (2006). doi:10.1117/12.669723
95. Takeda, H., Farsiu, S., Milanfar, P.: Kernel regression for image processing and reconstruction. IEEE Trans. Image Process. **16**(2), 349–366 (2007)
96. Milanfar, P.: A tour of modern image filtering. IEEE Signal Process. Mag. **30**(1), 106–128 (2013)
97. Sun, J., Ovsjanikov, M., Guibas, L.: A concise and provably informative multi-scale signature based on heat diffusion. In: Eurographics Symposium on Geometry Processing, vol. 28, no. 5, pp. 1383–1392 (2009)
98. de Goes, F., Goldenstein, S., Velho, L.: A hierarchical segmentation of articulated bodies. Comput. Graph. Forum **27**(5), 1349–1356 (2008)
99. Lieu, L., Saito, N.: High dimensional pattern recognition using diffusion maps and Earth Mover's distance (2008). https://www.math.ucdavis.edu/∼saito/publications/saito_prdmemd.pdf. Accessed 7 July 2017
100. Reuter, M., Wolter, F.-E., Peinecke, N.: Laplace-Beltrami spectra as "Shape-DNA" of surfaces and solids. Comput. Aided Des. **38**(4), 342–366 (2006)
101. Memoli, F.: A spectral notion of Gromov-Wasserstein distance and related methods. App. Comput. Harmon. Anal. **30**(3), 363–401 (2011)
102. Ahonen, T., Hadid, A., Pietikainen, M.: Face recognition with local binary patterns. In: Pajdla, T., Matas, J. (eds.) Computer Vision—ECCV 2004, Part 1, LNCS, vol. 3021, pp. 469–481. Springer, Heidelberg (2004)
103. Boykov, Y., Kolmogorov, V.: An experimental comparison of min-cut/max-flow algorithms for energy minimization in vision. IEEE Trans. Pattern Anal. Mach. Intell. **26**(9), 1124–1137 (2004)

Chapter 4
Methods of Filtering and Texture Segmentation of Multicomponent Images

E. Medvedeva, I. Trubin and E. Kurbatova

Abstract Some modern video systems, for example, remote sensing systems analyze the multicomponent images. Limitations of on-board technical and energy resources and video data transmission by low power and over long distance lead to strong image distortions. The filtering is used to recover the distorted by noise images for subsequent tasks of image processing, such as detection of texture regions and objects of interest, estimations of their parameters, classification, and recognition. Multicomponent images can be represented as the multidimensional signals and have significantly greater statistical redundancy than one-component images. This redundancy would be appropriate to improve a quality of image restoration. Special cases of multicomponent images are color RGB images, each color component of which is a g-bit digital halftone image. The nature of the statistical relationship between elements within the digital halftone image and among the elements of color components allows to suggest an approximation for 3D color images using a Markov chain with several states and for bit binary image applying a 3D Markov chain with two states. The proposed filtering method is based on an approximation the multicomponent images using a 3D Markov chain and on an efficient use of statistical redundancy of multicomponent images. This method requires small computational resources and is effective with signal-to-noise ratio at the input of receiver up to –9 dB. Real images have areas with varying degrees of detail and different statistical characteristics. The authors propose to improve the accuracy of the statistical characteristics of each local region within an image and between the color components to improve a quality of the reconstructed image. A sliding window is used to estimate the local statistical characteristics of an image. The proposed method allows to detect the small objects and contours of objects more accurately in image distorted by white Gaussian noise. A method of

E. Medvedeva (✉) · I. Trubin · E. Kurbatova
Vyatka State University, 36 Moskovskaya Street, Kirov 610000, Russian Federation
e-mail: emedv@mail.ru

I. Trubin
e-mail: trubin@vyatsu.ru

E. Kurbatova
e-mail: kurbatovae@gmail.com

© Springer International Publishing AG 2018   97
M.N. Favorskaya and L.C. Jain (eds.), *Computer Vision in Control Systems-3*,
Intelligent Systems Reference Library 135, https://doi.org/10.1007/978-3-319-67516-9_4

texture regions' detection on the reconstructed images based on Markov random fields is proposed. An estimate of the probability of a transition between image elements is used as the texture feature. The method efficiently detects the texture regions with different statistical characteristics and makes it possible to reduce the computational costs.

Keywords Multicomponent images · Nonlinear filtering
Texture segmentation · Markov chain · Halftone image · Noise redundancy

4.1 Introduction

Technique of multispectral image analysis can be significantly improved by the description of the observed physical processes and phenomena in many technical applications. Images obtained by the multispectral and hyperspectral remote sensing systems contain the dozens or even hundreds of spectral channels, in which there are noises of varying intensity. The sequence of images received in a radio channel can be significantly distorted. Restrictions of technical and energy resources for satellite or unmanned aerial vehicles, as well as a transfer of video data with low power at long distances, can be the reason of such distortions.

Image restoration (filtering) attracts many researchers because of ability to recover image from a distorted version. Many different filtering algorithms have been developed for a specific model of interferences (noises) [1–7]. Thus, for example, the well-known linear filtering algorithms based on the use of local operators are effective in large signal-to-noise ratio but with increasing power of noise cause the smoothing of fine detail and blurring the boundaries of objects [1–3]. Methods based on different versions of the median filter and surrounding elements filters including bilateral and nonlocal values filters are widely used the nonlinear filtering methods because of low computer costs [1–6]. All these filters insignificant distort the sharp edges of images and well suppress impulse noise but have low efficiency in the presence of White Gaussian Noise (WGN). At present time, the Block-Matching and 3D Filtering (BM3D) and Block-Matching and 4D Filtering (BM4D) filters based on the concepts of the grouping and collaborative filtering provide the most effective removing of Gaussian noise [7]. The main disadvantages of the filters BM3D, BM4D are low speed and blurring boundaries at low signal/noise relationship.

It should be noted that the majority of known filtering algorithms are 2D algorithms. They are applied to each separate component of the image and, as a result, cannot always provide an appropriate quality of an image especially in the conditions of high intensity noise action. Multispectral images are multidimensional signals and have much greater statistical redundancy than the single images that is useful for improving a quality of the restoration of noisy images.

Segmentation is widely applied after images' recovery to solve the subsequent tasks of image processing and analysis. This allows to detect the regions of interest in images. Images of the Earth's surface contain areas (for example, areas of forests and urban areas) that do not have distinctly pronounced borders and essential details. In this case, it is necessary to identify the image areas based on the analysis of their texture. The efficiency of separation based on the texture features determines the segmentation quality significantly.

A wide variety of texture segmentation methods can be divided into some categories, such as the statistical methods based on calculation of different statistical characteristics of textures (for example, absolute gradient, run length matrix, and gray level co-occurrence matrices), structural methods, which describe the texture regions as primitives with some properties (for example, average element intensities, perimeters, orientation, and elongation) and compare them with patterns for each texture, and methods using signal processing algorithms (for example, Fourier, Gabor, and wavelet transforms) for extraction texture features [1, 6, 8, 9]. Many of these methods require large computational resources and do not exactly indicative of shared space. Another approach for texture segmentation is the model based methods including autoregressive model, Markov random fields, and fractal methods. Computational complexity of such approach depends on a complexity of the used image model. Modern segmentation approaches usually combine several methods based on the extraction of several texture features. The statistical characteristics of extended regions in an image can be different significantly but they are uniform within certain local regions. In such cases, it is expedient to use the estimates of the statistical characteristics calculated within a window for each local region.

This chapter is devoted to the combined algorithm for detecting texture regions in noisy digital images, which makes it possible to recover the digital halftone (color) images with low signal/noise ratios at the first stage and detect the extended regions with the homogeneous statistical characteristics at the second stage. We used the causal multidimensional multi-valued random Markov processes (i.e. the multidimensional Markov chain with several states) as an approximation of multispectral images. To improve the quality of reconstructed image and efficiency of segmentation, the sliding window was used to evaluate the statistical characteristics of each local area within and between adjacent Bit Binary Image (BBI) frames for different color component.

The chapter is organized as follows. Method for nonlinear multidimensional filtering of images is represented in Sect. 4.2. Method of texture segmentation is discussed in Sect. 4.3. Section 4.4 concluded the chapter.

4.2 Method for Nonlinear Multidimensional Filtering of Images

Consider the proposed method for nonlinear multidimensional filtering of images. Mathematical model of multispectral images is developed in Sect. 4.2.1. The issues of nonlinear multidimensional filtering are investigated in Sect. 4.2.2.

4.2.1 Mathematical Model of Multispectral Images

The modern systems of remote sensing can generate the digital images in different spectral bands. The multispectral image may contain from three (like RGB color images) to hundreds of components, each of which can be considered as the monochrome (grayscale) image. The color RGB images can be represented as a special case of multispectral images. It is known that for some areas of the Digital Half Tone Images (DHTI) belonging to different spectral components there is a big statistical dependence between image elements. Given the nature of the statistical relationship between the elements within and between elements of the DHTI color components (RG, GB, BR), one can assume that RGB images allow an approximation of 3D multi-valued Markov chain.

During processing the DHTI with 2^g brightness levels, the problem of memory storage and working with the transition probability matrices $2^g \times 2^g$ in size is appeared. Such image processing requires large computational resources. The works [7–9] suggest to divide the DHTI represented by g bits binary numbers into g the BBI or bit planes. This allows to reduce the computational resources owing to working with 2×2 transition probability matrices.

We represent a sequence of elements in lth BBI as a Markov random field with a separable autocorrelation function. Then elements of the lth binary channel image can be represented as the superposition of three one-dimensional Markov chains with two equally probable states M_1, M_2 and transition probability matrices, such as the horizontal $^1\Pi = \left\|^1\pi_{ij}\right\|_{2\times2}$, vertical $^2\Pi = \left\|^2\pi_{ij}\right\|_{2\times2}$, and between spectral components (channel) $^4\Pi = \left\|^4\pi_{ij}\right\|_{2\times2}$, $\left(i,j = \overline{1,2}\right)$. The BBI of two color components lth bit DHTI divided by the area $F_i\left(i = \overline{1,4}\right)$, elements of which are the Markov chain of different dimensions, is depicted in Fig. 4.1.

The fragment of 3D Markov chain with two equiprobable states M_1 and M_2 is shown in Fig. 4.2, where the following designation are taken:

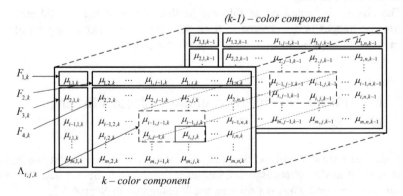

Fig. 4.1 The BBI of two color components lth bit DHTI

Fig. 4.2 The fragment of
BBI of two color components

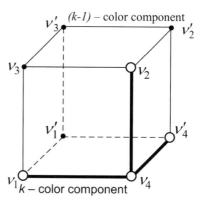

$$v_1 = \mu_{i,j-1,k} \qquad v_2 = \mu_{i-1,j,k} \qquad v_3 = \mu_{i-1,j-1,k} \qquad v_4 = \mu_{i,j,k}$$

$$v_1' = \mu_{i,j-1,k-1} \qquad v_2' = \mu_{i-1,j,k-1} \qquad v_3' = \mu_{i-1,j-1,k-1} \qquad v_4' = \mu_{i,j,k-1}$$

where i, j are the spatial coordinates, $k = \overline{1,3}$ is a number of color components in RGB image.

The conditional entropy of the element v_4 relative to the states of the elements neighborhood $\Lambda_{i,j,k} = \{v_1, v_2, v_4'\}$ defined as the difference between the unconditional entropy of element v_4 and mutual information received from the vicinity of elements $\Lambda_{i,j,k}$ [10, 11]. This is expressed by Eq. 4.1, where the products marked with Π are calculated for all possible combinations of different non-matching subscripts of 3D random field, $p(v_1, v_2, v_4', v_4)$, $p(v_i, v_j, v_k)$, $p(v_i, v_j)$, $i = j = k = \overline{1,4}$; $i \neq j \neq k$ are the joint probability density values of the elements, $p(v_i)$, $i = \overline{1,4}$ is a priori probability density values of the elements, $w(v_4|v_i)$, $i = \overline{1,3}$ is 1D density of probabilities of transitions, $w(v_4|v_i, v_j)$, $i = j = \overline{1,3}$; $i \neq j$ is a probability density of transitions in 2D Markov chains, $w(v_4|v_1, v_2, v_4')$ is a probability density of transitions in 3D Markov chains.

$$
\begin{aligned}
H(v_4|v_1, v_2, v_4') &= H(v_4) - I(v_1, v_2, v_4', v_4) \\
&= -\left[\log p(v_4) + \log \frac{\Pi p(v_i, v_j) p(v_1, v_2, v_4', v_4)}{\Pi_{i=1}^{4} p(v_i) \Pi p(v_i, v_j, v_k)} \right] = \\
&= -\log \frac{w(v_4|v_1) w(v_4|v_2) w(v_4|v_4') w(v_4|v_1, v_2, v_4')}{w(v_4|v_1, v_2) w(v_4|v_1, v_4') w(v_4|v_2, v_4')}
\end{aligned}
\tag{4.1}
$$

The transition probability density in the binary 3D Markov chain can be expressed by Eq. 4.2, where $\delta(\cdot)$ is the delta function.

$$w\left(v_4|\Lambda_{i,j,k}\right) = \sum_{i,\dots,r=1}^{2} \pi\left(v_4 = M_i|v_1 = M_j; v_2 = M_k; v_4' = M_r\right)$$
$$\times \delta\left(v_1 - M_j\right)\delta(v_2 - M_k)\delta\left(v_4' - M_r\right) \tag{4.2}$$

Taking into account Eq. 4.2, a transition probability matrix Π' for various combinations elements states of the neighborhood $\Lambda_{i,j,k}$ has the form of Eq. 4.3.

$$\Pi' = \begin{Vmatrix} \pi_{iiii} & \pi_{iiij} \\ \pi_{iiji} & \pi_{iijj} \\ \vdots & \vdots \\ \pi_{jjji} & \pi_{jjjj} \end{Vmatrix} = \begin{Vmatrix} \alpha_1 & \alpha_1' \\ \alpha_2 & \alpha_2' \\ \vdots & \vdots \\ \alpha_8 & \alpha_8' \end{Vmatrix} \quad i,j = \overline{1,2} \quad i \neq j \tag{4.3}$$

Elements of the matrix Eq. 4.3 satisfy the normalization requirement $\alpha_q + \alpha_q' = 1$. For instance, elements of the first row of the matrix Π' can be calculated by Eq. 4.4, where ${}^r\pi_{ii}\left(i = \overline{1,2}; r = \overline{1,7}\right)$ are the elements of transition probability matrix for 1D two equiprobable values Markov chain along the horizontal ${}^1\Pi$, vertical ${}^2\Pi$, and between spectral components (channel) ${}^4\Pi$, and additional matrices that calculated by Eq. 4.5.

$$\alpha_1 = \pi_{iiii} = \frac{{}^1\pi_{ii} \cdot {}^2\pi_{ii} \cdot {}^4\pi_{ii} \cdot {}^7\pi_{ii}}{{}^3\pi_{ii} \cdot {}^5\pi_{ii} \cdot {}^6\pi_{ii}} \quad \alpha_1' = \pi_{iiij} = \frac{{}^1\pi_{ij} \cdot {}^2\pi_{ij} \cdot {}^4\pi_{ij} \cdot {}^7\pi_{ii}}{{}^3\pi_{ii} \cdot {}^5\pi_{ii} \cdot {}^6\pi_{ii}} \tag{4.4}$$

$${}^3\Pi = {}^1\Pi \cdot {}^2\Pi \quad {}^5\Pi = {}^1\Pi \cdot {}^4\Pi \quad {}^6\Pi = {}^2\Pi \cdot {}^4\Pi \quad {}^7\Pi = {}^3\Pi \cdot {}^4\Pi \tag{4.5}$$

Other elements of the matrix Π' are determined in accordance with the vicinity $\Lambda_{i,j,k}$ element values.

4.2.2 Nonlinear Multidimensional Filtering

Assume that the elements of the BBI frames are transmitted by binary pulse signals independently of each other in the presence of additive WGN $n(t)$ with parameters $\left(0, \sigma_n^2\right)$.

Let $v_4 = \mu_{i,j,k}$ be the (i,j,k) element of lth BBI in kth component. Then the final a posteriori probability density of the elements can be written by Eq. 4.6, where $w\{\mu_{i,j,q}|\mu_{i,j-1,q}, \mu_{i-1,j-1,q}, \mu_{i-1,j,q}, \mu_{i,j-1,q-1}, \mu_{i-1,j-1,q-1}, \mu_{i-1,j,q-1}, \mu_{i,j,q-1}\}$ is a priori multivariate conditional probability density of transition from one combination of elements of the vicinity $\Lambda_{i,j,k}$ to next [12, 13].

$$P\{\mu_{i,j,k}\} = \prod_{q=1}^{k}\prod_{i=1}^{m}\prod_{j=1}^{n} P\{\mu_{i,j,q}\}$$
$$\times w\{\mu_{i,j,q}|\mu_{i,j-1,q},\mu_{i-1,j-1,q},\mu_{i-1,j,q},\mu_{i,j-1,q-1},\mu_{i-1,j-1,q-1},\mu_{i-1,j,q-1},\mu_{i,j,q-1}\}$$

$$(4.6)$$

If $i = j = q = 1$, then $P\{\mu_{i,j,q}\} = 0.5$

If $i = 1, j > 1, q = 1$, then $P\{\mu_{1,j,1}\} = P\{\mu_{1,1,1}\}\prod_{j=2}^{n} w\{\mu_{1,j,1}|\mu_{1,j-1,1}\}$

If $i > 1, j > 1, q = 1$, then $P\{\mu_{i,j,1}\} = P\{\mu_{1,1,1}\}\prod_{i=2}^{m}\prod_{j=2}^{n} w\{\mu_{i,j,1}|\mu_{i-1,j-1,1}\}$,

where $P\{\mu_{1,1,1}\}$ is an initial 3D probability density values of the Markov chain.

At presence of the additive WGN distorting the image, the likelihood function for the sequence of element values of lth BBI can be written by Eq. 4.7, where $f(\mu_{i,j,q})$ is a logarithm of the likelihood function of the element value $\mu_{i,j,q}$ in kth component and lth BBI.

$$F\{\mu_{i,j,k}\} = \exp\left\{\sum_{q=1}^{k}\sum_{i=1}^{m}\sum_{j=1}^{n} f(\mu_{i,j,q})\right\}$$

$$(4.7)$$

Taking into consideration Eqs. 4.6–4.7, a posteriori value distribution of 3D lth binary Markovian chain will have the form of Eq. 4.8, where c is a normalizing coefficient.

$$p_{as}\{\mu_{i,j,k}\} = c\exp\left\{\sum_{i=1}^{m}\sum_{j=1}^{n}\sum_{q=1}^{k} f(\mu_{i,j,q})\right\}\prod_{q=1}^{k}\prod_{i=1}^{m}\prod_{j=1}^{n} P\{\mu_{1,1,q}\}$$
$$\times w\{\mu_{i,j,q}|\mu_{i,j-1,q},\mu_{i-1,j-1,q},\mu_{i-1,j,q},\mu_{i,j-1,q-1},\mu_{i-1,j-1,q-1},\mu_{i-1,j,q-1},\mu_{i,j,q-1}\}$$

$$(4.8)$$

With account of designations taken in Fig. 4.2, an equation for a posteriori probability density $p_{as}(v_4)$ of the image element v_4 value $\{\mu_{i,j,k}\}$ will have the form of Eq. 4.9.

$$p_{as}(v_4) = c\exp\left\{\sum_{i=1}^{m}\sum_{j=1}^{n}\sum_{q=1}^{k} f(\mu_{i,j,q})\right\}\prod_{q=1}^{k}\prod_{i=1}^{m}\prod_{j=1}^{n} p_{as}(v_1, v_2, v_3, v_1', v_2', v_3', v_4')$$
$$\times w(v_4|v_1, v_2, v_3, v_1', v_2', v_3', v_4')$$

$$(4.9)$$

Taking into account properties of the complicated Markovian chain and assumptions made at construction of the model of multi-spectral images Sect. 4.2.1, we present (with account of Fig. 4.2) a priori probability density in Eq. 4.9 in the form of Eq. 4.10.

$$
\begin{aligned}
&p_{as}\left(v_1, v_2, v_3, v_1', v_2', v_3', v_4'\right) w\left(v_4 | v_1, v_2, v_3, v_1', v_2', v_3', v_4'\right) \\
&= \frac{p_{as}(v_1) w(v_4|v_1) p_{as}(v_2) w(v_4|v_2) p_{as}\left(v_4'\right) w\left(v_4|v_4'\right) p_{as}\left(v_3'\right) w\left(v_4|v_3'\right)}{p_{as}(v_3) w(v_4|v_3) p_{as}\left(v_1'\right) w\left(v_4|v_1'\right) p_{as}\left(v_2'\right) w\left(v_4|v_2'\right)}
\end{aligned}
\tag{4.10}
$$

Substituting Eq. 4.10 in Eq. 4.9 and integrating Eq. 4.9 over all values $\{\mu_{i,j,k}\}$, we obtain the final equation for the a posteriori probability density of the state element v_4 lth BBI expressed in terms of 1D posteriori probability density of states elements and the density of the transition probabilities of the states of the elements in the vicinity $\Lambda_{i,j,k} = \{v_1, v_2, v_3, v_1', v_2', v_3', v_4'\}$ to state of the element v_4 (Eq. 4.11).

$$
\begin{aligned}
P_{as}(v_4) = c \exp\{f(v_4)\} \int \cdots \int & \frac{p_{as}(v_1) w(v_4|v_1) p_{as}(v_2) w(v_4|v_2)}{p_{as}(v_3) w(v_4|v_3) p_{as}\left(v_1'\right) w\left(v_4|v_1'\right)} \\
\times & \frac{p_{as}\left(v_4'\right) w\left(v_4|v_4'\right) p_{as}\left(v_3'\right) w\left(v_4|v_3'\right)}{p_{as}\left(v_2'\right) w\left(v_4|v_2'\right)} dv_1 dv_2 dv_3 dv_1' dv_2' dv_3' dv_4'
\end{aligned}
\tag{4.11}
$$

Let us present a posteriori probability densities values and the transition probability densities in Eq. 4.11 in a view of Eqs. 4.12–4.13, where $p_i(v_q)$ is a posteriori probability of value M_i element v_q of lth BBI at kth-frame, $p_i\left(v_q'\right)$ is a posteriori probability of value M_i element v_q' of lth BBI at kth-frame, $^r\pi_{ij}$ is the elements of the rth $\left(r = \overline{1,7}\right)$ transition probabilities matrix, $\delta(\cdot)$ is the delta function.

$$
\begin{aligned}
p_{as}(v_q) &= \sum_{i=1}^{2} p_i(v_q) \delta(v_q - M_i) \\
p_{as}\left(v_q'\right) &= \sum_{i=1}^{2} p_i\left(v_q'\right) \delta\left(v_q' - M_i\right) \quad q = \overline{1,4}
\end{aligned}
\tag{4.12}
$$

$$
\begin{aligned}
w(v_4|v_q) &= \sum_{i=1}^{2} {}^r\pi_{ij} \delta(v_4 - M_j) \quad q = \overline{1,3} \quad r = \overline{1,3} \\
w\left(v_4|v_q'\right) &= \sum_{i=1}^{2} {}^r\pi_{ij} \delta(v_4 - M_j) \quad q = \overline{1,4} \quad r = \overline{4,7}
\end{aligned}
\tag{4.13}
$$

Substituting Eqs. 4.12 and 4.13 into Eq. 4.11, integrating with delta functions $\delta(v_q - M_i)$ and equating coefficients of the same delta functions $\delta(v_4 - M_i)$, we obtain an equation for the final a posteriori probability of state of element v_4 lth BBI (Eq. 4.14).

$$p_j(v_4) = c \exp\{f\left(M_j(v_4)\right)\} \frac{p_j(v_1)^1 \pi_{ij} p_j(v_2)^2 \pi_{ij}}{p_j\left(v_3^{(l)}\right)^3 \pi_{ij}^{(l)}} \times$$

$$\times \frac{p_j(v_4')^4 \pi_{ij} p_j(v_3')^7 \pi_{ij}}{p_j(v_1')^5 \pi_{ij} p_j(v_2')^6 \pi_{ij}} \quad i,j = \overline{1,2}$$

(4.14)

Dividing Eq. 4.14 at $j = 1$ by equation at $j = 2$ and taking the logarithm of the left and right, we come to an equation for nonlinear filtering of elements of lth BBI type [11–13] of Eq. 4.15, where $u(v_4) = \ln\frac{p_1(v_4)}{p_2(v_4)}$ is a logarithm of a posteriori probabilities ratio a posteriori state probabilities of the status of filtered element v_4, $[f(M_1(v_4)) - f(M_2(v_4))] = 4\rho_{in}^2\left[\pm 1 + \frac{\xi}{\sqrt{2}\rho_{in}}\right]$ is the difference of the logarithms of the likelihood functions in the output of the phase discriminator at $M_1 = 1$, $M_2 = -1$, $\rho_{in}^2 = \frac{A^2 T}{N_0}$ is a signal-to-noise ratio (A is an amplitude, T is a pulse duration, N_0 is a noise power spectral density), ξ is the noise components, H is a threshold in accordance with the ideal observer criterion (in our case $H = 0$) [14].

$$u(v_4) = [f(M_1(v_4)) - f(M_2(v_4))] + u(v_1) + z_1\left[u(v_1),^1 \pi_{ij}\right]$$
$$+ u(v_2) + z_2\left[u(v_2),^1 \pi_{ij}\right] + u(v_4') + z_4\left[u(v_4'),^4 \pi_{ij}\right]$$
$$+ u(v_3') + z_7\left[u(v_3'),^7 \pi_{ij}\right] - u(v_3) - z_3\left[u(v_3),^3 \pi_{ij}\right]$$
$$- u(v_1') - z_5\left[u(v_1'),^5 \pi_{ij}\right] - u(v_2') - z_6\left[u(v_2'),^6 \pi_{ij}\right] \geq H$$

(4.15)

$$z_r(\cdot) = \frac{\ln\left(^r\pi_{ii} + {}^r\pi_{ji}\exp(-u(v_r))\right)}{^r\pi_{jj} + {}^r\pi_{ij}\exp(u(v_r))} \quad r = \overline{1,7}$$

(4.16)

All a priori information about the degree of correlation between the image elements is contained in nonlinear function $z_r(\cdot)$.

Hence, it follows that the filtering efficiency will directly depend on the accuracy of the calculated estimates for elements of transition probabilities matrix. The real images contain regions with varying degrees of detail and different statistical characteristics. To improve a quality of the reconstructed image, we suggest to use a sliding window to evaluate the statistical characteristics of each local area of frames.

The estimates of transitions probabilities $^1\hat{\pi}_{ij}$ in row, $^2\hat{\pi}_{ij}$ in column in the image areas, and $^4\hat{\pi}_{ij}$ between adjacent spectral components (channel) for the lth BBI within a sliding window are calculated. Thereafter, the estimates of transitions probabilities are inserted into Eq. 4.2 to restore the central element in a window. In this study, we used the statistical characteristics of original non-noisy image to evaluate the effectiveness of filtering [15].

Figure 4.3 shows an example of the estimates $^1\hat{\pi}_{ij}$ and $^2\hat{\pi}_{ij}$ calculated for a single line of artificial BBI within the sliding window with size 21×21. We used 2D mathematical model with matrixes of transitions probabilities (Eq. 4.17) for each local area to synthesize the artificial BBI containing areas with different statistical characteristics [10, 11].

$$^1\Pi = {}^2\Pi = \begin{Vmatrix} 0.9 & 0.1 \\ 0.1 & 0.9 \end{Vmatrix} \text{ and } {}^1\Pi = {}^2\Pi = \begin{Vmatrix} 0.5 & 0.5 \\ 0.5 & 0.5 \end{Vmatrix} \quad (4.17)$$

Analysis of plots (Fig. 4.3b, c) shows that the estimation of transition probabilities within a window is close to the true probabilities (Eq. 4.17) in each texture area (Fig. 4.3a). Obviously, the larger the size of the local areas with similar statistical characteristics, the larger the window should be used. At the same time, the use of large size window for small local areas leads to averaging the statistical characteristics within a window and increase the error estimates.

Figure 4.4 shows real test RGB image with sizes 970×534. The result of a nonlinear filtering by different methods is shown in Fig. 4.5. The enlarged fragments of the image frame distorted by WGN for $\rho_{in}^2 = -6$ dB are presented in Fig. 4.5a. Figure 4.5b shows the same image recovered by the BM3D filter. The images recovered by 2D Non-Linear Filter (2DNF) algorithm without sliding window (independent color component filtering) and with the help 3D Non-Linear Filter (3DNF) without sliding window are represented in Fig. 4.5c, d, respectively. The same image reconstructed by 3D Non-Linear Filter with a Sliding Window (3DNF_SW) is shown in Fig. 4.5e. Figure 4.5f demonstrates the results of Combined Filtering (CF) that includes the developed 3D and median filters.

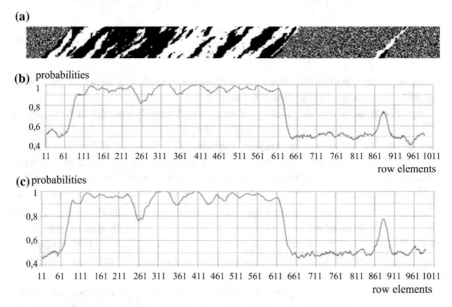

(a)

(b) probabilities

(c) probabilities

row elements

Fig. 4.3 Example of calculating estimates artificial BBI within the sliding window: **a** part of artificial BBI, **b** the estimates of transitions probabilities $^1\hat{\pi}_{ij}$ in row, **c** the estimates of transitions probabilities $^2\hat{\pi}_{ij}$ in column

Fig. 4.4 The real test RGB image

The results show that 3D filter based on the sliding window allow to reduce a quantity of artifacts like pulse disturbances and to ensure a more accurate selection of boundaries and small objects. To eliminate the pulse noise, we proposed to use a simple median filter in further implementation [16]. A processing of image distorted by the WGN becomes effective using the nonlinear and median filtering.

Generally, for comparison a quality of the recovered images, the criteria Peak Signal-to-Noise Ratio (PSNR), Feature SIMilarity (FSIM) index for color images, and Root Mean-Square Error (RMSE) are used [1, 17, 18]. Table 4.1 gives the estimates of the PSNR and the FSIM average according to 100 images for different filtering methods and different signal-to-noise ratio at the input of receiver ρ_{in}^2.

From Table 4.1, it is clear that the proposed method 3DNF_SW outperforms the known method BM3D in the filtering efficiency for small signal-to-noise ratio. The combined filter enables to improve the visual quality of the restored images reducing "salt and pepper-type" artifacts. At signal-to-noise ratio $\rho_{in}^2 = -6 \ldots 0$ dB, the CF provides a gain in the PSNR in comparison with the 3DNF_SW 20...30% and the FSIM 5...20%.

Figure 4.6 shows the dependence of the RMSE in the test image reconstructed by 2D non-linear filters without (2DNF) and considering a sliding window (2DNF_SW) as well as 3D nonlinear filters without (3DNF) and in a view of the sliding window (3DNF_SW) from bits of binary numbers of the DHTI. Signal-to-noise ratio at the input of the receiver in the simulation was $\rho_{in}^2 = -6$ dB. The 2DNF and the 3DNF filters without sliding window are represented in the plot by one line because such filtering provides similar results in the RMSE. 3D filter based on a sliding window allows to take the proper account of the statistical characteristics of each local region, both within and between adjacent BBI frames for different color component. It should be noted that a quantity of errors is smaller, when the high-order bits BBI are restoring that most severely affects on a visual image quality.

Figure 4.7 shows a gain in the RMSE for the developed 3D filtering algorithm (3DNF_SW) with respect to the prototype algorithm (3DNF) with different signal-to-noise ratio. For signal-to-noise ratio $\rho_{in}^2 = -9 \ldots -3$ dB gain in the RSME is in the range from 30 to 70%, respectively.

(a) (b)

(c) (d)

(e) (f)

Fig. 4.5 The result of nonlinear filtering: **a** noisy image ($\rho_{in}^2 = -6\,$dB), **b** image recovered by BM3D filter, **c** image recovered by 2DNF filter, **d** image recovered by 3DNF filter, **e** image recovered by 3DNF_SW filter, **f** image recovered by the CF filter

Table 4.1 Average characteristics of the restored images

	$p_{in}^2 = 0\,dB$ $\sigma = 41$		$p_{in}^2 = -3\,dB$ $\sigma = 59$		$p_{in}^2 = -6\,dB$ $\sigma = 70$	
	PSNR	FSIM	PSNR	FSIM	PSNR	FSIM
Without filtering	15.51	0.86	11.16	0.71	9.5	0.62
BM3D	25.1	0.95	17.43	0.83	14.32	0.69
3DNF_SW	24.42	0.99	19.74	0.95	17.26	0.90
CF	29.6	0.99	25.8	0.97	21.29	0.92

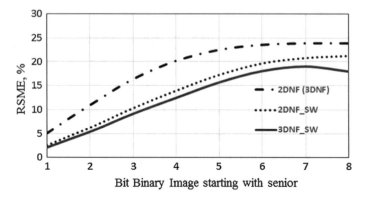

Fig. 4.6 RMSE in the restored test image

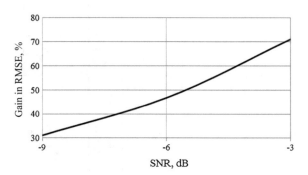

Fig. 4.7 Gain in RMSE for three-dimensional filter based on a sliding window

For unknown statistical characteristics of image, it is necessary to apply the adaptive processing algorithms [19]. To do this, it is required to calculate the evaluation of the element of the matrixes of transitions probabilities directly and perform the adaptation of filtering algorithm parameters.

4.3 Method of Texture Segmentation

Suppose that the images of Earth's surface contains the extensive areas with similar statistical characteristics within an area and have different values of statistical characteristics in neighboring areas. In this case, it is appropriate to use an estimate of statistical characteristics as a texture feature.

For texture regions detection in the images transmitted through a noisy channel, it is necessary, first, to filter an image from the presence of noise by algorithm Eq. 4.15. Second, after getting more accurate estimates of states of lth BBI elements and estimates of elements of the transition probability matrices, the regions with different textures are detected. Considering that the main details of an area are expressed in higher bits of the DHTI, it is proposed to process the BBI of higher bits of the DHTI with the most pronounced texture features. The sliding window method has been used to calculate the statistical characteristics.

Taking into account the set of elements $\psi = \{v_1, v_2, v_3, v_4\}$ (Fig. 4.2), the estimate of the transition probability along the horizontal $^1\widehat{\pi}_{ii}$, the estimate of the transition probability along the vertical $^2\widehat{\pi}_{ii}$ and estimate $\widehat{\pi}_{iii}$ of the transition probability in 2D Markov chain are calculated within the window using Eq. 4.18, where $^3\widehat{\pi}_{ii} = {}^1\widehat{\pi}_{ii}^2\widehat{\pi}_{ii} + {}^1\widehat{\pi}_{ij}^2\widehat{\pi}_{ij}$.

$$\widehat{\pi}_{iii} = \frac{{}^1\widehat{\pi}_{ii} \cdot {}^2\widehat{\pi}_{ii}}{{}^3\widehat{\pi}_{ii}} \tag{4.18}$$

Averaging the estimates $\widehat{\pi}_{iii}$ of transition probabilities within the window, the estimate of the mean transition probability $\widetilde{\pi}_{iii}$ for the element corresponding to the central element of the window has been calculated by Eq. 4.19.

$$\widetilde{\pi}_{iii}^{(r,k)} = \frac{1}{m \times n} \sum_{r=1}^{m} \sum_{k=1}^{n} \widehat{\pi}_{iii}^{(r,k)} \tag{4.19}$$

To detect regions with different textures, the estimate $\widetilde{\pi}_{iii}$ is compared by thresholding. The threshold value between two different texture regions is the estimate $\widetilde{\pi}_{iii}$, which can be chosen based on the analysis of image histograms for texture regions with different probability characteristics [20]. In the case, when it is known a priori that an image contains two textural regions with different statistical characteristics, two markers are sufficient to detect them (0 and 1). All elements, for which the estimate $\widetilde{\pi}_{iii}$ exceeds the threshold, are assigned with 1, while the remaining elements are denoted as 0. In the case, when the DHTI contains several textures, it is necessary to assign the unique marker to each region, as well as several threshold values ought to be used for various textures.

To estimate the quality of a method for detecting texture regions, the number of Erroneously Segmented Elements (ESEs) was calculated and compared with an ideal benchmark by Eq. 4.20, where h and w are the height and width of an image, respectively, F is a quantity taking a value of 0, when an image element is properly segmented, and 1, otherwise.

$$ESE = \frac{1}{h \cdot w} \sum_{i=1}^{h} \sum_{j=1}^{w} F(i,j) \tag{4.20}$$

In this case, the "ideal benchmark" means an artificial image divided into regions with different brightness, on which each brightness level corresponds to texture region with fixed probability characteristics in the real image.

The quality of detecting texture regions is determined by the size of a sliding window and size of the threshold value. For detection texture regions more exactly, it is required different window size for regions with different transition probabilities.

Table 4.2 The estimates of criterion ESE for the noisy artificial images

	$^1\Pi_1 = {}^2\Pi_1$	$^1\Pi_1 = {}^2\Pi_1$	Estimate of ESE criterion (%)			
			Initial image	Noisy image		
				$\rho_{in}^2 = 0$ dB	$\rho_{in}^2 = -3$ dB	$\rho_{in}^2 = -6$ dB
1	$\begin{Vmatrix} 0.95 & 0.05 \\ 0.05 & 0.95 \end{Vmatrix}$	$\begin{Vmatrix} 0.5 & 0.5 \\ 0.5 & 0.5 \end{Vmatrix}$	1.49	1.93	2.41	4.02
2	$\begin{Vmatrix} 0.95 & 0.05 \\ 0.05 & 0.95 \end{Vmatrix}$	$\begin{Vmatrix} 0.7 & 0.3 \\ 0.3 & 0.7 \end{Vmatrix}$	2.96	3.39	4.26	5.85
3	$\begin{Vmatrix} 0.8 & 0.2 \\ 0.2 & 0.8 \end{Vmatrix}$	$\begin{Vmatrix} 0.6 & 0.4 \\ 0.4 & 0.6 \end{Vmatrix}$	2.37	2.77	4.00	7.94
4	$\begin{Vmatrix} 0.9 & 0.9 \\ 0.1 & 0.1 \end{Vmatrix}$	$\begin{Vmatrix} 0.8 & 0.2 \\ 0.2 & 0.8 \end{Vmatrix}$	9.56	11.2	14.29	42.30

The larger transition probabilities the textures have, the larger window size is required for their detection. Therefore, it is necessary to select such sizes of sliding window that would be efficient for different textures. Large windows ensure the homogeneity of the detected regions. However, it is not possible to determine the region borders exactly. Application of too small window sizes leads to significant heterogeneity of the segmented regions. However, small window sizes permit to localize the region borders exactly. In the works [21, 22], it was shown that a window with sizes 21×21 is the most efficient from a viewpoint of quality/processing time ratio for the majority of textures. When a window with sizes 21×21 is used, the method permits to separate an image into the textural regions, in which the probability of transitions between elements differs by 0.15 and the segmentation error does not exceed 6%.

The quality of detecting texture regions depends also on a noise level in radio channel. Table 4.2 shows the estimates of ESE criterion for the noisy artificial images, which contain two texture regions formed according to the transition probability matrices $^1\Pi_1 = {}^2\Pi_1, {}^1\Pi_2 = {}^2\Pi_2$ for each texture region. Texture segmentation has been produced after image pre-processing by nonlinear 2D filter.

Figure 4.8 shows an example of segmentation of 1024×1024 artificial image with two texture regions. Figure 4.8a is an artificial image generated by benchmark (Fig. 4.8b). Figure 4.8c is a segmented initial image. Figure 4.8d is a texture feature histogram. Figure 4.8e, g are the noisy images with signal-to-noise ratio -3 dB and -6 dB, respectively. Figure 4.8f, h are the segmented noisy images. The artificial image in Fig. 4.8a contains the texture regions with the transition probability matrices equal $^1\Pi_1 = {}^2\Pi_1 = \begin{Vmatrix} 0.95 & 0.05 \\ 0.05 & 0.95 \end{Vmatrix}$ and $^1\Pi_2 = {}^2\Pi_2 = \begin{Vmatrix} 0.5 & 0.5 \\ 0.5 & 0.5 \end{Vmatrix}$.

Fig. 4.8 Segmentation of artificial image with two texture regions: **a** initial image, **b** benchmark, ▶ **c** segmented initial image, ESE = 1.49%, **d** texture feature histogram, **e** noisy image with signal-to-noise ratio SNR = -3 dB, **f** segmented noisy image, ESE = 2.41%, **g** noisy image with signal-to-noise ratio SNR = -6 dB, **h** segmented noisy image, ESE = 4.05%

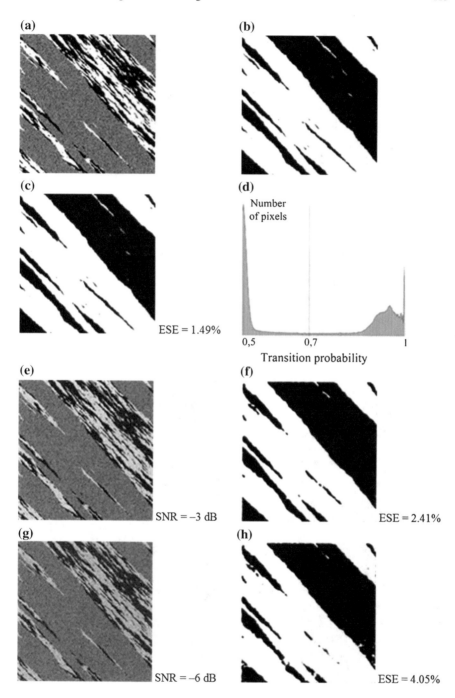

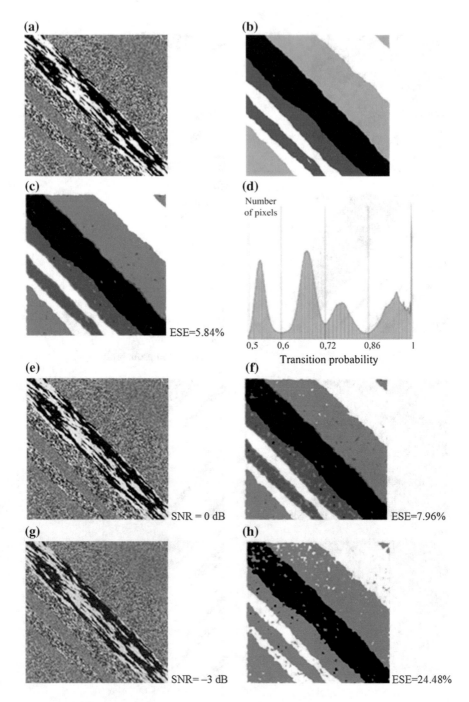

(a)

(b)

(c)

ESE=5.84%

(d)

Number of pixels

Transition probability

(e)

SNR = 0 dB

(f)

ESE=7.96%

(g)

SNR= −3 dB

(h)

ESE=24.48%

◄**Fig. 4.9** Segmentation of artificial image with four texture regions: **a** initial image, **b** benchmark, **c** segmented initial image, ESE = 5.84%, **d** texture feature histogram, **e** noisy image with signal-to-noise ratio SNR = 0 dB, **g** noisy image with signal-to-noise ratio SNR = –3 dB, **f** segmented noisy image, ESE = 7.96%, **h** segmented noisy image, ESE = 24.48%

Figure 4.9 shows an example of segmentation of 1024×1024 artificial image with four texture regions. Figure 4.9a is an artificial image generated by benchmark (Fig. 4.9b). Figure 4.9c is a segmented initial image. Figure 4.9d is a texture feature histogram. Figure 4.9e, g are the noisy images with signal-to-noise ratio equal 0 dB and −3 dB, respectively. Figure 4.9f, h are the segmented noisy images. The artificial image in Fig. 4.9a contains the texture regions with the transition probability matrices equal $^{1}\Pi_{1} = {}^{2}\Pi_{1} = \left\| \begin{matrix} 0.95 & 0.05 \\ 0.05 & 0.95 \end{matrix} \right\|$, $^{1}\Pi_{2} = {}^{2}\Pi_{2} = \left\| \begin{matrix} 0.7 & 0.3 \\ 0.3 & 0.7 \end{matrix} \right\|$, $^{1}\Pi_{3} = {}^{2}\Pi_{3} = \left\| \begin{matrix} 0.8 & 0.2 \\ 0.2 & 0.8 \end{matrix} \right\|$, $^{1}\Pi_{4} = {}^{2}\Pi_{4} = \left\| \begin{matrix} 0.55 & 0.45 \\ 0.45 & 0.55 \end{matrix} \right\|$.

The presented results (Table 4.2 and Figs. 4.8 and 4.9) show that developed segmentation method allows efficiently to divide a noisy image into the textural regions (with signal-to-noise ratio until to –6 dB) if the transition probability between elements in areas does not exceed 0.15. In this case, the segmentation error is less than 8%.

The number of peaks in texture feature histogram corresponds to the number of textures in the image (Figs. 4.8d and 4.9d). The threshold value is chosen as the minimal value between two neighboring peaks in the histogram. After the threshold is chosen from the test set of images with different statistical characteristics, the given threshold value can be used to segment other images containing regions with the same statistical characteristics. The BBIs, which have the most pronounced texture regions, are selected for texture segmentation of the DHTI.

Figure 4.10 shows an example of segmentation of satellite image with sizes 1143×844, which contains two of the most pronounced areas. Result of segmentation of initial noiseless image is used as an ideal benchmark (Fig. 4.10d). Segmentation was carried out over the BBI of 5th bit of the DHTI (Fig. 4.10b). The texture feature histogram (Fig. 4.10c) contains two main peaks, which correspond to the two areas in the image—the forest area ($\tilde{\pi}_{iii} = 0.67$) and field area ($\tilde{\pi}_{iii} = 0.83$). The threshold was chosen as the minimum value between two peaks in the histogram. The image distorted by the WGN at $\rho_{in}^{2} = -6$ dB is displayed in Fig. 4.10e. The filtered DHTI is shown in Fig. 4.10f. The segmentation of noisy DHTI is demonstrated in Fig. 4.10g. The presented results show that the developed method efficiently detects the textural regions of forest and field in a noisy image.

Texture segmentation of color images can be performed applying the developed method to each color components. The results of texture segmentation of color components are combined to one color segmented image, in which the regions with different textures are allocated by the different colors.

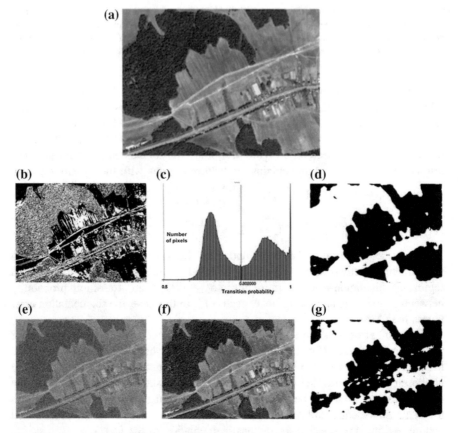

Fig. 4.10 Result of segmentation of satellite image: **a** initial satellite image, **b** BBI, **c** texture feature histogram, **d** result of initial image segmentation, **e** noisy image ($\rho_{in}^2 = -6$ dB), **f** filtered image, **g** result of noisy image segmentation

4.4 Conclusions

In this chapter, a method of 3D nonlinear filtering of multicomponent images under the WGN using a detection of extensive texture regions is proposed. The method is based on theory of conditional Markov processes with discrete arguments and applies a sliding window. The method of 3D nonlinear filtering permits to improve a quality of multispectral images distorted by the WGN owing to improve an accuracy of statistical characteristics calculation. The algorithm is effective at low signal-to-noisy ratio. The artifacts similar to "salt-and-pepper" noise may occur during 3D nonlinear filtering as the sudden changes in correlation between elements of the DHTI. Such "salt-and-pepper" noise can be removed using the combined nonlinear filter that includes the developed 3D non-linear and median filters. For detection the extensive regions with similar statistical characteristics in the restored Images, the texture segmentation method is developed. This method allows to segment a noisy image (with signal-to-noise ratio –6 dB) into textural regions efficiently if the transition probability between elements in the areas does not exceed 0.15. In this case, a segmentation error is less than 8%.

References

1. Gonzalez, R.C., Woods, R.E.: Digital Image Processing. Prentice Hall, Upper Saddle River (2008)
2. Schowengerdt, R.A.: Remote Sensing. Models and Methods for Image Processing. Elsevier, USA (2007)
3. Pratt, W.K.: Digital Image Processing: Concepts, Algorithms, and Scientific Applications. Springer, Berlin (2001)
4. Chobanu, M.K.: Multidimensional Multirate Systems for Signal Processing. Technosphera, Moscow (in Russian) (2009)
5. Rubel, A., Lukin, V., Pogrebniak, O.: Efficiency of DCT-based denoising techniques applied to texture images. In: Proceedings of MCPR, Cancun, Mexico, pp. 111–120 (2014)
6. Behtin, Y.S., Emelyanov, S.G., Titov, D.V.: Theoretical Foundations of Digital Image Processing Embedded Optoelectronic Systems. Argamak-Media, Moscow (in Russian) (2016)
7. Dabov, K., Foi, A., Katkovnik, V., Egiazarian, K.: Image denoising by sparse 3-D transform-domain collaborative filtering. IEEE Trans. Image Process. **16**(8), 2080–2095 (2007)
8. Haralick, R.M.: Statistical and structural approaches to texture. Proc. IEEE **67**(5), 786–804 (1979)
9. Zhang, J., Tan, T.: Brief review of invariant texture analysis methods. Pattern Recognit. **35**, 735–747 (2002)
10. Petrov, E.P., Trubin, I.S., Medvedeva, E.V., Smolskiy, S.M.: Mathematical models of video-sequences of digital half-tone images. In: Atayero, A.A., Sheluhin, O.I. (eds.) Integrated Models for Information Communication System and Networks: Design and Development, pp. 207–241. IGI Global, Hershey (2013)
11. Petrov, E.P., Medvedeva, E.V., Harina, N.L., Kurbatova, E.E.: Methods and Algorithms of Digital Image Processing in Remote Sensing Systems on the Small-Size Platforms. LAP LAMBER Academic Publishing, Germany (in Russian) (2016)
12. Petrov, E.P., Trubin, I.S., Medvedeva, E.V., Smolskiy, S.M.: Development of nonlinear filtering algorithms of digital half-tone images. In: Atayero, A.A., Sheluhin, O.I. (eds.) Integrated Models for Information Communication System and Networks: Design and Development, pp. 278–304. IGI Global, Hershey (2013)
13. Petrov, E.P., Medvedeva, E.V.: Nonlinear filtering of statistically connected video sequences based on hidden Markov chains. J. Commun. Technol. Electr. **55**(3), 307–315 (2010)
14. Tikhonov, V.I.: Statistical Radio Engineering. Sov. Radio, Moscow (1966) (in Russian)
15. Medvedeva, E.V., Trubin, I.S.: Improving the noise immunity of receiving video distorted white Gaussian noise. In: International Siberian Conference on Control and Communications (SIBCON'2016) (2016)
16. Petrov, E.P., Medvedeva, E.V., Metelyov, A.P.: Method of combined nonlinear filtration of correlated videoimages. Nonlinear World **11**, 677–684 (2010) (in Russian)
17. Lin, Zhang, Lei, Zhang, Mou, X., Zhang, D.: FSIM: a feature similarity index for image quality assessment. IEEE Trans. Image Process. **20**(8), 2378–2386 (2011)

18. Zemliachenko, A.N., Kolganova, E.O., Lukin, V.V., Tchobanou, M.K.: Pre-filtering and image compression corrupted by additive noise. Aviat. Space Technol. **6**, 109–117 (2013) (in Russian)
19. Petrov, E.P., Trubin, I.S., Medvedeva, E.V., Chastikov, I.A.: Adaptive nonlinear filtration of video sequences affected by heavy noise. Informatics **2**, 49–56 (2009) (in Russian)
20. Shapiro, L.G., Stockman, G.C.: Computer Vision. Prentice-Hall, Upper Saddle River, NJ (2001)
21. Kurbatova, E.E., Medvedeva, E.V., Okulova, A.A.: Method of isolating texture areas in images. Pattern Recognit. Image Anal. **25**(1), 47–52 (2015)
22. Medvedeva, E.V., Kurbatova, E.E.: Image segmentation based on two-dimensional Markov chains. In: Favorskaya, M.N., Jain, L.C. (eds.) Computer Vision in Control Systems-2. Innovations in Practice, vol. 75, pp. 277–295. Springer International Publishing, Switzerland (2015)

Chapter 5
Extraction and Selection of Objects in Digital Images by the Use of Straight Edges Segments

V.Yu. Volkov

Abstract New method for finding geometric structures in digital gray-level images is proposed. The method is based on grouping straight line segments, which correspond to the edges of the object. It includes extraction of straight line segments by oriented filtering of gradient image and gives the ordered list of segments with the endpoints' coordinates for each segment. Adaptive algorithm for straight edge segments extraction is developed that uses angle adjustment of oriented filter in order to extract the line corresponding to the real edges accurately. This algorithm permits the extraction and localization of artificial objects with the rectangular or polygonal shape in digital images. Perceptual grouping approach is applied to extracted segments in order to obtain the simple and complex structures of lines using their crossings. Proposed approach uses the points of intersection of ordered segments as the main property of object structure and also takes into account some specific properties of grouped lines, such as the anti-parallelism, proximity, and adjacency. At the first step, the simple structures are obtained by lines grouping taking into consideration all crossing lines or only part of them. At the second step, these simple structures are joined allowing for restrictions. Initial image is transformed to a collection of closed rectangular or polygonal structures with their locations and orientations. Structures obtained by this method represent an intermediate-level description of interesting objects, which have polygonal view (buildings, parts of roads, bridges, and some natural places of landscape). Application with real aerial and satellite images shows a good ability to separate and extract the specific objects like buildings and other line-segment-rich structures.

V.Yu. Volkov (✉)
The Bonch-Bruevich State Telecommunications University, 61, Moika.,
St. Petersburg 191186, Russian Federation
e-mail: vl_volk@mail.ru

V.Yu. Volkov
State University of Aerospace Instrumentation, 67, Bolshaya Morskaya,
St. Petersburg 190000, Russian Federation

© Springer International Publishing AG 2018
M.N. Favorskaya and L.C. Jain (eds.), *Computer Vision in Control Systems-3*,
Intelligent Systems Reference Library 135, https://doi.org/10.1007/978-3-319-67516-9_5

Keywords Line-segment detector · Local descriptors · Geometric primitives
Edge-based feature detector · Perceptual contour grouping · Object recognition
Content-based image retrieval · Building and road extraction · Feature-based image
matching

5.1 Introduction

The extraction of objects, their selection, and classification are the most studied
problems of image processing and computer vision. They have important applications for segmentation, visual tracking, image matching, indexing, and retrieval
of images [1–8]. Model-based approaches instead of view-based ones are generally
used for man-made object recognition. Techniques of this type analyze the
semantic information, which is contained in an object shape. The usual method is to
extract the contours and investigate their properties.

The crucial moment of object recognition is to describe the category of a recognized object. This is usually done through a set of reliable and repeatable features, which may be obtained from an object model or given from a reference image
by a universal method. Local features and descriptors are very successful in providing a compact representation for image matching, with applications to registration, wide baseline matching, image retrieval, object recognition, and
categorization. Perceptual grouping is the process, by which primitive image elements are aggregated into larger and more meaningful collections without prior
knowledge of the image content [1, 2]. The grouping principles embodied such
concepts as grouping by the proximity, similarity, continuation, closure, and
symmetry [1]. Grouping of contours is a natural way to get more complex structures
[4–6].

There are several levels of feature description. Low-level descriptors may be
broadly classified into three main types: the point-based, edge-based or linear, and
region-based [9–11]. Corners and edges are two of the most important geometrical
primitives in image processing. Intermediate-level or mid-level descriptors can be
obtained by perceptual grouping of the geometrical primitives to get simple
structures. High-level descriptors result from the comparative analysis of obtained
structures and can get enough information for the image interpretation, understanding, and matching with other image or template.

Detecting and matching specific features across different images typically
involves three distinct steps. First, a feature detector identifies a set of image
locations, presenting rich visual information and whose spatial location is well
defined. Second, a description as a vector characterizing local visual appearance is
computed from the image near the nominal location of the feature. Third, a given
feature is associated with one or more features in other images that is called
matching [9–13]. There are the intensity-based and geometrically-based methods of
feature extraction. If images are obtained from different sources or one image is a
sketch, it leads to problems in the use of intensity-based methods.

Very important property of every approach in object recognition is the scale invariance and affinity. This allows to recognize an object viewed from a different distance or with different camera settings. Another problem is occlusion and background clutter, which can significantly change the appearance of features localized in object boundaries. There are two main approaches to performance assessment of feature detection and extraction algorithms [13, 14]. The first approach deals with stability and localization accuracy of obtained features. The second approach uses the final effect of solving the problem [11]. The performance of the different combinations of detectors and descriptors was evaluated for a feature matching problem [9–16]. This study relates to the design of useful structures for intermediate-level descriptions of the objects in an image. It includes the extraction of straight edge segments and perceptual grouping of geometric primitives taking into account their intrinsic and relative properties [1–12, 15–24].

There are some very important issues, such as the land-use detection and classification, automatic detection of buildings and roads, location of rivers and streams, detection of landscape changes, image fusion and multi-image feature-based matching, which require the development and investigation of specific object models and object descriptions using the straight line segments [19–31].

Though plenty of works were devoted to theoretical aspects of grouping Problems, there are not so much practically effective algorithms for man-made object selection in real images. In addition, it is often difficult to obtain the performance characteristics for such algorithms, choose criteria, and make a comparative analysis [3, 22–40].

The chapter is organized as follows. Section 5.2 presents related works. Problem statement and method of solution are given in Sect. 5.3. Modelling of straight edge segments extraction is discussed in Sect. 5.4. Modelling of segment grouping and object detection, selection and localization is considered in Sect. 5.5. Section 5.6 provides the experimental results for the aerial, satellite, and radar images. Conclusions are mentioned in Sect. 5.7.

5.2   Related Works

Straight edge segments play an important role in features description because almost all contours of real objects are locally straight [3, 12, 19–31]. There are many objects, whose distinctive features are edges with the geometrical relations between them. In addition, many real objects of interest in air imagery have locally straight edges. They are the buildings, towers, bridges and other architectural objects, roads, rivers, landscape boundaries, and so on. There are many approaches of getting straight edge segments as lines from a gradient image. Most of them are discussed in [12, 19, 40]. Recent publication [20] presents another interesting algorithm based on the aligned points and *a contrario* approach with a false detection control.

There are plenty well-known operators for the extraction of straight lines, which operate in the gradient images. In general, the line segment detection methods can be divided into two categories: the gradient-orientation-based and gradient-magnitude-based. Burns' straight lines detector [1, 41] uses the orientation of the local gradient in the neighborhood of the pixel in order to obtain the "support line" regions of similar orientation of the gradient. The structure of the associated intensity surface is studied to extract the location and properties of the region.

The gradient-magnitude-based methods are applied as an edge detector to extract the edge or contour map in an input image and then detect the line segments based on the extracted map. Hough Transform (HT) is a traditional line detector based on an edge map, which extracts all lines containing a number of edge points exceeding a threshold. A lot of variants of the HT have been proposed, e.g. the elliptical Gaussian kernel-based HT, but they usually extract infinitely long lines instead of line segments and easily cause many false detections in richly-textured regions with strong edges [41].

Hough algorithm draws the straight lines in the image in the parametric space without considering the spatial relationships between points. As a result, such detectors often give a fragmentation that cannot provide the entirely smooth straight edges in short segments, which virtually destroy the geometric structure of an object. A longer line segment can be broken into several short ones; also the weak gradient parts of a line segment may be lost. The key problem of the edge segment based methods lies on that the line segment detection result suffers from the deficiency of the edge segment detection algorithm to a great extent. As a rule, these algorithms are very sensitive to values of many working parameters, which ought to be settled manually. Attempts are made to make the algorithms less sensitive to parameter set and obtain the robust structures. Robust line segment detector called as *CannyLines*, was proposed in [41], which is based on the edge map obtained by applying a parameter-free Canny operator in the input image. However, even the perfect contours defined by Canny detector contain jitter that has a negative impact on the result of the HT.

Recently proposed Line Segment Detector (LSD) [20] produces accurate line segments and controls a number of false detections in a low-level by efficiently combing the gradient orientations and line validation according to the Helmholtz principle. The LSD states clearly, what is a line segment, how to detect it and how to verify it, but the gradient magnitude threshold eliminates some useful line information. Decreasing the threshold value leads to detecting more but coarser line segments, which is the problem of the internal LSD because the gradient orientation is unstable, when the gradient magnitude has small value.

A new method proposed in [12, 19] uses the oriented filtering (slope line filter) and forming a gradient profile in the chosen direction. It has very important advantage over other methods. It allows to get the crossing points between extracted lines. The second important property of this method is ordering of line segments with respect to the output of the slope line filter. These properties are essential here for grouping lines into simple structures, which relate to the intermediate-level description of an image.

Matching images using linear features was discussed in [21]. The authors used the term *anti-parallel* lines, which was defined by Nevatia and Babu, and Anti-PARallel lineS (APARS). Such parallel segments were represented by a single line through the middle of two segments, whose orientations differ by approximately 180°, the tolerance being inversely proportional to their length, and separated by a width lower than a given threshold. The anti-parallel properties along with crossings present the correspondences between line segments.

Idea of straight line grouping for features description was theoretically developed in [17]. An image was interpreted as a collection of objects and relationships between these objects. They used the hierarchy of line segments and relations between these segments to describe the geometric structures. The proposed intermediate-level relational graph describes image structure and contains points (pixels) at the lower level. At the first level, points combines to get segments, which can form the ribbons, junctions, and curves at higher levels. Grouping at each level is based on some geometric constraints, such as the continuity, parallelism, symmetry, overlapping, and coincidence [1, 18, 21–23, 33, 34]. The information embedded in the graph is useful for a variety of tasks. Object recognition is often mapped into a graph matching problem.

In [30], the authors developed new structural features called *consistent line clusters*, which are useful in recognizing and locating man-made objects in images. The idea of adjacency was exploited in [5] for construction *k-AS adjacent segments*. An important question for content-based image retrieval is how to use the extracted segments to form more advanced features that can be used to recognize various objects.

Coordinates of straight line segments together with angles and magnitudes form the first level for object description [12, 19, 40–43]. We can find the points of cross-section to detect the corners and junctions on the basis of extracted lines. The simplest features are represented by collections of lines, which have some relations. Better extraction of straight line segments allows to detect the corners and junctions of edges. We can further develop the known matching algorithms [3, 17] through the use of additional features. Some new ideas were discussed in [34], though without considering the sign of edge gradient.

Searching for related line pairs was implemented by comparing the relation of angles. In [3], a weighed matching measure model of straight lines, which simultaneously use various linear features, was constructed and the values of weights of different features were proposed. The method adopted a hierarchical straight line matching strategy, which uses the matching result of the first step as a restriction to reduce the searching range, and, thus, to finish the complete matching in a whole imagery. However, this method cannot overcome the incorrect matching caused by parallel straight lines. Other descriptors, which are based on the active contours, snakes, graph/trees, and evaluation of the convex hull and the minimum bounding rectangle, have been proposed in [27, 34–37].

Adjacent fields for this research are the extraction the shapes and meaningful curves from images [38], Scale-Invariant Feature Transform (SIFT) local descriptors and interest point extraction [7–11, 13–15, 23], detection of multi-part objects [33], region-based features, and shape representation [4–10, 35–39].

5.3 Problem Statement and Method of Solution

Hereinafter, the problem statement and tasks, image processing structure, advanced algorithm of straight edge segments extraction, as well as the lines grouping algorithms for object description and selection are considered in Sects. 5.3.1–5.3.4, respectively.

5.3.1 Problem Statement and Tasks

The goal of this investigation is to develop a practical algorithm to extract the straight line segments of edges, which are grouping to describe the man-made objects in real grayscale images. The extracted straight lines may have a polygonal configuration (in most interesting cases, they are rectangular in shape).

A detailed description of the new method for straight line segments grouping is to get the structures for intermediate-level object description. Novel method, which was proposed and described in [12, 19, 40, 43] for extraction of straight edge segments, is developed here to obtain the correctly localized edges and their crossings. Advanced algorithm includes a line angle adaptation loop to get a precise estimate of edge orientation. Straight line segments are ordered with respect to the mean gradient magnitudes of edges. Additional features are the orientation, intensity, and width of the edge.

The problem is how to construct the object description on the basis of straight line segments and set of low-level additional features. A novel method uses the crossings in the ordered segments as the main property for grouping. The next problem is the practical application and evaluation of this method to real aerial and satellite images for object extraction and recognition, as well as for image matching tasks.

5.3.2 Image Processing Structure

Image processing structure is shown in Fig. 5.1. The pre-filtering and straight line segments extraction form a low-level description of an image content.

A gray-level image X is obtained from the registered initial image after some pre-filtering in order to smooth the initial image. Algorithm for extraction of straight

Fig. 5.1 Image processing structure for object selection on the base of extracted straight edge segments

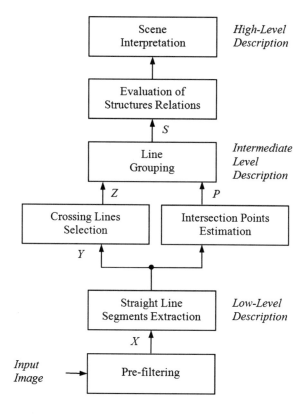

edge segment was described in details in [12, 19, 40]. The goal is to show the advantages of the proposed algorithm for the detection and extraction of the straight edge segments. This extraction is the first problem because well-known detectors do not permit the obtaining of surely localized edges and their intersections. Straight edge segments are ordered with respect to the mean gradient magnitudes of edges. Additional features are the orientation, width, and intensity.

The second problem is how to construct the object description on the basis of straight line segments obtained and use a set of low-level additional features and characteristics. A hierarchical structure for object representation is developed, which includes a line combining at four levels corresponding to their intersections and orientations. The third problem is a modelling of corresponding situations for the performance evaluation and practical application of these algorithms for real aerial and satellite images for object extraction and recognition, as well as for image matching tasks.

5.3.3 Advanced Algorithm of Straight Edge Segments Extraction

There is some reserve to enhance edge extraction algorithms for noisy imagery. The first possibility is the use of spatial contiguity of points in lines. The second possibility is to use a linear filtering before any non-linear transformations, which can decrease a Signal-to-Noise Ratio (SNR). These ideas were included in the gradient-based method proposed in [12, 19, 40].

A gray-level image X is obtained from the registered initial image after some pre-filtering, which smooths an initial image. Gaussian smoothing is used with the scale parameter $\sigma = 1.5$. Differentiation in local window results in several gradient images. Four Prewitt masks are used to obtain the gradient images. Since it needs to separate edges with ascending and descending intensity, these gradient images are inverted that is resulted into eight gradient images. Initially, there are eight angle sectors with 45° for coarse searching the edge slope, as shown in Fig. 5.2. The rough estimate then should be improved by adaptive tuning algorithm.

Filtering along the chosen line slope allows the average gradient profiles A to be obtained. The output of the Slope Line Filter (SLF) estimates the average gradient magnitude and is used to rank the line directions according to their maximal mean average profile values. The slope line filter mask should be matched to the length of straight segment to be extracted. It was initially chosen by 16 × 9 for horizontal direction that results in matching with a segment length of 16 pixels. This mask presents a support region of pixels for the initial extraction of a line segment.

To avoid the scale-space problems in the case of an unknown scale of initial image, it is necessary to search the optimum filter mask size, which is matched to edge length in this direction. The optimum SLF mask size is chosen automatically among different masks and it gives the maximum normalized filter output in the chosen direction.

The number of directions for searching line segments depends on angular resolution, which is provided by the pre-filtering and slope line filter mask. It also depends on the minimal length of the edge to be extracted and filter mask size [36].

Fig. 5.2 Eight angle sectors for coarse searching of edge slope

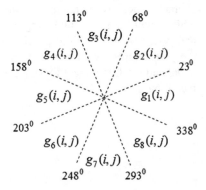

During processing, an angular drift of 2° was chosen for sparse searching and it is settled at 0.5° for fine searching. Gradient image Y is rotated by this value so that we can use the same SLF mask. It is worth noting that the sign of intensity variation at the edge is kept in order to discriminate edges from black to white and from white to black in the same direction.

The next processing step is the selection of a gradient profile Z along the line with a given slope and making estimates of the start and end points of a line segment. The gradient profile Z in a given direction looks like a one-dimensional noisy process with random impulses within it. The task of segment localization is reduced to the estimation of the start and end points of random impulse in noise.

The part of extraction processing is modified. The previous algorithm in [40] gave straight lines ranked with magnitude in each direction. In this method, all directions are investigated to choose the most valuable line segment. Its profile is formed from pixels in the support region by the Profile Line Filter (PLF). This support region is then deleted from the gradient image Y and from the SLF output A. After that, the searching process starts again and the second valuable line segment is extracted, which may have either the same or a different direction. Extracted profiles are compared with the threshold to obtain the start and end points of the segment. We used a constant threshold, though in complex tasks adaptive solutions like t-Detector or non-parametric rank algorithms may also be used.

In comparison with previous algorithm, several improvements have been made to get better edge locations and to decrease the calculation time. Instead of rotating gradient images, the oriented filtering was obtained using a bank of rectangular filter masks. Every mask has small width (about 3–5 pixels) and a length $lmask$. This length is a filter parameter, which affected on the resulting lines ordering. It is also related to image sizes and determined edges, which were extracted at the first step.

Eight local gradients are calculated using Prewitt masks in corresponding sectors. Directional filter masks have different angles of orientation in these sectors with spacing of 6°. The first extracted point has maximum value among all oriented filter outputs. Direction of this filter defines rough estimation of the first line orientation angle.

To obtain the endpoints' coordinates of a segment, the gradient profile along the rough direction is formed, which has to be averaged among several adjacent lines. In Fig. 5.3, a gradient ridge is represented, which has a small positive angle of orientation. Dashed line relates to the oriented filter mask, which got maximum output filter value. It has a horizontal direction, which is a rough direction of the line. The corresponding gradient profile is described below. There is an angle error a between rough direction and the ridge slope. This error results in bad endpoints' estimation, which was a drawback of the previous algorithm.

The structure of the advanced algorithm for getting endpoints' coordinates for each straight edge segment is presented in Fig. 5.4. At the first step, a gradient image is filtered by four oriental filters in order to obtain eight output images (four positive and four negative). Maximum value of output intensity is found among all these images and its place is taken as the region for the first straight edge segment. At the second step, this region is improved by adaptive tuning algorithm, as was

---

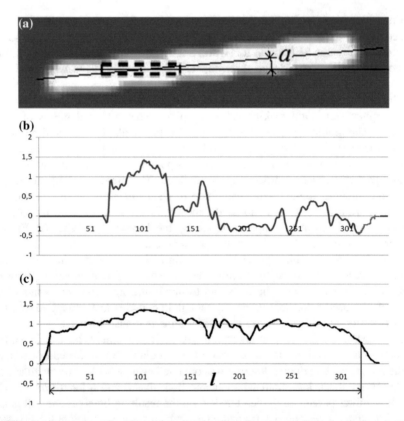

Fig. 5.3 Gradient image of a ridge and gradient profiles of a segment along the rough and precise directions: **a** initial image, **b** rigde profile, **c** gradient profile

Fig. 5.4 Structure for getting endpoint coordinates

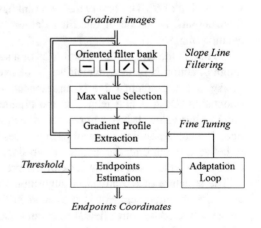

described earlier. Sign of gradient in this place enables to estimate the start and end points of the segment together with width of segment region.

Endpoint coordinates obtained are stored in the list of segments, and corresponding region is deleted from the initial gradient image. After that, all operations are repeated to get the following straight edge segment. In contrast to early algorithmic version, the orientation angle of a line is adapted by maximizing the estimated length of a segment at this point. Oriented mask is rotated within the bounds of 6° with the step of 0.5°.

At every step, a length of segment is calculated through the threshold circuit and precise angle *phi* is set, which corresponds to the maximum of the segment length. This procedure prevents a fragmentation of lengthy lines in the image. The resulting profile is presented in Fig. 5.3c.

Threshold value *lengthtresh* is another parameter of the algorithm, which determines the resulting line lengths. All gradients inside the segment have to exceed the *lengthtresh*. Thus, high values of *lengthtresh* may cause a line fragmentation. Too small values may result in connection of different lines in the same direction.

Number of lines *nlines* is the last parameter, which has to be chosen. It determines the maximal number of extracted segments in an image. In practice, we may set additional threshold *gradthresh*, which restricts minimal gradient values for lines extraction. In this case, the exact number of lines may be less than *nlines*. The task of setting the value of *gradthresh* relates to the problem of noisy lines cancellation.

At the beginning of processing, a number *nlines* of extracted lines should be restricted using some a priori information about number of objects in an image. Together with coordinates of endpoints' *coordinates* and the length of each line segment *length*, the algorithm gives its angle *phi*, the maximal output of a directional filter *mfilter* and width of the ridge *width*. In order to calculate this width, the second threshold value *widththresh* is required. Algorithm forms several cross-sections of the ridge gradients and makes estimates of width, which are averaged to form the resulting *width*. All segments obtained are ranked with respect to the value *mfilter* of filter output.

The main operation for the next level of feature description is a detection of lines crossings. Every line may be crossed by several lines, and a final table Z contains the rows with ordered numbers of all lines, which cross (or adjoin) the chosen line. Corners and junctions are also included in this table. Crossing points' coordinates P are geometrically calculated and may be also used for final structure description.

5.3.4 Lines Grouping Algorithms for Object Description and Selection

The problem is to construct the feature descriptors on the base of extracted ordered straight line segments for object recognition and image matching. A hierarchical set of features was developed in [12, 19, 43]. Here, we present the detailed description and evaluation the performance of the method. At the intermediate level of description, the straight edge segments are grouped getting a simple structure for a given segment line. Here, we define simple structure $C_k = \{L_k, L_m, L_n\}$ as a set of lines, which may include up to two crossing lines for a given main line L_k, $k = 1$, ..., $nlines$, $k < m, n$.

At the first step of grouping lines, several restrictions may be applied to select the most interesting simple structures:

- The contiguity describes touching or bordering of two lines. Here, a crossing of lines is a type of contiguity.
- The anti-parallelism of two lines, which cross the main line, means that they have absolute difference in orientations near 180°. In practice, we may define some angle *gamma* in degrees (the half of possible error) as a measure of anti-parallelism; anti-parallel lines are called APARS [21].
- The proximity is being to or near. It can be evaluated by the distance d between lines.
- The adjacency is being enough so as to touch. For example, adjacent line results from the road boarders' extraction. It characterizes by the shift *delta* of one of anti-parallel line with respect to another.

These parameters are illustrated in Fig. 5.5. It needs to normalized values d and *delta* with respect to the minimal length of anti-parallel lines. Application of these restrictions results in selection of simple structures with desired properties among all possible structures.

Complex structure $S_k = \{C_k, C_m, C_n, C_p, \ldots\}$ represents a collection of simple structures for a given line and crossing lines with the mentioned restrictions. Some of simple structures may also be excluded from S_k if the corresponding line has small magnitude M_m with respect to the magnitude M_k of the main line. Resulting complex structure is used for object description along with properties of segments contained in this complex structure.

In this study, the compound objects with closed parallelogram structures are of primary interest. They may be considered as the salient regions. Then a complex structure may consist of two simple structures with mutual lines. This method can be generalized to form more complex collections of straight segments with corresponding descriptors.

Fig. 5.5 Parameters of adjacency and proximity of anti-parallel lines

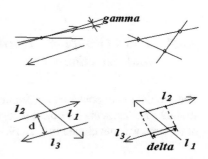

5.4   Modelling of Straight Edge Segments Extraction

Several noisy models were used to demonstrate the ability of proposed algorithm compared to other known methods. Suppose a square is corrupted by Gaussian noise that is shown in Fig. 5.6a. The SNR is $d = 2.326$. Image in Fig. 5.6b contains all line segments and image in Fig. 5.6c shows all possible crossing points. Among all lines, there are the isolated lines, simple structures consist of two, three and four lines, and only two closed structures with four lines. After line grouping and closed structure selection with geometrical constraints square object is extracted and it is shown in Fig. 5.7a. The same task is solved using the LSD algorithm [20], which is available on the site. Resulting extraction is shown in Fig. 5.7b. It gives only five segments and fails to select square object. The result of Canny edge detector with aperture 3×3 followed by the HT is shown in Fig. 5.7c. Canny-Hough algorithm (from Matlab) gives exact contour of square object but it consists of   21 line segments, thus, it cannot be recognized as a square contour.

Consider another model of a noisy image, which contains four equal horizontal strip regions. Each strip region contains signals represented by four rectangular boxes with different sizes in pixels: 8×16, 16×16, 32×16, and 64×16. The image size is 256×256 pixels. A level of a signal is constant at each strip and rises from the top strip to the bottom (Fig. 5.8).

The noisy image contains standard Gaussian noise, which is the same in all regions. SNR are equal $d = 0.58, 1.16, 2.33, 4.65$ in sequential strips from top to bottom, respectively. There are Canny contours in Fig. 5.8c obtained by Matlab with $\sigma = 1.5$.

Results of straight edge segment extraction are represented on Fig. 5.9. Proposed algorithm accurately extracts the straight edges and uses the crossing points to get closed structures with SNR $d = 1.16$.

Results for line segment extraction by the LSD algorithm are displayed in Fig. 5.10a. It can be seen that the LSD detector cannot extract closed structures at all. However, it does not produce false segments. The output of Canny-Hough

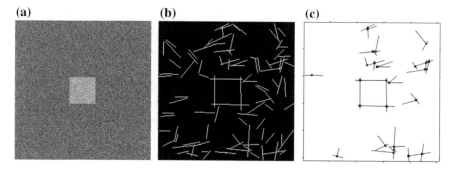

Fig. 5.6 Straight edge segment extraction in noisy image: **a** image corrupted by Gaussian noise, **b** detected line segments, **c** lines' grouping

(a)   **(b)**   **(c)**

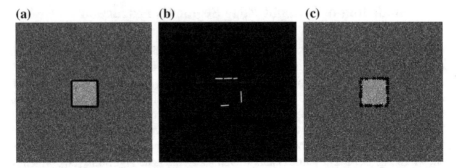

Fig. 5.7 Square object extraction by different algorithms: **a** proposed algorithm, **b** the LSD algorithm [20], **c** Canny edge detector followed by the Hough transformation

(a)   **(b)**   **(c)**

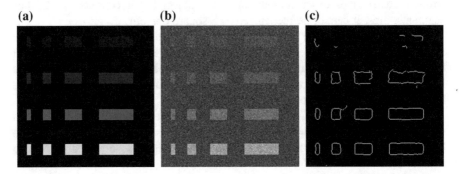

Fig. 5.8 Results of the Canny edge detector: **a** test image, **b** noisy image, **c** detected edges

(a)   **(b)**

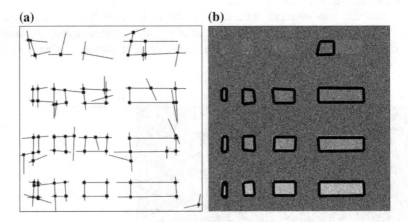

Fig. 5.9 Extraction of closed structures: **a** proposed straight line segment detector, **b** extracted closed structures

(a)   (b)

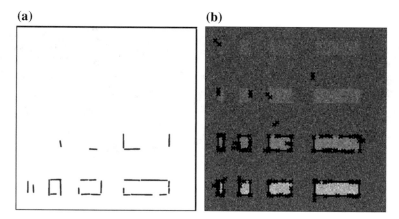

Fig. 5.10 Extraction of line segments: **a** proposed straight line segment detector, **b** extracted closed structures

straight line detector is shown in Fig. 5.10b. Algorithm allows to get the closed structures at SNR more than $d = 4.65$, but lines are very fragmented and consist a lot of shot fragments. Thus, it needs great efforts to select closed structures in noisy image.

Both of competitive algorithms cannot localize edges correctly, even for high SNR. Their outputs also contain fragmented lines. In addition, the information about signs of gradient is lost, thus, we cannot use anti-parallel features of edges. Two necessary Canny threshold values were manually settled every time to get the best view of lines. It is impossible to set lower threshold values because it results in lots of noisy contours of edges.

The proposed line segments extraction algorithm also needs in threshold values for maximal gradient value selection and for the endpoints' extraction [19, 40]. First threshold is not very critical because it only restricts the number of extracted lines. The second threshold is used to get the endpoints from gradient profile. It determines the length of the extracted line and may course to disappearing of crossing point in some cases.

The algorithm extracts the edges and orders the line segments with respect to decreasing filter outputs. We can show the corners and junctions with certainty, even for rather low SNR. The additional advantage is the information about gradient signs. It permits the combination of anti-parallel lines and the combination of a set of line segments for the constructed feature description model.

Figure 5.11 shows the corner extraction results for the Harris corner detector. Matlab algorithm was applied with sensitive factor $k = 0.1$ and different level values. The left image gives Cornermetric, and other images show results of thresholding with low and high values. The Harris and the Shi-Tomasi algorithms were inferior the proposed scheme in terms of corner detection ability. Detection characteristics are shown in Fig. 5.12. The false alarm level was manually settled to

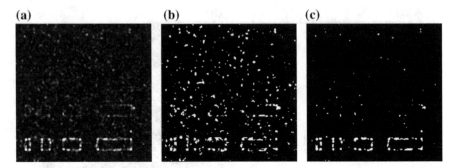

Fig. 5.11 Corner extraction results for the Harris detector: **a** Cornermetric, **b** thresolding with low value, **c** thresolding with high value

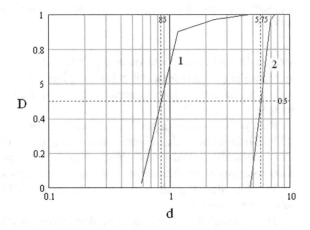

Fig. 5.12 Detection of corners characteristics for the advanced algorithm (*line 1*) and for the Harris detector (*line 2*)

ensure negligible false responses (less than 0.01). The advanced algorithm (line 1) extracts edges with less SNR than both contenders (line 2). Its errors do not include false corner points but are represented by displacements and misses.

5.5 Modelling of Segment Grouping and Object Detection, Selection and Localization

Consider a model of a noisy image, which contains ten equal horizontal stripes. Each stripe contains ten square objects of size 16 × 16 with Gaussian noise background (Fig. 5.13a). Thus, there are totally 100 squares in noise. The SNR is different in stripes. Its values vary from the top stripe to the bottom: 0.58, 1.16, 2.32, 3.49, 4.65, 5.81, 6.98, 8.14, 9.3, and 11.6. The task is to detect and select the square objects in the image.

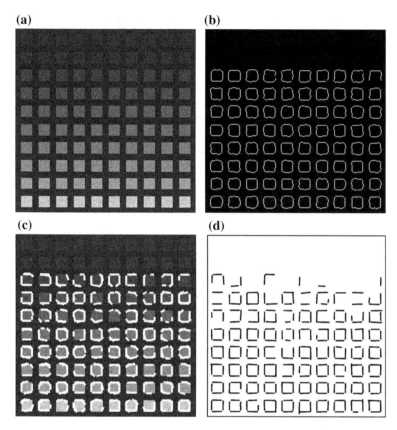

Fig. 5.13 Detection of straight line segments: **a** noisy image of 100 squares with different SNR values, **b** contours extracted by Canny detector, **c** straight line segments extracted by Canny-Hough processing, **d** the LSD algorithm

Well-known Canny detector gives excellent extracted edge presentation (Fig. 5.13b) for object localization but it is difficult to test square shapes of the objects. It needs getting straight segments to solve the selection problem. The Hough transform can get straight line segments on the base of Canny edges. They are represented in Fig. 5.13c. It is evident that few square objects may be extracted here even at high SNR.

One of modern algorithms is the LSD [20], which obtains the presumed false alarms of noisy lines in the image. The result of lines extraction is shown in Fig. 5.13d. The drawback of the LSD is the lack of crossing points but it is possible to construct closed objects using the lines' fusion. By this way the algorithm can detect and localize the closed square objects but with low quality even at high SNR (above $d = 10$). Finally, both of Canny-Hough and LSD algorithms have low possibility to extract square objects.

The proposed algorithm gives the closed square objects as complex structures. It means that simple structures are obtained, which consist of three lines. Then these simple structures are grouped to the complex closed structures, which satisfied geometrical restrictions.

Results of detection and localization are represented in Fig. 5.14a. Parameters for detection were chosen to set rather weak restrictions on lines: *lmask* = 16, *gamma* = 30, *dl* = 0.1, *deltal* = 0.1–10. Here *dl* and *deltal* are normalized distances (see Fig. 5.5) to the minimal length among two anti-parallel lines. Threshold *gradthresh* was set so as to obtain not more than one noise line in an image. The value of *lengthresh* was chosen to get stable crossing points. We can see good detection for SNR more than *d* = 3.5. Algorithm has better detection ability than LSD [20], which cannot get closed structures even for high SNR. The detected square objects are depicted in Fig. 5.14b.

Crossing points can be easily extracted from straight lines and they all correspond to corners. It is possible to compare results with Harris corner detector. This is shown in Fig. 5.15 for sensitivity factor *k* = 0.01. Cornermetric is displayed in Fig. 5.15a and extracted corners are shown in Fig. 5.15b.

Comparative analysis with Harris detector shows that the proposed algorithm can better extract corners. This fact had been pointed out earlier (see Figs. 5.9 and 5.11). The gain in the threshold SNR is around 3.

Consider again a noisy image, which contains four stripes and different rectangular objects in each stripe (Fig. 5.11). Objects have different sizes in each stripe and SNR takes increasing values 0.58, 1.16, 2.33, and 4.65 from top stripe to the bottom. The task is to select rectangular objects with different shapes and estimate their location and orientation parameters.

It is possible to control the selection process by varying the parameters *gamma*, *dl*, and *deltal*. When we do not restrict the shape of objects, the algorithm extracts every rectangular object for SNR more than 1.16. This is presented in Fig. 5.16c, d, where images show the locations and orientations of the two objects from the

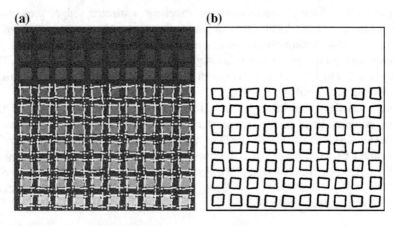

Fig. 5.14 Results of proposed algorithm: **a** straight line segments, **b** square objects

(a)   **(b)**

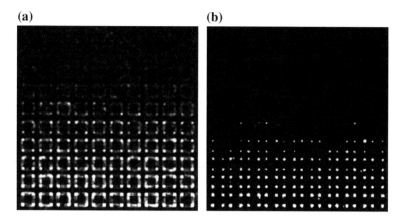

Fig. 5.15 Corner extraction results for the Harris detector: **a** Cornermetric, **b** extracted corners

bottom stripe, which has high SNR. If minimal *dl* equals to 0.5, then algorithm gives six objects in the centre of noisy image (Fig. 5.16e). In this case, the longest rectangulars have lower values of this parameter. Reducing angle *gamma* to 1 but allowing *dl* = 0.1 and *delta* = 0.1–10, it is possible to get only well-shaped four objects (Fig. 5.16f).

5.6   Experimental Results for Aerial, Satellite and Radar Images

Original aerial and satellite images are shown in Fig. 5.17. They contain buildings, which have straight edges. The aerial image (Fig. 5.17a) has a better resolution than the satellite one (Fig. 5.17b).

Extracted straight line segments (*nlines* = 500) and their crossing points are shown in Figs. 5.18 and 5.19. The closed structures are represented in Fig. 5.20 after applying the geometrical restrictions. Here we get 154 different closed structures in Fig. 5.20a and 107 structures in Fig. 5.20b, respectively. These structures represent whole objects and also different fragments of them and all have locations and orientations. We can initially distinguish ten objects, which are the same in both images. Six of them can be extracted as a whole or partly. Locations and orientations of one of the objects, which have larger sizes, are shown in Fig. 5.21.

Another pair of images, which describes the same scene, is represented in Fig. 5.22. Aerial image in Fig. 5.22a has high resolution but it depends on atmospheric vision conditions. Synthetic Aperture Radar (SAR) image in Fig. 5.22b is independent on it but has low resolution and specific speckle noise. The task is to select the same objects in these two images.

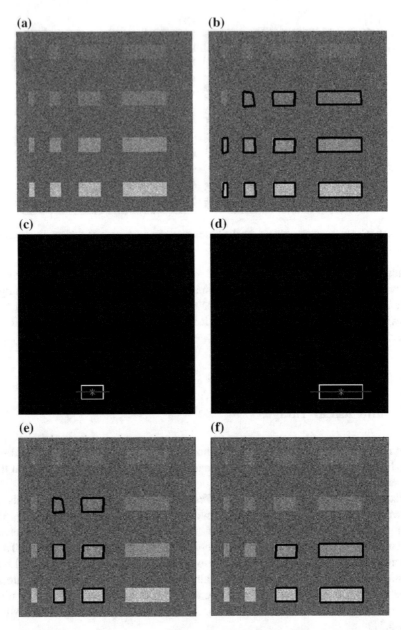

Fig. 5.16 Rectangular object shapes selection in noisy image: **a** initial noisy image, **b** extracted objects with any restriction of parameters, **c** longest rectangular, $dl = 0.5$, **d** longest rectangular, $gamma = 1$, $dl = 0.1$, $deltal = 0.1–10$, **e** six extracted objects, $dl = 0.5$, **f** four extracted objects, $gamma = 1$, $dl = 0.1$, $deltal = 0.1–10$

(a)   **(b)**

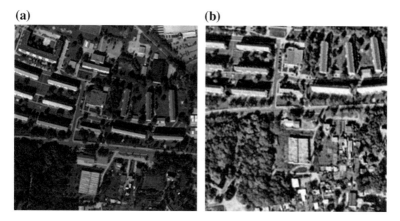

Fig. 5.17 Examples of original images: **a** aerial image, **b** satellite image

(a)   **(b)**

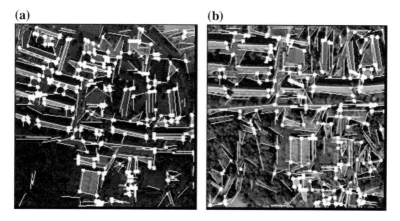

Fig. 5.18 Straight edge segments in the images: **a** aerial image, **b** satellite image

Extracted straight line segments (*nlines* = 500) are shown in Fig. 5.23 for both images. Crossing points are represented in Fig. 5.24 and allow to consruct the simple and complex structures. Different structures are obtained by means of grouping the crossing straight lines subject to geometrical constraints.

Aerial image in Fig. 5.24a contains 187 structures and SAR image in Fig. 5.24b contains 190 structures for objects and their parts. After selection of structures with four corners subject to geometrical constraints, the closed structures are extracted and they are shown in Fig. 5.25. Despite of poor quality of both images, about a half of objects can be selected correctly. Every object can be localized and their orientations may be estimated as it is shown in Fig. 5.26.

The perspective investigations may relate to application of region-based methods of object extraction and recognition after previous segmentation is made using a lines grouping.

(a)　　　　　　　　　　　　　　　　　(b)

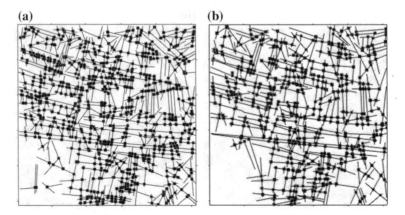

Fig. 5.19 Crossing points of straight line segments: **a** aerial image, **b** satellite image

(a)　　　　　　　　　　　　　　　　　(b)

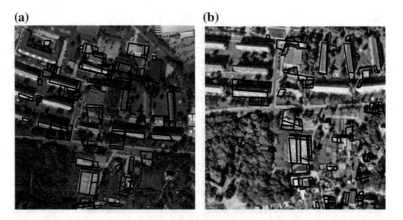

Fig. 5.20 All closed structures in images: **a** aerial image, **b** satellite image

(a)　　　　　　　　　　　　　　　　　(b)

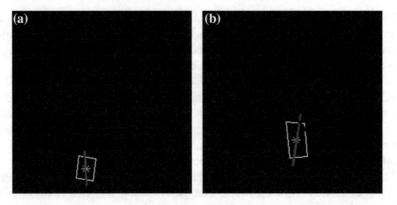

Fig. 5.21 Object extraction and localization in both images: **a** aerial image, **b** satellite image

(a)   (b)

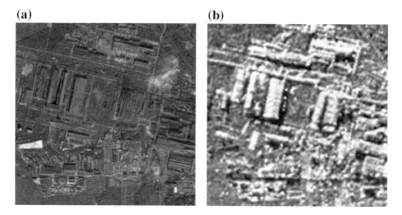

Fig. 5.22 Original images: **a** aerial image, **b** SAR image

(a)   (b)

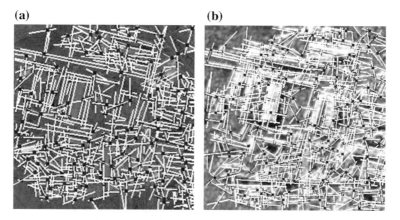

Fig. 5.23 Straight edge segments in images: **a** aerial image, **b** SAR image

(a)   (b)

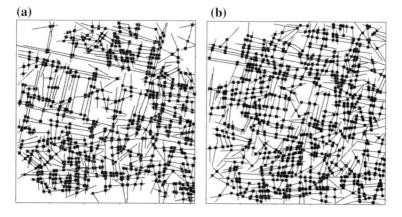

Fig. 5.24 Straight line segments with crossing points: **a** aerial image, **b** SAR image

(a)   (b)

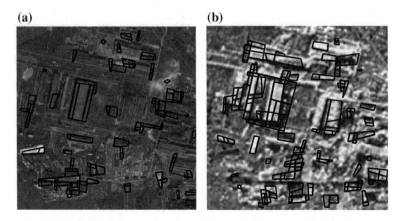

Fig. 5.25 Closed polygonal structures with four corners extracted in both images: **a** aerial image, **b** SAR image

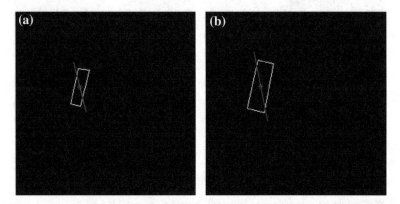

Fig. 5.26 Localized rectangular object with estimated orientation: **a** aerial image, **b** SAR image

Consider aerial image in Fig. 5.27a from [27], where active contour (snake) procedure was performed, and the result is repeated in Fig. 5.27b. Such procedure needs to set the initial points for successive object selection.

The proposed algorithm works without initial setting of object centers. Extracted straight line segments with crossing points for *nlines* = 500 are shown in Fig. 5.28. Perceptual grouping gives a set of closed structures shown in Fig. 5.28b. Algorithm can select almost all rectangular objects but gives several surplus structures. These structures do not relate to false or noisy objects so that a selection of useful objects requires the additional analysis.

Extracted straight line segments with crossing lines can be used for selection objects, which are not rectangular, for example polygonal structures. Thus, the

(a)

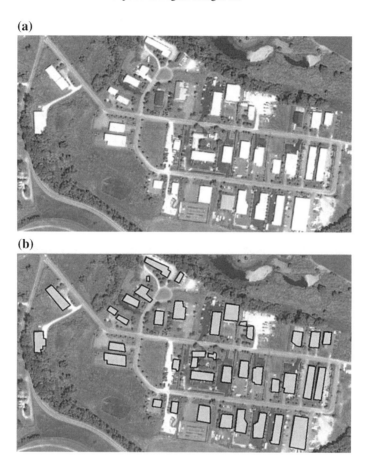

(b)

Fig. 5.27 Example of processing: **a** aerial image from [27], **b** result of building selection using active contours

extracted triangular objects are depicted in Fig. 5.29a, while the extracted roads, which are a partial case of rectangular objects, are shown in Fig. 5.29b.

5.7 Conclusions

The problem of feature construction for the object description and extraction has been discussed in this chapter. An advanced edge-based method for automatic detection and localization of straight edge segments is developed, which gives an ordered set of straight line segments with their orientations and magnitudes. The algorithm automatically adjusts an appropriate mask size for the slope line filter to give the maximum normalized filter output in order to reliable detection of the local edges and edge intersection points. The comparative analysis of the noisy image

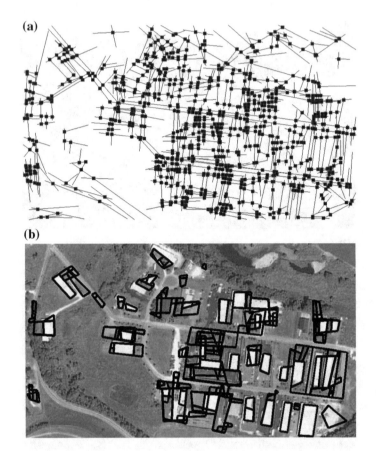

Fig. 5.28 Selection of closed structures in aerial image by proposed algorithm: **a** extracted straight line segments, **b** grouping of closed structures

shows that the advanced algorithm is inferior to others in feature detection performance. A hierarchical set of features is developed for object description subject to the proposed feature detector. This set contains four levels of line combination, each level relating to the number of combined line segments.

The proposed method was applied to the processing of real aerial and satellite images. Features of the first four levels of the hierarchical description were extracted, and the results show that the quality of extraction is high enough to recognize similar objects. Evaluation of the recall-precision characteristics for image matching will be carried out in future work. It is also important to generalize these feature descriptions to the processing of color images. Also the problem of object selection· using the straight line segments extraction and grouping has been discussed. The experiments with noisy images were conducted. Applications to real aerial, satellite, and SAR images show a good ability to separate and extract rectangular objects like buildings and other line-segment-rich structures. Most of

(a)

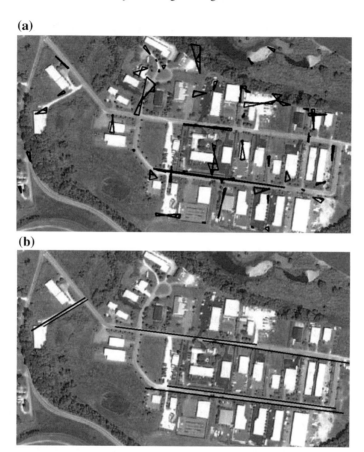

(b)

Fig. 5.29 Objects' extraction in aerial image: **a** extracted triangular objects, **b** roads' extraction

objects are selected somehow or other and the following problem is how to improve the grouping process.

Acknowledgements Author thanks to Prof. Rudolf Germer from TU Berlin for collaboration, Dr. J. Wernicke from EMT (Penzberg) for picture material and HTW Berlin and DAAD for support of the work.

References

1. Iqbal, Q., Aggarwal, J.K.: Retrieval by classification of images containing large manmade objects using perceptual grouping. Pattern Recognit. **35**(7), 1463–1479 (2002)
2. Movahedi, V.: Contour Grouping. Department of Computer Science and Engineering & Centre for Vision Research, Qual Exam, York University (2009)

3. Sohn, G.: Extraction of buildings from high-resolution satellite data and Lidar. In: XXth ISPRS Congress, vol. XXXV, Part B3, pp. 1036–1042 (2004)
4. Srinivasan,P., Wang, L., Shi, J.: Grouping contours via a related image. In: 21st International Conference on Neural Information Processing Systems (NIPS'2008), pp. 1553–1560 (2008)
5. Ferrari, V., Fevrier, L., Jurie, F., Schmid, C.: Groups of adjacent contour segments for object detection. IEEE Trans. Pattern Anal. Mach. Intell. **30**(1), 36–51 (2008)
6. Lu, Ch., Latecki, L.J., Adluru, N., Yang, X., Ling, H.: Shape guided contour grouping with particle filters. In: IEEE 12th International Conference on Computer Vision (ICCV'2009), pp. 2288–2295 (2009)
7. Hedau, V., Arora, H., Ahuja, N.: Matching images under unstable segmentations. In: IEEE Conference on Computer Vision and Pattern Recognition (CVPR'2008), pp. 551–563 (2008)
8. Shao, J., Mohr, R., Fraser, C.: Multi-image matching using segment features. Int. Arch. Photogramm. Remote Sens. **XXXIII**(Part B3), 837–844 (2000)
9. Mikolajczyk, K., Zisserman, A., Schmid, C.: Shape recognition with edge-based features. In: British Machine Vision Conference (BMVC'2003), pp. 779–788 (2003)
10. Kadir, T., Brady, M.: Saliency, scale and image description. Int. J. Comput. Vis. **45**(2), 83–105 (2001)
11. Tuytelaars, T., Mikolajczyk, K.: Local invariant feature detectors: a survey. Comput. Graph. Vis. **3**(3), 177–280 (2007)
12. Volkov, V., Germer, R., Oneshko, A., Oralov, D.: Object description and extraction by the use of straight line segments in digital images. In: International Conference on Image Processing, Computer Vision and Pattern Recognition (IPCV'2011), pp. 588–594 (2011)
13. Moreels, P., Perona, P.: Evaluation of features detectors and descriptors based on 3D objects. Int. J. Comput. Vis. **7**(3), 263–284 (2006)
14. Mikolajczyk, K., Schmid, C.: A performance evaluation of local descriptors. IEEE Trans. Pattern Anal. Mach. Intell. **27**(10), 1615–1630 (2005)
15. Sohn, G., Dowman, I.J.: Extraction of buildings from high resolution satellite data. In: Baltsavias, E.P., Gruen, A., VanGool, L. (eds.) Automatic Extraction of Man-Made Objects from Aerial and Space Images (III), pp. 345–355. CRC Press (2001)
16. Brown, M., Hua, G., Winder, S.: Discriminative learning of local image descriptors. IEEE Trans. Pattern Anal. Mach. Intell. **33**(1), 1–14 (2011)
17. Horaud, R., Veillon, F., Skordas, T.: Finding geometric and relational structures in an image. In: Faugeras, O. (ed.) Computer Vision—ECCV 90, LNCS, vol. 427, pp. 374–384. Springer (1990)
18. Kim, S.K., Ranganah, H.S.: Efficient algorithms to extract geometric features of edge image. In: International Conference on Image Process, Computer Vision, and Pattern Recognition (IPCV'2010), vol. 2, pp. 519–525 (2010)
19. Volkov, V., Germer, R., Oneshko, A., Oralov, D.: Object description and finding of geometric structures on the base of extracted straight edge segments in digital images In: International Conference on Image Process, Computer Vision and Pattern Recognition (IPCV'2012), Part II, pp. 805–812 (2012)
20. Grompone von Gioi, R., Jakubovich, J., Morel, J.M., Randall, G.: LSD: a line segment detector. IEEE Trans. Pattern Anal. Mach. Intell. **32**(4), 722–732 (2010)
21. Medioni, G., Nevatia, R.: Matching images using linear features. IEEE Trans. Pattern Anal. Mach. Intell. **6**(6), 675–685 (1984)
22. Fu, Z., Sun, Z.: An algorithm of straight line features matching on aerial imagery. Int. Arch. Photogramm. Remote Sens. Spat. Inf. Sci. **XXXVII**(Part B3b), 97–102 (2008)
23. Zhao, Y., Chen, Y.Q.: Connected equi-length line segments for curve and structure matching. J. Pattern Recognit. Artif. Intell. **18**(6), 1019–1037 (2004)
24. Lavigne, D.A., Saeedi, P., Dlugan, A., Goldstein, N., Zwick, H.: Automatic building detection and 3D shape recovery from single monocular electro-optic imagery. In: Kadar, I. (ed.) Signal Processing, Sensor Fusion, and Target Recognition XVI, SPIE Defence & Security Symposium, vol. 6567, Article id. 656716 (2007)

25. Magli, E., Olmo, G., Presti, L.L.: On-board selection of relevant images: an application to linear feature recognition. IEEE Trans. Image Process. **10**(4), 543–553 (2001)
26. Tretyak, E., Barinova, O., Kohli, P., Lempitsky, V.: Geometric image parsing in man-made environments. Int. J. Comput. Vis. **97**(3), 305–321 (2012)
27. Theng, L.B.: Automatic building extraction from satellite imagery. Eng. Lett. **13**(3), EL_13_3_5 (2006)
28. Jin, X., Davis, C.H.: Automated building extraction from high-resolution satellite imagery in urban areas using structural, contextual, and spectral information. EURASIP J. Appl. Signal Process. **14**, 2196–2206 (2005)
29. Ettarid, M., Rouchdi, M., Labouab, L.: Automatic extraction of buildings from high resolution satellite images. In: XXIst ISPRS Congress, vol. XXXVII, Part B8, pp. 61–65 (2008)
30. Li, Y., Shapiro, L.G.: Consistent line clusters for building recognition in CBIR. In: 16th International Conference on Pattern Recognition (ICPR'2002), vol. 3, pp. 952–956 (2002)
31. Jia, W., Zhang, J., Yang, J.: SAR image and optical image registration based on contour and similarity measures. In: Geo-spatial Solutions for Emergency Management (GSEM'2009), pp. 1–5 (2009)
32. Song, Y., Yuan, X., Xu, H.: A multi-temporal image registration method based on edge matching and maximum likelihood estimation sample consensus. Int. Arch. PRSSI Sci. **XXXVII**(Part B3b), pp. 61–66 (2008)
33. Bergevin, R., Bernier, J.F.: Detection of unexpected multi-part objects from segmented contour maps. Pattern Recognit. **42**(11), 2403–2420 (2009)
34. Kim, S.K., Ranganah, H.S.: Efficient algorithms to extract geometric features of edge images. In: Image Process, Computer Vision, and Pattern Recognition (IPCV'2010), Part II, pp. 519–525 (2010)
35. Brown, M., Hua, G., Winder, S.: Discriminative learning of local image descriptors. IEEE Trans. Pattern Anal. Mach. Intell. **33**(1), 1–14 (2011)
36. Venkateswar, V., Chellappa, R.: Extraction of straight lines in aerial images. IEEE Trans. Pattern Anal. Mach. Intell. **14**(11), 1111–1114 (1992)
37. Belongie, S., Malik, J., Puzicha, J.: Shape matching and object recognition using shape contexts. IEEE Trans. Pattern Anal. Mach. Intell. **24**(24), 509–521 (2002)
38. Cao, F., Muse, P., Sur, F.: Extracting meaningful curves from images. J. Math. Imaging Vis. **22**(2–3), 159–181 (2005)
39. Bernstein, E.J., Amit, Y.: Part-based statistical models for object classification and detection. In: IEEE Computer Society Conference on Computer Vision and Pattern Recognition (CVPR'2005), vol. 2, pp. 734–740 (2005)
40. Volkov, V., Germer, R.: Straight edge segments localization on noisy images. In: International Conference on Image Process, Computer Vision and Pattern Recognition (IPCV'2010), vol. II, pp. 512–518 (2010)
41. Lu, X., Yaoy, J., Li, K., Li, L.: Cannylines: a parameter-free line segment detector. In: IEEE International Conference on Image Processing (ICIP'2015), pp. 507–511 (2015)
42. Liu, Zh., Wang, J., Liu, W.P.: Building extraction from high resolution imagery based on multi-scale object oriented classification and probabilistic Hough transform. In: IEEE International Geoscience and Remote Sensing Symposium (IGARSS'2005), pp. 2250–2253 (2005)
43. Volkov, V., Germer, R., Oneshko, A., Oralov, D.: Object selection by grouping of straight edge segments in digital images. In: International Conference on Image Process, Computer Vision and Pattern Recognition (IPCV'2013), pp. 321–327 (2013)

Chapter 6
Automated Decision Making in Road Traffic Monitoring by On-board Unmanned Aerial Vehicle System

Nikolay Kim and Nikolay Bodunkov

Abstract The study is dedicated to solving the target issues of the ground traffic monitoring aided by the Unmanned Aerial Vehicles (UAV) based on applying the on-board computer vision systems. The classification of the road situations using images obtained after Traffic Accident (TA) is based on the feature set, facts, and attributes specified directly and/or indirectly on a possible situation class. The hierarchical structure of description of a road situation observable after the TA event is developed. For decision making, the production model of knowledge representation and corresponding Knowledge Base (KB) is offered to use. The issues related to decision making for recognition of the occurring traffic situations have been considered. The analysis of the strategies have been carried out based on the principles of minimizing the overall losses, limiting the admissible UAV flight altitude, and ensuring the required class recognition reliability. The models describing the functional criteria of the losses, flight safety of the UAV, and reliability of class recognition have been proposed. It has been shown that applying the minimum loss criterion ensures considerable savings of resources under different ratio of the loss quotients. The example for classification of a road incident using the real images is given.

Keywords Road traffic monitoring · Unmanned aerial vehicle · Situation classes
Objects recognition · Functional criteria · Losses · Flight safety
Recognition reliability

N. Kim (✉) · N. Bodunkov
Moscow Aviation Institute, 4 Volokolamskoe Shosse,
Moscow 12599, Russian Federation
e-mail: nkim2011@list.ru

N. Bodunkov
e-mail: boduncov63@hotmail.com

© Springer International Publishing AG 2018   149
M.N. Favorskaya and L.C. Jain (eds.), *Computer Vision in Control Systems-3*,
Intelligent Systems Reference Library 135, https://doi.org/10.1007/978-3-319-67516-9_6

6.1 Introduction

Currently, great attention has been paid to automating processes of traffic control and traffic management. In particular, the use of the UAVs equipped with the built-in video cameras acquires a wide application. In these and other investigations, the researchers try to increase the reliability for detection of objects of interest [1–4] and productivity of calculations [5]. In the long term, a use of the UAV for automatic traffic monitoring can provide the essential improvement of the capacity of traffic, optimization of actions for elimination of consequences of road incidents, and reduction of the losses from incidents.

The UAV is equipped by the observation systems including video cameras that provide the images (video sequences) and on-board computing system that can allocate the Objects of Interest (OI) like roads, vehicles, and people. Estimation of the movement parameters of the OI (in the case of traffic monitoring—vehicles) will allow automatically (without the assistance of the operator) to allocate segments of the road with the lowered capacities or with the happened TAs.

The on-board programs use the techniques of computer vision, as well as methods of image processing and analysis. For example, various methods of image processing provided the OI detection using templates and features of the movement are presented in researches [6, 7]. Problems of extension of the observation conditions are considered in work [8]. Based on similar techniques, large number of researches, in particular, of allocation and detection of cars on the roads is carried out [9–11]. At the same time, the only detection of the objects, which are present in the observed scene, and determination of their movement parameters do not allow to estimate a current road situation comprehensively.

For the estimation and forecast of development of the present road situations, it is required to classify these situations [12]. In particular, such classification allows to define the necessary actions for mitigation of consequences of a current situation. Image recognition (classification) itself is extremely difficult process based on the estimation of accessory of the allocated features to the corresponding classes of images (in the considerable case, the situation classes) [6, 13]. At this, if classification is based on identification and analysis of visual characteristics, one should distinguish and detect all OIs, as well as relationships between them [14–16]. The OIs include vehicles, drivers, pedestrians, and environmental conditions, under which the particular road traffic situations may occur. There are some works connected with the detection and identification of different vehicles on the roads or detection of a traffic accident, for example, in the case of cars' collision. Thus, in work [12] it was demonstrated that it is possible to examine a current situation class during tracking two vehicles before a collision.

However, a traffic accident with simultaneous UAV monitoring of such road segment is very unlikely. That is why in the vast majority of real cases, the UAV is not a direct testifier of a specific road traffic situation, for example, collision of cars. In such a way, it is necessary to consider a possibility of situations' classification based on the observed scene after the situation occurrence. At this, the characteristics of situation classes include the relative positions of cars, their damages, and position and behaviour of people. All these data must be contained in an automatically formed description of the observed scene (situation) on the UAV board [17, 18].

As it is noted above, numerous problems connected with the detection, recognition, and position examination of the OIs are effectively solved. At the same time, the problems related to detailing some attributes of an observed scene are relatively novel. Nevertheless, a significant success was achieved in this direction. The normal and abnormal behaviour detection was considered, for example, in works [18, 19]. A possibility of separating out people with abnormal behaviour at the observed scene was demonstrated in these ones and other researches.

It has been shown in the study [3] that a difficulty in detection and classification of the occurring special traffic situations is associated with the high information workload on an operator, as well as with a lack of useful information in the obtained images. In that study, the automation issues, such as the detection of special situations, preliminary classification of the special situations, and collection video information that is relevant to a preliminary identified class and required for the operator to make informed decisions, are discussed.

However, the available studies do not consider the issues, when several tasks are need to be solved simultaneously, such as the following ones:

- Ensuring the required traffic capacity of a road based on the detected traffic impediments, pre-congestion, congestion, and other special accident cases that eliminate the consequences of special situations and prevent the road accidents.
- Ensuring the safety of the people by implementing actions (measures) aimed at preserving their life and health.
- Securing the safety of the UAV together with its equipment.

A complexity of the decisions increases, when the efforts that minimize the potential material damages are undertaken simultaneously to avoid expenses for false alarms in rescue services, technical assistance services, or expenses related to downtime of transport vehicles. Mistakes of the operator or delay in a decision making on classification of the special situations can result in the considerable material losses and insufficient safety of the people. Thus, a solving the these issues is an urgent and practically important problem.

The purpose of the chapter is to develop a classification (recognition) method for road traffic situation based on the analysis of an observed scene obtained in the UAV board after occurrence of an accident. Among the problems being solved (ensuring the required capacity of a controlled road section), we are interested in classification connected with a mitigation of situation consequences. The classification of occurred road traffic situations is required for the examination and forecasting of their progressing [13]. Particularly, it is required for determining the response measures, which are necessary in such cases.

The chapter is organized as follows. The algorithm for classification of a road traffic situation is discussed in Sect. 6.2. Section 6.3 provides an alphabet forming the situation class. Detection of scene objects is considered in Sect. 6.4. Forming the observed scene descriptions are situated in Sect. 6.5. Decision making procedure is represented in Sect. 6.6, while a use of the statistical criteria for decision making is considered in Sect. 6.7. Section 6.8 includes a discussion about the functional criteria model. Examples are located in Sect. 6.9. Section 6.10 concluded the chapter.

6.2 Algorithm for Classification of Road Traffic Situation

Consider the algorithm for classification of road traffic situation. The observed scene image (or video sequence) obtained from the on-board surveillance system (video camera) is sent to the algorithm input. Decision on the suggested class of the observed situation is the algorithm output. Hereinafter, a situation analysis is represented by the detection of the OIs, identification of their parameters, and detection of the inter-object relationships in the context of classification. Flow-chart of this algorithm is given in Fig. 6.1.

Video sequences of the observed scene are passed to Block 1. The detection, recognition, and estimation of positions and parameters of the OIs are carried out on the basis of processing and analysis of the obtained images. A list of the OIs must be determined in Block 4 (based on the ontology). The descriptions of the observed

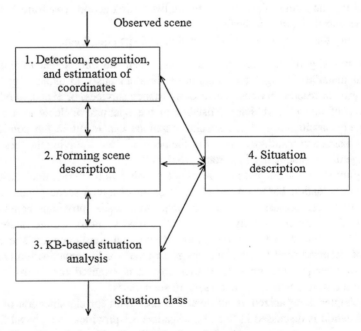

Fig. 6.1 Flow-chart of algorithm for classification of situation

scene are formed in Block 2 on the basis of obtained data. At this, general description structure is determined in Block 4. As a result, a description containing a set of facts required for situation classification is formed. Situation analysis is carried out in Block 3. The determination of the target class of a road traffic situation is a result of such analysis. It is suggested that classification should be carried out on the basis of causal logic [11] connecting causes and effects of examined events by means of pre-arranged KB. Facts obtained from a description formed in Block 2 are used for a decision making. Block 4 contains the general description of situations determining the structure of the OIs descriptions, their properties (attributes), and inter-object relationship.

Accept that a classification method includes solution of the following tasks (similarly to [12, 13]):

Task 1. The selection of classification type and an alphabet forming of situation class.

Task 2. The forming descriptions in the observed scene containing attributes of recognized situations (close to the feature dictionary [13]).

Task 3. The choice of the decision-making algorithms that allows ranging the examined situation to corresponding class.

Consider in details the steps for classification of traffic situations.

6.3 Alphabet Forming for Description of Situation Class

Situation class facilitates a selection of control actions aimed to mitigate the consequences of specific road traffic situations. It means that the selected situation class alphabet must ensure a possibility of such situations' detection and comply with the set of possible control actions. Assume that during mitigation of consequences one should ensure:

- Timely delivery of healthcare to the injured.
- Minimization of normal traffic recovery time.
- Reduction of possible losses due to a calling the aids that do not correspond to the occurred situation.

In such a way, the classification of the occurred situation must facilitate decision making regarding selection of aids for mitigation of road traffic situation consequences.

In the simplest cases, the traffic situations are described by a set of static quantities (statistics) calculated by a vehicle scalar speeds' set. To convert a vector velocity into scalar form, it is possible to project a velocity on the road centreline (the scalar product of the velocity vector and directrix vector of the road centreline). In particular, the velocity distribution diagrams, number of the vehicles, average and maximum speed, and mean square deviations (Sect. 6.6) can be selected as static quantities. Figure 6.2 shows the examples of vehicle velocity distribution histograms on the segments of a road in various standard situations. The examples were obtained by modelling of traffic processes. The first histogram in Fig. 6.2a shows an example of traffic with some vehicles moving in both directions at a normal speed, some at a low speed, and some being stationary. The second histogram in Fig. 6.2b shows a standard situation with two-way traffic (a number of vehicles is displayed on the vertical axis; a velocity, km/h, is mentioned on the horizontal axis).

More complex options include a location of vehicles in relation to one another and to the lines of traffic, as well as the direction of the velocity vector in relation to the road centre line.

In general, the process of classification or recognition (or detection in some cases) within a scope of tasks solved by Support Vector Machine (SVM) consists of specifying the class of the observed traffic situation on the basis of video information and relating a set of features to the corresponding class. Denote a set of situation classes as $X = \{x_1, x_2, \ldots, x_m, \ldots, x_M\}$, where M is a total number of situation classes. Vectors $Y = \{y_1, y_2, \ldots, y_n, \ldots, y_N\}$ characterizing different classes of situations in the feature space are referred to as the implementation vectors, N is a total number of features used to describe a situation. The complete set of features is called the feature dictionary. Conventionally, it is assumed that all traffic situations are divided into five classes ($M = 5$) that comprise the source classes alphabet.

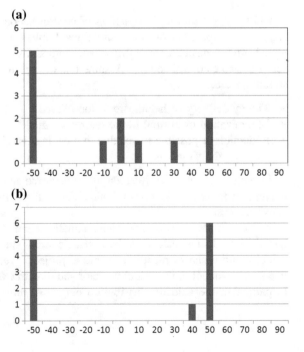

Fig. 6.2 Histograms of the vehicle speed: **a** traffic with some vehicles moving in both directions with different speed, **b** two-way traffic

Herewith, class x_1 corresponds to a standard traffic situation, where the traffic capacity of a specific road segment in a given season and at a given time of day or night lies within the specified tolerance limits. Assume that classes x_2, x_3, x_4, x_5 refer to the abnormal situations that push current values of the road traffic capacity outside the specified (for a normal situation) tolerance limits. Class x_2 is an abnormal situation that does not result in direct financial losses but disturbs the traffic capacity and push it outside the tolerance limits specified for a standard situation as well. The abnormal situations that belong to the classes x_3, x_4 are emergencies followed by financial losses including but not limited to vehicle damage of varying severity. Situation x_5 is catastrophic and results in human losses. Moreover, we assume that special situations x_2, x_3, x_4, x_5 occur because of vehicles collisions. The higher the class of the situation, the more severe are the consequences.

For detection and classification of traffic situations, we use the statistical technique of recognition. When a particular situation (regardless of its class) occurs, it must be detected (step 2).

6.4 Detection of Scene Objects

In accordance with Fig. 6.1, the detection, recognition, and measurement of coordinates of the objects of interest are carried out in Block 1. Recognition/detection of vehicles is of paramount importance for detection of road traffic situations. It is required to solve the tasks of the vehicle position and speed definition based on image/videos obtained from the UAV. There are two basic approaches to vehicle detection: the detection of moving vehicles using video sequences or vehicle detection irrespective of their speed based on separate images. Methods for detection of moving vehicles are more accurate [20]. During recognition of specific situations, a detection of moving vehicles is enough. For recognition of a wide range of road traffic situations (called as traffic jams, consequences of road accidents, no active traffic, etc.), a detection of immovable or slowly moving objects is also important.

Methods for vehicle detection based on the edge detection (such as Canny edge detector [21]) are not always successful: the false tripping and merging of vehicle contours with shadows, road cracks, wayside, road marking, or other vehicles take place. When the contours appear to be unstable, the separate vehicle sides may be not detected. Segmentation and/or extraction of characteristics are more appropriate methods. For this case, a vehicle detection scheme is given below:

1. Zooming and determination of a Region Of Interest (ROI), usually one or several rectangular covering a road together with a wayside. At this, navigation data are used.
2. Image segmentation in the ROI.

3. Initial segment filtering (called as filtering out very large of small segments) and merging some adjacent segments. A set of regions is obtained.
4. Filtering of regions.
5. Removal of irrelevant region attributes.
6. Vehicle recognition based on the region attributes.
7. Vehicle recognition based on the speed fields.

As a result of segmentation, all processed images are divided into a set of connected non-overlapping segments:

$$X = U_i S_i,$$ (6.1)

where X is a multitude of segmented image pixels, S_i is a segment, i is an integer index and

$$S_i \cap S_j = \varnothing \quad i \neq j.$$

Segmentation is arranged in such a way that a separate segment would be homogenous enough and different but the adjacent segments would differ in their colour or brightness. In order to choose the most effective method for the following segmentation, some methods were examined, such as Felzenszwalb-Huttenlocher Segmentation [22], Quick Shift [24], Simple Linear Iterative Clustering (SLIC) [23], Maximally Stable Extremal Regions (MSER) [24], and Model Based Clustering (MBC) [25]. At present, only Fast Hierarchical Segmentation (FHS) method ensures the operating speed required for on-board data processing [26]. At this, the FHS method ensures adequate segmentation quality and allows to merge large homogenous areas into one region (for example, it allows combining the major part of the road in one segment instead of many similar segments). All small segments (under limits) are combined with the adjacent large segments. The FHS method, as well as the majority of mentioned methods, uses colour information.

In work [26], as well as in our research, it is demonstrated that the vehicle imagery is close to target by one, two, or three segments using the FHS method. That is why, all separate segments combined pairs of adjoining segments and combined triplets of segments, where at least one segment is adjoining to two other ones, are considered for vehicle detection. The obtained sets of points (hereinafter, regions) are filtered by the limit area (on the top and on the bottom) and number of geometric parameters like the area, occupancy, and sizes and ratio of the sides of minimally inclusive turned rectangular (the effective algorithm for rectangular forming is represented in work [12]).

Now, the task of vehicle imagery detection comes down to the task of region binary classification. To solve such task, a set of attributes based on the geometrical and textural analysis is formed. The vehicle attributes can be generated on the basis of Histogram of Oriented Gradients (HOG) [27], Haar-feature [28], and AdaBoost [29]. The HOG and Haar-feature are turn-sensitive; that is why it is necessary to reproduce such features by turning angle with some step. Some approaches deal with the search of invariant attributes. Nevertheless, such methods of object classification are replaced by the unsupervised learning algorithms [29]. Another way to solve the classification task based on the training of multilayer auto-associative neural networks (deep learning) was suggested in [30]. The advantage of this method is that it is resistant to noise contamination and, in the case of quite representative sampling, to a wide range of distortions. The main idea of this method is that the initial attribute vector is transformed through a neural network cascade into a small vector and then this vector is transformed through the other neural network cascade into initial vector. Each class has its own training scheme. During detection, the class, for which transformation was the most accurate, is selected. For example, the task of pedestrian detection using video sequence can be solved by different methods like Viola–Jones classifier, correlation filter, and the HOG method.

In Fig. 6.3, the example of pedestrian detection using the HOG descriptors [27] based on calculation of the number of gradient directions in local image regions is given. Detection errors are the following: a number of the false targets is equal to 2 and a number of the missed targets is equal to 4.

This method supports the invariance of geometric and photometric transformations, except an object orientation. The final step of object detection is a descriptor classification using supervised systems, for example SVM.

Detection of moving vehicles. There are a lot of the works dedicated to detection of moving vehicles [20, 23, 30]. The offered detection methods are mainly reduced to the following ones:

Fig. 6.3 Example of pedestrian detection

- Variant I. Detection of static objects with their following tracking.
- Variant II. Detection of moving regions or points and their filtering.
- Variant III. Creation of supervoxels (the voxels segmented in time and space) [23]. (This variant does not consider in the chapter, it is a topic for separate research.)

Variant I is substantially keeping a track of the object; its implementation is considered further. For variant II we have examined the following movement detection methods:

- Correlation method (fining the maximum of correlation function between the adjacent frames).
- Lucas-Kanade method for detection of optical flow in the image pyramid and fixed grid.
- Farneback method based on quadratic (in relation to coordinates and time) models of image intensity representation [31].

Correlation method with a fine grid works very slowly but it cannot provide the necessary precision using a rare grid. Lucas-Kanade method allows to create a speed field in each point in the image Pyramid. However, it is not as fast and accurate as other methods. On the coarse grid, Lucas-Kanade method works fast and is suitable for determination of total wipe. Farneback method creates a list of moving points and keeps track of them autonomously. Farneback method works fast (for example, it well processes the images with resolution 1080×760 pixels using a laptop with 120 FPS, a single core mode, Intel i7 3300 MHz processor). Farneback method is suitable for detection of moving vehicles.

All movement detection methods give a false tripping, which is rarely observed on the road and more frequently—on the sideway or in bushes. That is why, the results of moving object detection should be filtered or combined with other methods. The output of Farneback method is represented by a set of points that are determined to be moving. For each such point, a region containing it (not all such regions correspond to actual vehicles) may be determined. Sets of such regions are also classified based on supervised techniques similar to classification of regions in separate images.

Detected vehicle tracking and speed estimation. In order to create the descriptions of road traffic situations, it is important not only to detect vehicles but also to estimate their speed and, in some cases, a path. The highest accuracy of movement estimation is also reached using the tracking algorithms of moving objects.

Sets of descriptions in each frame (in fact, image regions that correspond to vehicles) represent a source data for the tracking system. Separate vehicle detected in the current frame is described by attribute vector $\vec{Z} = (z_1, z_2, \ldots, z_l, \ldots, z_L)$, where L is a total number of attributes. This set of attributes includes data of the object location and its geometric characteristics, as well as initial estimations of its speed (carried out on the basis of optic flows creation). Using vector $\vec{Z}$, one can obtain the general estimation of 2D coordinates of the object $\vec{X}(\vec{Z})$ (for example, coordinates of centre of mass or centre of inclusive turned rectangular in practice may coincide with two $\vec{Z}$ coordinates). Set of vehicle descriptors $U = (\vec{Z}_1, \vec{Z}_2, \ldots, \vec{Z}_k, \ldots, \vec{Z}_K)$ is a result of frame processing, where K is a total number of vehicles detected in the frame. For frame with number t, a set of object descriptors U_t is formed.

Two levels are suggested for vehicles' tracking: the level of hypothesis generation using the separate vehicle movement and level of separate vehicle tracking based on Kalman filtering [32–34].

Hypotheses are generated on the basis of analyzing f sequential frames (f must be a small number, for example 3 and 4, in order to determine a period of time, within which one can neglect acceleration of objects). In these frames, all pairs of objects, where one object is on the current frame, and another ones is on the other frame but in limited distance are selected. Limitation is calculated using the limit speed of a vehicle, scope of image, and frame rate. The hypothesis is that the found pair of objects corresponds to one vehicle (and this vehicle movement is with some accuracy sustainable and linear on the last f frames).

Then a set of the obtained hypotheses is compared with the hypotheses accepted earlier regarding consistency. If there is no hypothesis consistent to it that was found earlier, it is added to the Hypotheses List (HL), to which Kalman filtering is applied. In order to confirm or decline the hypotheses from the HL, search of the best confirmations of Kalman filtering prognosis is used. If a hypothesis is not confirmed within f_1 frames, it will be deleted. Algorithm output at every step is a set $S = ((\vec{Z}_1 \vec{V}_1), (\vec{Z}_2 \vec{V}_2), \ldots, (\vec{Z}_R \vec{V}_R))$, where R is a number of Kalman filters, $\vec{V}_l$ is a speed calculated by Kalman filter:

$$\vec{V}_l = \frac{\vec{F}_{K_i} - \vec{X}_i}{\Delta}, \tag{6.2}$$

where Δ is a time interval between frames, $i = \overline{1, R}$ is an object index, K_i is Kalman filter for the ith object, $\vec{X}_i$ is the current coordinates of the ith object, $\vec{F}_{K_i}$ is a prediction of Kalman filter for the next step for the ith object.

Thus, there is a wide range of algorithms that allow to solve the problems of the OIs coordinates' detection, recognition, and estimation.

6.5 Forming Observed Scene Descriptions

Consider a description of structures for various possible situations occurring after traffic accidents and containing useful information that allows to classify these situations (Fig. 6.1, Block 2). The semantic description is used for convenience of its further analysis. Such description is formed based on the unknown in a priori information that causes a necessity to apply the expert estimations.

Assume that a description of the observed scene includes the following data:

1. General state of an observed scene (Class "Observed Scene").
2. Descriptions of objects of interest, i.e. those that could be a part of occurrence and progression of the TA (Class "Objects of TA").
3. Description of external conditions that could influence on the occurrence and progression of the TA (Class "External Condition").
4. Description of objects that could indirectly influence on the occurrence and progression of the TA (Class "Additional Objects").

This description has a hierarchic structure and is divided into classes, subclasses, and divisions of various levels depending on a hierarchic level. In general, various types of descriptions, such as spatial, spatiotemporal, temporal, and causal, may be used [14]. If a putative solution is based on the analysis of the OIs' positions, i.e. the spatial relationships between the objects, it is appropriate to use the spatial descriptions. During classification of road traffic situations (by their consequences), this type of descriptions is the most important. Hereafter, 2D representation of scene description involves the direction and distance between the OIs.

When a process is connected with moving object tracking, the spatiotemporal description is used [12]. The temporal descriptions are irrelevant for the considered task and not examined in this research. The causal descriptions are used during situation analysis (Fig. 6.1, Block 3). In Fig. 6.4, the ontology fragment for traffic accident description is shown.

In Fig. 6.4, class "Observed Scene" contains general information about the observed scene required for its identification, localization, and determination of the OIs. Class "Objects of TA", subclass "General Description" must contain a basic list of OIs, such as the vehicles, people, and objects, including roads and obstacles that can be objects in a traffic accident. In general, this subclass includes descriptions of the objects: their attributes (types, properties, state, and position) and relations between the objects. Subclass "People" of this class contains the descriptions of people, including properties, states, and spatial relations. Subclass "Vehicle" includes all vehicles directly involved in the TA. A description ought to include the information about the properties, condition of vehicles, and their relations to other objects in a scene. Subclass "Objects" may include the following objects: the road, buildings, or constructions, against which a vehicle crashed. Class "External Condition" contains information about the external conditions that may influence on the occurrence and progression of the TA. Examples of subclasses of this class are the following: "Weather", "Road Conditionals", and "Features of the Road". Such parameters as illumination of this road segment or vehicle density may be contained in subclass "Road Conditionals", while the subclass "Features of the Road" may contain the road design features that may cause by the TA. Class "Additional Objects" contains a description of objects that could influence on the occurrence and progression of the TA. The scheme shown in Fig. 6.3 determines a description of the observed scene after traffic accident.

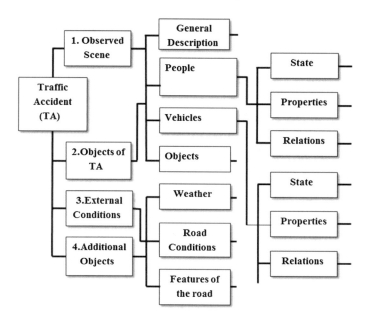

Fig. 6.4 Ontology fragment

Several vehicles have stopped on one of the traffic lanes, which greatly restricted a traffic capability in this road segment. For convenience of analysis, the precise coordinates of the objects (C) obtained in Block 1 (Fig. 6.1) are transformed into descriptions of spatial relations between the objects. The objects are numbered from left to right and from top to bottom. In descriptions, we mark directions (D) that change in steps from 1 to 8 counterclockwise in 45 angular degrees. Position of the object on the road is marked as 0. Distance between the objects (L), considering their sizes, is marked as mentioned below:

- "In close contact"—0.
- "Close"—1.
- "Neither close nor far"—2.
- "Far"—3.
- "Very far"—4.

Let us form a description of situation by classes in accordance with the given ontology structure (Fig. 6.4). Fragment of observed scene description is given below.

Let us describe a situation shown in Fig. 6.5 in s following manner: tram («$V(1)$»), vehicles («$V(2)$», «$V(3)$», «$V(4)$»), groups of people («$GP(1)$», «$GP(2)$»), and some of their properties (type and state of the object). Then a fragment of description of the observed scene is given below:

"Observed Scene":

$(FN(5777), TD(16-00), RN(3444), FD(7))$

"Objects of TA":

General Description: State: $(SP(8)^\wedge SV(6)^\wedge R(1))$

People: $\qquad P(1) : (\text{State} : PC(x : \ldots, y : \ldots)\ldots, SP(0)^\wedge B(N)^\wedge \ldots)$

$\qquad\qquad\quad P(2) : (\text{State} : PC(x : \ldots, y : \ldots)\ldots, SP(0)^\wedge B(N)^\wedge \ldots)$

Relations: $\quad GP(P(1)^\wedge P(2)) : (B(N)^\wedge St())$

$\qquad\qquad\quad GP(P(3)^\wedge P(4)^\wedge P(5)) : (B(N)^\wedge St())$

Vehicles: $\quad V(1) : (\text{State} : PC(x : \ldots, y : \ldots)\ldots, SV(7)^\wedge B(N)^\wedge D(R : 1))$

$\qquad\qquad\quad V(2) : (\text{State} : PC(x : \ldots, y : \ldots)\ldots, SV(1)^\wedge B(Un)^\wedge D(R : 7))$

Relations: $GV(V(1)^\wedge V(2)): (D(1)^\wedge L(0))$

"External Conditions":

Weather: (Wm), Road Conditions: $((Dr), Visibility(3), \ldots)$

where $FN(\cdot)$ is a frame (file) number, $TD(\cdot)$ is a time, $RN(\cdot)$ is a number of observed road section, $FD(\cdot)$ is a direction to the north (from the center of the image), $SP(8)$, $SV(6)$, $R(1)$ are the numbers of objects belonging to a specific class (people, vehicles and road correspondingly), St and W are the states: "standing" and "walking", respectively, $B(N)$ is a state (N normal, Un abnormal, and Vr not specified), $GP(P(1)^\wedge P(2))$ is a group of people, $GV(V(1)^\wedge V(2))$ is a group of objects, objects of groups are specified in brackets (one object may be ranged in several groups), $D(7)$, $L(1)$ is a course to the object (in the eight line system) including a distance to the object, $PC(x:\ldots, y:\ldots)$, $VC(x:\ldots, y:\ldots)$ are the coordinates of the objects, $Weather: (Wm)$ is the weather conditions ("warm"), Dr is a road conditions ("dry"), $Visibility(5)$ is a visibility on a 5-point scale (0 is zero visibility).

Fig. 6.5 Observed scene after traffic accident

Presented example demonstrates that offered description form is universal and ensures the possibility to form descriptions of various traffic accidents.

6.6 Decision Making Procedure

The obtained descriptions are passed to Block 3 "Analysis of the situation" (Fig. 6.1). Block 3 contains the KB, where the information about conditions of the TAs occurrence is kept. Different models of knowledge presentation are used during creation of such systems: logic, production, and network models. In order to solve these tasks, we use the production system technology based on the production rules. Description of current scene is passed to the block input, and in its output a conclusion about the most reliable situation class is given.

As some description elements are presented in imprecise form (for example, object state), the imprecise systems based on the imprecise type rules are used for creating the knowledge bases:

$$\text{if } \mu_{A_1} \cap \mu_{A_2} \ldots \mu_{A_N} \quad \text{then} \quad C = x_i$$

where $A_1, \ldots, A_N$ is the imprecise statements (for example, "normal behavior of the people", "no visible damages of the vehicles"), μ_{A_N} is a reliability of an imprecise statement, C is an output rule (for example, a conclusion about the possible situation class).

Based on the available facts obtained from the situation description, a reliability of rules is determined. Examples of rules from the KB are given in Table 6.1.

For simplicity, accept the order of sequential (in ascending order of numbers) use of conditions coinciding with available facts. For convenience of corresponding condition search $A(\ldots)$ in "Description of Observed Scene", the conditions must be described in accordance with the rules of the observed scene description. For example, item 5 from Table 6.1 will correspond:

Table 6.1 Examples of rules from knowledge base

No.	A	B
1	Vehicle is immovable on the roadway, the rest of the road is clear	TA, situation class x_2 or x_3 or x_4 or x_5
2	Vehicle is standing in close contact on the roadway	Situation class x_1 or x_2, or x_3 or x_4 or x_5
3	Vehicle condition is normal (no visible damage)	Situation class x_1 or x_2 or x_3
4	Vehicle condition is abnormal (visible damage)	Situation class x_3 or x_4 or x_5
5	Two vehicles are standing in close contact and perpendicularly to each other	TA, situation class x_4 or x_5
6	TA and increased risk to one vehicle (tram)	Situation class x_4 or x_5
7	Reliabilities x_n and x_m close and $m \geq n$	Situation class x_m

$$\text{if } (D(1)^{\wedge}L(0)) \text{ then } x_4, x_5.$$

For the scene shown in Fig. 6.5, the reliabilities of statements $D(1)$ and $L(0)$ take values "1" and "0.9", respectively. A reliability of a rule is a minimal value of reliability included in the imprecise statements rule. Thus, a reliability of rule 5 will be "0.9". Reliability of situation class is calculated as a sum of reliabilities of rules belonging to this class:

$$p_{x_i} = \sum_{i=1}^{L} \min\left(\mu_{A_1}, \mu_{A_2}, \dots, \mu_{A_N}\right),$$

where p_{xi} is a reliability of the ith class, L is a number of rules relating to the ith class.

There are four vehicles in close contact in the examined scene (rule reliability 2 = 1.0). Two of those vehicles $(V(1)^{\wedge}V(2))$ are perpendicular to each other (rule reliability 5 = 0.9). One of the vehicles participating in the TA of an increased risk is a tram (rule reliability 6 = 0.9). Thus, current situation is with the most reliability ranged to class x_4 (with significant material losses) or x_5 (with the injured). The reliabilities of the obtained results are similar. In such case, the final decision is a rule 7, i.e. the worst variant. Thus, the current situation class is x_5.

6.7   Use of Statistical Criteria for Decision Making

In previous discussion, it was demonstrated that several rules with similar reliabilities can be applied for a decision making. Approach based on the statistical criteria is used in such cases. Denote a standard situation occurrence probability as $P(x_1)$ and an abnormal situation occurrence probability as $P(x_m)$, $m = \overline{2, M}$.

Note that at this point the abnormal situations are not classified. Object recognition task can be considered as a special case of recognition if a decision contains only two outcomes: whether an abnormal situation is detected or not. This approach simplifies the solution that allows to apply various statistical detection criteria without calculating the posterior probabilities by Bayes formula [35]. To determine the decisive boundaries, it is suitable to use likelihood factor (ratio) λ [36]

$$\lambda = p(Y|x_m)/p(Y|x_1) \tag{6.3}$$

and threshold (critical) likelihood factor λ_0

$$\lambda_0 = P(x_1)\,(R_{m1} - R_{11})/P(x_m)\,(R_{1m} - R_{mm}), \tag{6.4}$$

where $P(x_n)$ is a priori probability of the situation x_n, $P(Y|x_n)$ is a conditional probability density of the feature Y upon condition that the source of information is a situation x_n, $n = \overline{1, M}$, R_{mm} is the losses caused by identification errors (situation x_n is identified as x_m).

The decision as to whether a situation is standard or an abnormal is made based on values of λ and λ_0. Assume that the losses in case of a right decision are $R_{11} = R_{mm} = 0$. In addition, in given Eqs. 6.1–6.2 the condition (*Bayes criterion*) is obtained in a view of Eq. 6.5.

$$
\begin{aligned}
&\text{if} \quad \lambda = p(Y|x_m)/p(Y|x_1) > \lambda_0 = P(x_1)\,(R_{m1})/P(x_m)\,(R_{1m}) \\
&\quad\text{then} \quad \text{the situation observed is abnormal} \quad X = x_m \\
&\quad\text{else} \quad \text{the situation observed is standard} \quad X = x_1
\end{aligned}
\tag{6.5}
$$

For a detection process, it is important to consider the errors of object recognition, when an abnormal situation is not identified (α is a type one of errors) or mistakenly identified (β is a type two of errors). If there is a possibility of one of the abnormal situation identification errors leading to unacceptable consequences, Neyman-Pearson criterion must be used to limit the tolerable values of these errors. According to the criterion, the detection algorithm conditions are provided by Eq. 6.6, where α_0, β_0 are the given maximum limits for type one and type two of errors, respectively.

$$
\min \alpha \text{ provided } \beta \le \beta_0 \quad \text{or} \quad \min \beta \text{ provided } \alpha \le \alpha_0
\tag{6.6}
$$

Likelihood factor λ is compared with a threshold value λ_0 calculated from following the formulae [36]:

$$
\alpha_0 = \int_{\lambda_0}^{\infty} p(\lambda|x_m)d\lambda \quad \text{or} \quad \beta_0 = \int_{-\infty}^{\lambda_0} p(\lambda|x_m)d\lambda.
$$

Assume that the average speed of vehicles is measured as a feature, while detecting the abnormal situations:

$$
Y = \frac{1}{K}\sum_{k=1}^{K} V_k,
$$

where K is a number of vehicles, V_k is a speed of vehicle k.

To simplify the calculations, assume that a feature distribution for each $n = 1$ mth situation follows the normal law in a view of Eq. 6.7.

$$
p(y|x_n) = \frac{1}{\sigma\sqrt{2\pi}} e^{-\frac{[y - M(y|x_n)]^2}{2\sigma^2}}
\tag{6.7}
$$

Consider the examples of dangerous situations' detection based on Bayes and Neumann-Pearson criteria. For a specific controlled road section (discarding the feature measurement, km/h), it is assumed that

$$E(y|x_1) = 60 \quad \sigma(y|x_1) = 15$$
$$E(y|x_1) = 20 \quad \sigma(y|x_1) = 5$$

Here and hereafter, $E(y|x_m)$ is a mathematical expectation of y and $\sigma(y|x_m)$ is a conditional Root-Mean-Square (RMS) deviation of y, $m = \overline{1, M}$.

Option 1. Bayes criterion
Priori probabilities:

$$P(x_1) = 0.99 \quad P(x_m) = 0.01$$
$$\text{Tolerable loss}: \quad R_{m1} = 0.9 \quad R_{1m} = 0.1$$

Option 2. Bayes criterion

$$P(x_1) = 0.99 \quad P(x_m) = 0.01$$
$$\text{Tolerable loss}: \quad R_{m1} = 0.99 \quad R_{1m} = 0.01$$

Option 3. Neumann-Pirson criterion
Tolerable error (failure to identify the Abnormal Situation (AS))

$$\beta_0 = 0.001.$$

Assume that the following values of the feature (average speed) were taken as 20, 30, 40, 50. It is required to determine, what speed corresponds to the AS and what to the Standard Situation (SS). The computational results using Eqs. 6.3–6.4 are represented in Table 6.2.

Thus, the results of the decisions made are largely dependent on the prescribed initial data. For example, on reducing the assigned losses caused by the failure to identify the AS (Option 1 in relation to Option 2) at the average speed of 40 km/h, it is decided that the situation is standard (Option 2), but in Option 1 the situation is identified as an abnormal. Similar results can be obtained using various algorithms. Option 3 shows that Neumann-Pearson criterion can provide the same solutions as Bayes criterion but without loss evaluation R_{mn}. After detecting an abnormal situation, it is necessary to classify the situation at the next step of the scenario (step 3).

Table 6.2 Computational results

Speed	20	30	40	50
Option 1	AS	AS	AS	SS
Option 2	AS	AS	AS	SS
Option 3	AS	AS	AS	SS

6.8 Use of Functional Criteria Model

In order to analyze the specified problem, the criteria models of the UAV flight safety and classes x_3, x_4 recognition reliability should be formulated. To make the investigation of these criteria simpler, the sigmoid curves, which describe a behavior of the criteria alteration quite plausibly, are selected. The heuristic models for describing the UAV safety and class recognition (identification) errors can be represented as follows.

Criterion 1. At low flight altitudes (for example, lower than 50 m), the flight safety is considerably influenced by the objects (the OIs) located close to or on the trajectory of the UAV flight path. These objects are buildings and structures, trees and bushes, power transmission towers, etc.

Assume that a relative safety W_{su} of the UAV alters a dependence of the flight altitude in a range of 0–1 and it can be calculated according to Eq. 6.8, where k_{su} is an empirical quotient that is determined based on the conditions of the flight, h and h_{su} are the altitude of the UAV flight and the altitude of a flight, when a safety of the flight equals 0.5, respectively, su is the index of the flight safety criterion.

$$W_{su} = \frac{1}{1 + e^{-k_{su}(h - h_{su})}}, \tag{6.8}$$

At $W_{su} = 0$, the UAV is crashed and destroyed. At $W_{su} = 1$, the conditions of the flight are absolutely safe.

Criterion 2. The reliability of classes x_3 and x_4 recognition depend on the altitude, at which the consequences of the accident are investigated (the UAV flight altitude). In some cases, the type of damages will be determined reliably only at the altitude of several meters (for example, an investigation of a human in a car through the car windows). Then the reliability of identification W_a (for such cases) is calculated by Eq. 6.9, where k_a is an empirical quotient depending on the surveillance conditions and location of the OIs, h and h_a are the UAV flight altitude and the altitude of a flight, when a safety of the flight equals 0.5, respectively, a is the index of the flight safety criterion.

$$W_a = \frac{1}{1 + e^{-k_a(h - h_a)}} \tag{6.9}$$

At $W_a = 0$, the reliability of identification W_a is minimal. Assuming the equal probability of the outcomes, it can be considered that a probability (reliability) of false identification is equal to 0.5. At $W_a = 1$, the reliability is maximum and the probability of the identification error equals 0. Quotients k_{su}, k_a (dimensionality 1/m) and the values of altitude h_{su}, h_a (dimensionality m) can be determined based on the previous experience of investigating similar special situations.

As an example in Fig. 6.6, the plots of altering the relative safety W_{su} (dashed line) and identification reliability W_a of class x_3 (solid line) are represented. Their

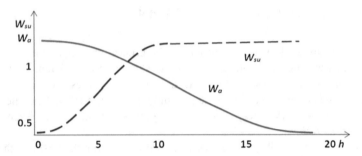

Fig. 6.6 Plots of altering the relative safety W_{su} (*dashed line*) and identification reliability W_a of class x_3 (*solid line*)

values are indicated on the vertical scale and on the horizontal scale the UAV flight altitude in meters is indicated. Here, the following values were assumed: $k_{su} = 1$ m^{-1}, $h_{su} = 6$ m, $k_a = 0.6$ m^{-1}, $h_a = 11$ m.

From the plot of W_{su} it is clear that at the UAV flight altitudes lower than 2 m, the flight safety is approximating to zero. The safety W_{su} is close to 0.5 at the flight altitude close to 6 m, and the flight higher than 10 m is practically safe. Analysis of plot W_a shows that a sufficiently high reliability of surveillance can be achieved at the flight altitude lower than 4 m. The reliability close to 0.5 is obtained at the flight altitude near 10–12 m. Thus, if it is required to improve a reliability of class identification and to ensure the UAV safety simultaneously, the corresponding criteria 1 and 2 can come to a conflict.

Basic option for reconciling such conflicts is to minimize the potential general (total) losses (Option 1 from Sect. 6.7), which can be put by Eq. 6.10:

$$R_\Sigma = R_{su} + R_a, \tag{6.10}$$

where

$$R_{su} = R_{su}^0 (1 - W_{su}) \tag{6.11}$$

is a loss associated with the UAV safety, R_{su}^0 is a quotient of losses depending on the UAV price and on its maintenance expenses and

$$R_a = R_a^0 (1 - W_a) \tag{6.12}$$

is a false alarm loss, when the rapid action team of the rescue service is called, which is associated with class x_3 identification error, R_a^0 is a quotient of losses depending on the expenses for the rapid action team false alarm.

Assume, for example, $R_{su}^0 = 100$ units and $R_a^0 = 15$ units, where units identify some conditional units of value. The plots of losses for the previous example are depicted in Fig. 6.7, where R_Σ, R_{su}, and R_a are shown in the dash-dot, dashed, and solid lines, respectively.

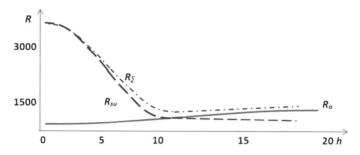

Fig. 6.7 The plots of losses, R_Σ, R_{su}, and R_a are shown in the *dash-dot*, *dashed*, and *solid lines*, respectively

It is clear from the plots that the minimal losses R_Σ is achieved at the flight altitude close to 10–11 m. The reliability of the situation class identification for these flight altitudes is close to 0.5 but an attempt to increase this reliability value by the expense of decreasing h will be resulted in serious potential losses associated with the UAV crash.

The following versions of the decision making strategies (Options 2, 3) are realized by limiting the UAV safety and identification reliability in the form of the conditions provided by Eq. 6.13, where W_{0su}, W_{0a} are the preset limitations.

$$W_{su} \geq W_{0su} \quad W_a \geq W_{0a} \tag{6.13}$$

Table 6.3 shows the options for calculating the losses and the parameters W_{su}, W_a at different correlations of the loss quotients.

Example 1 (Table 6.3) considers the case, when the loss quotients are substantially different. Due to the high relative value of $R_{su}^0 = 100$, the losses R_{su} determine the position of min $R_\Sigma = 8$ (Criterion 1.1) at altitude $h = 11$ m. This value of altitude is close to the maximum values of Criterion 1.2 $W_{su} \geq 0.99$ but at a relatively low value of $W_a = 0.5$. If the precondition $W_a \geq 0.99$ (Criterion 1.3) is given, then the minimum of losses is shifted to altitude $h = 6$ m. At that, the total losses increase considerably up to $R_\Sigma = 51$, and the UAV safety decreases $W_{su} = 0.5$.

Example 2 (Table 6.3) is different in that the loss quotients are equal, which results in shifting the minimum of losses for Criterion 2.1, to $R_\Sigma = 26$ at the flight altitude of $h = 8$ m. Limitations $W_{su} \geq 0.99$ and $W_a \geq 0.99$ give the same total losses $R_\Sigma = 51$ at the flight altitude values $h = 11$ m and $h = 6$ m, respectively. In all, according to the calculations, criterion min R_Σ is the most economically feasible but in a number of cases it does not allow to ensure the sufficiently high values of functional criteria W_{su} and W_a.

Table 6.3 Options for calculating the losses and the parameters W_{su}, W_a at different correlations of the loss quotients

Examples	Criterion	h	R_Σ	R_{su}	R_a	W_{su}	W_a
1. $R_{su}^0 = 100$ $R_a^0 = 15$	1.1 min R_Σ 1.2 $W_{su} \geq 0.99$	11	8	1	7	0.99	0.5
	1.3 $W_a \geq 0.99$	6	51	50	1	0.5	0.99
2. $R_{su}^0 = 100$ $R_a^0 = 100$	2.1 min R_Σ	8	26	13	13	0.87	0.87
	2.2 $W_{su} \geq 0.99$	11	51	1	50	0.99	0.5
	2.3 $W_a \geq 0.99$	6	51	50	1	0.5	0.99

6.9 Examples

Assume that a collision of two vehicles occurred in vision of the UAV, i.e. the AS development was observed in dynamics. The collision condition is provided by Eq. 6.14, where D_{qr} is a distance between the vehicles with indices q and r.

$$D_{qr} = 0 \tag{6.14}$$

In such case, the closing speed (dimension m/s) before the collision can be taken as a feature characterizing the AS classes

$$Y = \dot{D}_{qr}. \tag{6.15}$$

Connection between the AS class and closing speed is random in nature, thus statistical methods will be applied to the AS classification (recognition), as well as to their detection.

In general, the recognition or classification (including the AS) using the statistical methods are based on the calculation of the posterior probability by Bayes formula (Eq. 6.16), where $P(x_m)$ and $P(x_k)$ are the priori probabilities of situations x_m and x_k, respectively, $P(Y|x_m)$ is a conditional probability density of a feature vector Y upon condition that the source of information is a situation x_m, $k = 2, 3, 4$, M are the situation classes indices, $M = 5$.

$$P(x_m|Y) = \frac{P(x_m)P(Y|x_m)}{\sum_{k=2}^{M} P(x_k)P(Y|x_k)} \tag{6.16}$$

In particular, the simplest classification is to use the ideal observer criterion. We also accept as a condition that all the losses caused by identification failures are equal. The method is based on hypothesis selection corresponding to the maximum posteriori probability $P(x_m|Y)$.

Consider the AS classification procedure applied to the traffic monitoring task. Assume that $P(Y|x_m)$ is known, Y is an univariate vector ($N = 1$, $Y = (y_1)$, suppose

$y = y_1$). Assume that a feature distribution for class x_2 follows the exponential law and a conditional probability density is calculated by Eq. 6.17.

$$p(Y|x_2) = p(y|x_2) = \begin{cases} \lambda e^{-\lambda y} & y \geq 0 \\ 0 & y < 0 \end{cases} \tag{6.17}$$

Feature distribution for classes x_3, x_4, x_5 corresponds to the normal law.

Let us assume the following initial data: parameter $\lambda = 4$ (expert analysis for the class x_2) and for classes x_3, x_4, x_5:

$$\begin{aligned} E(y|x_3 &= 6) & \sigma = (y|x_3) &= 1.5 \\ E(y|x_4 &= 12) & \sigma = (y|x_4) &= 2.3 \\ E(y|x_5 &= 25) & \sigma = (y|x_5) &= 3.5 \end{aligned} \tag{6.18}$$

The plots of probability densities are represented in Fig. 6.8 (vertical axis is densities and horizontal axis is speed in m/s).

Figure 6.8 indicates 4 probability densities for different AS classes. Herewith, the probability density plots on the y-axis intervals (0–2.5), (5–10), (15–20) are mutually intersected.

These intersections are the most difficult for the AS to be classified. However, densities duality allows (in this example) to apply the detection algorithms (with two possible outcomes). Detection is not necessary for the remaining areas of the plot, thus the following rules are applied:

1. If $2.5 \leq y \leq 5$ then $X = x_3$.
2. If $10 \leq y \leq 5$ then $X = x_4$.
3. If $20 \leq y$ then $X = x_5$.

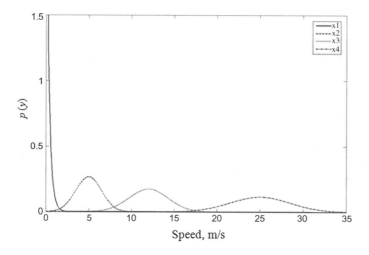

Fig. 6.8 Vehicles closing speeds probability densities for different situation classes

Consider two examples of the abnormal situations' classification, such as the AS on the straight-line section (Fig. 6.9) and the AS on the intersection (Fig. 6.10).

Figure 6.9 demonstrates two (non-adjacent) frames from video sequence registering the development of the AS. In Fig. 6.9a, a motor car marked by square 1 was moving in the left lane. Van marked by square 2 started left lane change. In the other frame (Fig. 6.9b), we can see the collision, which results in the motor car going to the roadside. The closing speed (based on the auto-tracking analysis) by the time of collision was equal $Y = 1.25$.

Let us assume as an example:
Option 1

$$P(x_2) = 0.1 \quad P(x_3) = 0.9 \quad R_{m1} = R_{1m} = 1$$

and in accordance with Bayes criterion, $\lambda_0 = 0.111$.
Option 2

$$P(x_2) = 0.8 \quad P(x_3) = 0.2 \quad R_{m1} = R_{1m} = 1 \quad \lambda_0 = 0.25.$$

Hence, it is decided that for Option 1 the resulting characteristic value corresponds to the class x_3, and for Option 2 it corresponds to the class x_2.

Figure 6.10a demonstrates the AS, where a motor car and a bus collide at the intersection, and Fig. 6.10b depicts the vehicles motion paths.

The paths of the vehicles that passed the intersection before the AS are highlighted in white, whereas the collision participants are highlighted in red. The calculated closing speed was equal to $Y = 12$ m/s. In accordance with the probability densities plots (Fig. 6.8), this characteristic value falls within the range corresponding to class x_3 and requires no further classification.

(a)   **(b)**

Fig. 6.9 The AS development on the straight-line section: **a** initial situation, **b** collision

(a)   (b)

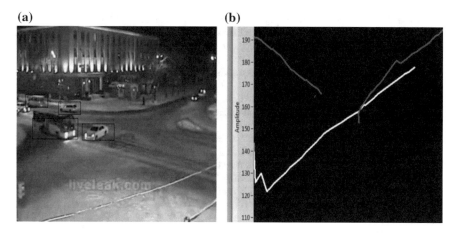

Fig. 6.10 The vehicles collision at the crossroads: **a** frame before the collision, **b** selected vehicles motion paths

6.10 Conclusions

Novelty of the presented research deals with the TAs classification based on the images obtained from the on-board UAV. Such classification of situations is often difficult due to the absence of direct attributes of a situation class, for example, visible signs of vehicle damage or clear signs indicating the injured people. The corresponding descriptive structure (a description of the observed scene) is developed in order to indicate the situations containing maximum data for a decision making. Also, a production system is offered to be used as a decision making method.

References

1. Türmer, S., Leitloff, J., Reinartz, P., Stilla, U.: Evaluation of selected features for car detection in aerial images. In: International Archives of the Photogrammetry, Remote Sensing and Spatial Information Sciences, vol. XXXVIII-4/W19, pp. 341–346 (2011)
2. Qadir, A., Semke, W., Neubert, J.: Implementation of an onboard visual tracking system with small unmanned aerial vehicle (UAV). Int. J. Innov. Technol. Creat. Eng. 1(10), 1–9 (2011)
3. Kim, N., Chervonenkis, M.: Situational control unmanned aerial vehicles for traffic monitoring. Mod. Appl. Sci. 9(5), 1913–1852 (2015)
4. Rainer, L., Maydt, J.: An extended set of Haar-like features for rapid object detection. In: International Conference on Image Processing (ICIP'2002), vol. 1, pp. 900–903 (2002)
5. Zhang, J., Liu, L., Wang, B., Chen, X., Wang, Q., Zheng, T.: High speed automatic power line detection and tracking for a UAV-based inspection. In: International Conference on Industrial Control and Electronics Engineering (ICICEE'2012), pp. 266–269 (2012)
6. Forssyth, D.A., Ponce, J.: Computer Vision: A Modern Approach. Prentice Hall, Ptr. Copyright by Pearson Education, Inc. (2003)

7. Bernd, J.: Digital Image Processing. Springer, Berlin, Heidelberg (2005)
8. Kim, N., Bodunkov, N.: Adaptive surveillance algorithms based on the situation analysis. In: Favorskaya, M., Jain, L.C. (eds.) Computer Vision in Control Systems-2, ISRL, vol. 75, pp. 169–200 (2015)
9. Yilmaz, A., Javed, O., Shah, M.: Object tracking: a survey. ACM Comput. Surv. **38**(4), 13.1–13.45 (2006)
10. Lin, F., Lum, K.Y., Chen, B.M., Lee, T.H.: Development of a vision-based ground target detection and tracking system for a small unmanned helicopter. Sci. China Ser. F: Inf. Sci. **52**, 2201–2215 (2009)
11. Lin, Y., Yu, Q., Medioni, G.: Efficient detection and tracking of moving objects in geo-coordinates. Mach. Vis. Appl. **22**(3), 505–520 (2011)
12. Kim, N.: Automated decision making in road traffic monitoring by on-board unmanned aerial vehicle system. Ind. J. Sci. Technol. **8**(S10), 1–6 (2015)
13. Gorelik, A.L.: Recognition Methods. Vishaya shkola, Moscow (in Russian) (2004)
14. Pospelov, D.A.: Situational Control: Theory and Practice. Nauka, Moscow (in Russian) (1986)
15. Li, L., Jiang, S., Huang, Q.: Learning hierarchical semantic description via mixed-norm regularization for image understanding. IEEE Trans. Multimedia **14**(5), 1401–1413 (2012)
16. Guarino, N., Oberle, D., Staab, S.: What is an ontology? In: Staab, S., Studer, R. (eds.) Handbook on Ontologies. International Handbooks on Information Systems, pp. 1–17 (2009)
17. Mizoguchi, R.: Tutorial on ontological engineering: Part 3: Advanced course of ontological engineering. New Gener. Comput. **22**(2), 193–220 (2004)
18. Yu, T.H., Moon, Y.S.: Unsupervised abnormal behavior detection for real-time surveillance using observed history. Adv. Biomet. 1019–1029 (2009)
19. Zhu, Y.Y., Zhu, Y.Y., Zhen-Kun, W., Chen, W.S., Huang, Q.: Detection and recognition of abnormal running behavior in surveillance video. Math. Probl. Eng. **296407**, 1–14 (2012)
20. Yang, K., Cai, Z., Zhao, L.: Algorithm research on moving object detection of surveillance video sequence. Sci. Res. **3**(28), 308–312 (2013)
21. Canny, J.: A computational approach to edge detection. IEEE Trans. Pattern Anal. Mach. Intell. **8**(6), 679–698 (1986)
22. Felzenszwalb, P.F., Huttenlocher, D.P.: Efficient graph-based image segmentation. Int. J. Comput. Vis. **59**(2), 167–181 (2004)
23. Implementation of the SLIC superpixel algorithm to work with OpenCV. http://docs.opencv.org/trunk/df/d6c/group__ximgproc__superpixel.html. Accessed 12 June 2017
24. Matas, J., Chum, O., Urban, M., Pajdla, T.: Robust wide-baseline stereo from maximally stable extremal regions. Image Vis. Comput. **22**(10), 761–767 (2004)
25. Zhong, S., Ghosh, J.: A unified framework for model-based clustering. J. Mach. Learn. Res. **4**, 1001–1037 (2003)
26. Meuel, H., Reso, M., Jachalsky, J., Ostermann, J.: Superpixel-based segmentation of moving objects for low bitrate ROI coding systems. In: 10th IEEE International Conference on Advanced Video and Signal Based Surveillance (AVSS'2013), pp. 27–30 (2013)
27. Dalal, N., Triggs, B.: Histograms of oriented gradients for human detection. In: IEEE Conference on Computer Vision and Pattern Recognition (CVPR'2005), vol. 1, pp. 886–893 (2005)
28. Polikar, R.: Ensemble based systems in decision making. IEEE Circuits Syst. Mag. **6**(3), 21–45 (2006)
29. Deng, L., Yu, D.: Deep learning: methods and applications. Found. Trends Signal Process. **7**(3–4), 197–387 (2014)
30. Bishop, C.M.: Neural Networks for Pattern Recognition. Oxford University Press, Inc., New York, NY, USA (1995)
31. Farnebäck, G.: Very high accuracy velocity estimation using orientation tensors, parametric motion, and simultaneous segmentation of the motion field. In: 8th IEEE International Conference on Computer Vision (ICCV'2001), vol. 1, pp. 171–177 (2001)

32. Welch, G., Bishop, G.: An Introduction to the Kalman Filter. Technical report, University of North Carolina at Chapel Hill Chapel Hill, NC, USA (1995)
33. Kleeman, L.: Understanding and applying Kalman filtering. In: 2nd Workshop on Perceptive Systems, pp. 1–4 (1996)
34. Kelly, A.: A 3D State Space Formulation of a Navigation Kalman Filter for Autonomous Vehicles. CMU-RI-TR-94-19-REV 2.0, Carnegie Mellon University (1994)
35. Hazewinkel, M. (ed.): Encyclopedia of Mathematics (set). Kluwer Academic Publishers, Dordrechr, Holland (1988)
36. Neyman, J., Pearson, E.S.: On the problem of the most efficient tests of statistical hypotheses. Philos. Trans. R. Soc. A: Math. Phys. Eng. Sci. **231**, 289–337 (1933)

Chapter 7
Warping Techniques in Video Stabilization

Margarita N. Favorskaya and Vladimir V. Buryachenko

Abstract Digital image and video stabilization are crucial issues in many surveillance systems. Good stabilization of the raw data provides a successful processing of visual materials. At present, the main approach directs on the search of the trade-offs between 2D and 3D stabilization methods in order to derive the benefits of both techniques. Our contribution is twofold. First, the multi-layered motion fields are applied in the warping during stabilization. For this purpose, the term "Structure-From-Layered-Motion" was introduced. Second, the warping and inpainting of the frame boundaries are executed using a pseudo-panoramic key frame and the multi-layered motion fields. Such inpainting permits to restore fast the cropped stabilized frames up to the sizes of the original non-stabilized frames. The dataset Sports Videos in the Wild, as well as the additional non-stationary video sequences, were used in experiments, which demonstrated good visibility results with a preserving of the frame sizes.

Keywords Video stabilization · Warping · Dynamic scene · Scene depth
3D motion · Camera trajectory · Multi-layered motion field
Structure-from-layered-motion

7.1 Introduction

The warping techniques were developed since 1990s as a form of the geometric transformations [1, 2] and the improving tool for the optical flow methods [3, 4]. Thereafter, Brox et al. [5] provided a variational formulation of the optical flow

M.N. Favorskaya (✉) · V.V. Buryachenko
Institute of Informatics and Telecommunications, Reshetnev Siberian State
University of Science and Technology, 31, Krasnoyarsky Rabochy Avenue,
Krasnoyarsk 660037, Russian Federation
e-mail: favorskaya@sibsau.ru

V.V. Buryachenko
e-mail: buryachenko@sibsau.ru

© Springer International Publishing AG 2018   177
M.N. Favorskaya and L.C. Jain (eds.), *Computer Vision in Control Systems-3*,
Intelligent Systems Reference Library 135, https://doi.org/10.1007/978-3-319-67516-9_7

method with the high accuracy. In the following publications [6], these authors mentioned that the warping techniques implemented the minimization of the linear data terms in the case of small motion displacements [7] and the use of the non-linear data terms under the large motion displacements [8].

Four approaches, including mechanical, optical, electronic, and digital stabilization, are well-known [9]. In recent years, the Digital Image Stabilization (DIS) that only depends on the image processing attracts many researchers as a reasonable tool for improvement of the non-stationary video sequences obtained from the moving vehicles (robotic platforms, ships, and unmanned aerial vehicles), as well as the home hand-held videos. All existing stabilization algorithms in the conventional DIS framework can be roughly divided into two categories, such as 2D and 3D stabilizations. During 2D stabilization, unwanted motions (jitters and shakes) are separated from the camera motion from frame to frame. Then the compensation procedures like a low-pass filtering or pixels' positions recalculation are implemented in order to smooth a camera path and provide a natural motion. Some 2D stabilization methods use the Block Matching Algorithm (BMA) for motion estimation [10], while the others applied the feature matching methods [11] or the optical flow methods [12], overcoming the main limitations of the BMA, such as the absence of occlusions and the homogeneous regions in a frame. The 2D stabilization methods might be successful in removing of the small camera jitters but they cannot build 3D camera trajectory that corresponds to 3D stabilization in the wild. Also these methods cannot model the parallax effects and map the 2D warps into 3D camera motion. The limitations of 2D stabilization provoked the appearance and development of 3D stabilization methods [13–16].

The pioneer research in 3D stabilization appeared in 2009 as a state-of-the-art method devoting to the Content-Preserving Warping (CPW)-based Stabilization (CPWS) algorithm that was developed by Liu et al. [13]. The CPWS algorithm includes three following stages:

• The creation of the Structure-From-Motion (SFM) is based on 3D camera motion analysis and a sparse set of 3D feature points.
• The desired 3D camera path is specified as the linear, parabolic, or smoothed version of the original trajectory. Note that the CPWS algorithm fits automatically a camera path to the input path.
• The least squares optimization algorithm computes a spatially varying warp from each input frame into an output frame, using a homography matrix.

At present, many 3D warping techniques are proposed for solving of the stabilization tasks. This is one of the crucial subtasks, and many reasonable warping techniques will be considered in the following discussions. However, other stages will also be highlighted in this chapter.

The contributions of this chapter are twofold. First, the Multi-Layered Motion Fields (MLMF) are implemented in the CPWS algorithm. For this purpose, the term "Structure-From-Layered-Motion" was introduced. Second, a building of pseudo-panoramic key frame helps to fill the missing boundary areas after stabilization. This inpainting stage permits to restore the cropped stabilized frames up to the sizes of the original non-stabilized frames, also using the MLMF technique.

The remainder of this chapter is organized as follows. The development of the warping techniques is discussed in Sect. 7.2. Section 7.3 reviews the related works. The scenes' classification is considered in Sect. 7.4. The proposed 3D warping during stabilization presents in Sect. 7.5, while the warping during video inpainting is described in Sect. 7.6. Section 7.7 presents the experimental results. Section 7.8 concludes the chapter summarizing our contribution.

7.2 Development of Warping Techniques

All warping techniques for 2D signals, in a mathematical sense, are classified as the parametric and non-parametric methods. Translation, translation and dilation, translation and rotation, procrustes, affine, perspective, bilinear, and polynomial transformations are concerned to the parametric methods. They provide the compact representation and fast computation of a warping but cannot perform well the local distortions. Opposite to them, the non-parametric methods like elastic deformations, thin-plate splines, and Bayesian approach demonstrate a heavy computational load and the presence of local optima.

Hereinafter, the researchers did not follow this concept strictly and many modifications suitable for 2D and 3D image analysis were proposed. Thus, Milliron et al. [17] suggested the framework represented by the geometric transformations that are parameterized by some functions but cannot be considered as the general approach for the wishful by the end-user warps. The framework proposed by Bechmann [2] is based on the feature points and generalizes the free-form deformations. The classification of the image warps suitable for stabilization task is depicted in Fig. 7.1.

Some of the warping techniques are shortly described in Table 7.1. A variety of their modifications can be found in literature. Note that the warping techniques have been widely used in 3D solid object modelling, 3D object reconstruction based on separate photographs, panorama construction, and other classical tasks of computer vision. More, the recent challenges require the fast implementation of the warping techniques that comes into collision with their heavy computations.

The initial methods of image warping and deformation were built on the earlier work of Alexa et al. [18] devoting to As-Rigid-As-Possible (ARAP) shape interpolation that minimizes the perceivable distortion of a shape in order to avoid the superfluous global or local deformations. These authors proposed the least-distorting triangle-to-triangle morphing based on the affine transformation model. The hybrid model of the ARAP parameterization that combines a local mapping of each 3D triangle to the plane with a global "stitch" operation of all triangles, involving a sparse linear system, was proposed by Liu et al. [19]. Another approach for the ARAP modification is focused on an interactive image

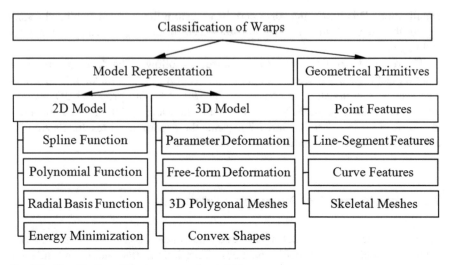

Fig. 7.1 Classification of the image warping technique

Table 7.1 Description of some warping techniques

Warping technique	Assignment	Short description
Free-form deformation	Solid modelling	This is a feature-based warping technique. The objects are deformed in the lattice. The movement of the feature points from the source positions to the target positions are approximated (not interpolated)
Line-segment features	Image warping and metamorphosis	This feature-based warp operates in 2D images as a part of an image metamorphosis algorithm. The deformation is controlled using the source and target line segments, which are defined by the end-user. This warp involves an inverse mapping in order to determine, which source-image pixels should contribute to the color of the destination pixel
Parameter-based deformation	Solid modelling	The deformations are defined by the geometric parameters, for example, along an axis. For each point, a single transformation is applied from the parameterized transformation's continuous spectrum
Curve-based deformation	Surface modelling and animation	This feature-based deformation is called "wires" that uses 3D curve features to manipulate the smooth surfaces for 3D modelling and animation

deformation controlled by the end-user [20, 21]. Thus, the shape is represented by a triangle mesh and the end-user moves several vertices of the mesh as the constrained handles in [20]. The main challenge is to find the appropriate definition for the distortion of an individual triangle. These authors designed an error metric and solved the minimization problem as a simple matrix computation. The research [21] is an extension of Igarashi et al. investigations [20]. This image deformation method is based on the moving least squares, using various classes of linear functions including affine, similarity, and rigid transformations. The end-user specifies the affine deformations, using the sets of points, line segments, curves, or profiles in the image.

Thereafter, some techniques in 3D meshes were proposed by Zhou et al. [22] as the techniques for large deformations on 3D meshes using the volumetric graph Laplacian. Sheffer and Kraevoy [23] proposed a local shape representation referred to as the pyramid coordinates, which are invariant to the rigid transformations. The pyramid coordinates capture the local shape (lengths and angles) of the mesh around each vertex that permits to provide the mesh deformation and morphing based on a small number of the user-specified control vertices.

Weng et al. [24] offered a 2D shape deformation algorithm based on the nonlinear least squares optimization. This algorithm preserves the Laplacian coordinates of the boundary curve of the shape and local areas inside the shape. The problem of a nonlinear least squares minimization was solved by an iterative method and provided better physically plausible results that are hard to achieve with the linear least squares methods.

Gal et al. [25] suggested an inhomogeneous 2D texture mapping method that preserves the shape of masked regions of the texture according to the end-user specifications during an image warping. The end-user provides a feature mask, marking the parts of the image whose shape should be preserved. The goal is to find a mapping of the original grid that is as close as possible to the input warp and respects the shape of the features specified by the mask. The algorithmic core of such feature-sensitive texture mapping is a solution of the least-squares optimization. These authors clarified that their technique is useful for texture mapping, image re-scaling, and warping.

7.3   Related Works

The perception of the warping techniques in the stabilization tasks had been formed during the recent decades. The overviews of the warping for 2D and 3D stabilization tasks are represented in Sects. 7.3.1–7.3.2, respectively. The video completion methods that are applied after stabilization are discussed in Sect. 7.3.3.

7.3.1   Overview of Warping Techniques in 2D Stabilization

As Litvin et al. mentioned in [26], the overall 2D stabilization algorithm involves two main stages:

- The stabilization of video sequence (unwanted motion compensation) with three sub-steps like as:

  – The estimation of the pair-wise transformations between the adjacent frames.
  – The evaluation of the intentional motion parameters.
  – The compensation of each frame for removal of the unwanted motion (frame warping).

- The reconstruction of missing boundary regions using mosaicking.

  – The estimation of the transformation between the distant frames.
  – The warping distant frames and mosaic construction for the undefined regions in each frame.

At that time, a frame warping implied a Global Warping (GW) in a full frame using one of the interpolation functions, such as bilinear, cubic, or Bézier curve. The warping distant frame signified the alignment of the neighboring frames with respect to the warped current frame in order to reduce the visual artifacts at the boundaries of the undefined regions.

Wang et al. [27] suggested the robust technique that permits to reconstruct the long feature trajectories as a Bezier curve and maintains the spatial relations between these trajectories by preserving the original offsets of neighboring curves. A regular grid mesh with cell sizes 10×10 pixels is imposed on each frame. The frame warping is executed as a 2D homography transformation of each cell. The features of the foreground moving objects do not taken into consideration. A Bezier curve cannot fit well the shape of a long and twisting trajectory. First, the authors computed the best fit Bezier curve for each trajectory without considering spatial rigidity. Second, they portioned the complicated trajectory into sub-trajectories as the iteration in the optimization process.

It is interesting that after some outstanding investigations in 3D stabilization task (Sect. 7.3.2) Liu et al. returned to 2D stabilization in the research [28]. These authors introduced the term "bundled camera paths", which presents the multiple and spatially-variant camera paths of different locations in video. Such approach allows to estimate a nonlinear motion caused by the parallax and rolling shutter effects, at the same time providing the robustness and simplicity of 2D methods. An image warping model represents a motion between the consecutive frames, adopting the warping model [13, 20] though general models, such as "moving-least-square" algorithms [21]. In each frame, a uniform grid mesh is imposed and a motion is represented by an unknown warping of the grid mesh in order to receive the same bilinear interpolation of the four corners of the enclosing grid cell after warping. In other words, the warping-based motion model is a set of spatially-variant homographies on 2D grid. The motion is minimized in two energy terms: a data term for matching features and a shape-preserving term for enforcing regularization. However, a problem of stitching of the warping frame parts is an actual sub-task in such solutions.

In 2014, Liu et al. [29] introduced the concept of pixel profiles. A pixel profile is a set of motion vectors collected at the same pixel location. (In comparison, a feature trajectory follows the motion of a scene point.) Video sequence can be stabilized by smoothing all pixel profiles that are collected through the raw optical flow in order to get so called SteadyFlow. The authors clarified that their results are free from many artifacts. In spatial domain, the gradient magnitude of raw optical flow is restricted to identify the discontinuous regions. The temporal analysis examines the accumulation motion vectors on a pixel and determines if this pixel is "outlier" or not. The pixels, belonging to the foreground moving objects, are considered as the "outlier" pixels. The authors claimed that their algorithm fails, when the dominant foreground objects consistently occupy more than a half area of a frame and exist for a long time. However, the same can be spoken for many stabilization algorithms because in this case the foreground objects are accepted as a background.

7.3.2 Overview of Warping Techniques in 3D Stabilization

The conventional stabilization techniques based on 2D warps cannot stabilize a shaky camera motion through a 3D scene. The conventional approach for 3D stabilization contains three major stages [30]:

- The extraction of the 3D image features in each frame and the features' matching between the neighboring frames.
- The estimation of the original 3D camera path and the sparse 3D scene geometry by the SFM algorithm on the matched features.
- The generation of a desired camera path and synthesis of the images along the path using the warping techniques.

Also the reconstruction stage of the obtained stabilized frames is often required. Smith et al. [30] proposed a novel method to exploit the multi-view video streams for stabilization based on the special light field cameras. This approach does not require the reconstruction of the desired 3D camera motion over a long video sequence because the light field cameras capture multiple images of a scene from different viewpoints at the same time instant. Such technical decision suitable, for example, for robotic platforms can be successfully applied for a dynamic scenes' rendering.

One of the popular approaches for 3D video stabilization uses the epipolar geometry, when, first, the feature points in a scene are tracking and, second, the fundamental matrices are computed in order to model the stabilized camera motion [31]. The classic epipolar point transfer can be applied only to points that correspond to the static objects in 3D space but do not belong to the non-stationary objects. Goldstein and Fattal [32] overcome this limitation using a time-view point re-projection of the moving objects under assumption of a smooth inertial object motion in 3D space. They introduced the time-differentiation matrices in order to obtain the trajectories, resulting from the minimal forces such that every point lies on its corresponding epipolar line.

Liu et al. [13] developed the CPW technique under assumption that the desired camera path will not be very far from the original camera path. This means that the local content in the original image should not need to be distorted significantly. The CPW method spreads the error near occlusions across the rest of frame in such manner that a human eye will not notice them. The edges of the foreground objects as well as the texture content of the frame are preserved. Liu et al. offered two solutions. The first solution applies a generic sparse data interpolation to yield a dense warp. The original video frame is divided into a uniform grid mesh and the warped version of this grid is computed. Each quadrangle mesh is split into two triangles and the displacements inside the triangles are interpolated using the barycentric coordinates. This solution is flexible but has some disadvantages. The scene structures near occlusions may be distorted. The problems occur near the frame boundaries during extrapolation. Also the displacements as the hard constraints lead to the temporal incoherence that causes a necessity of the preserving temporal coherence based on the displacements as the soft constraints. Liu et al. minimized an energy function of two weighted energy terms: a data term for each sparse displacement and a similarity transformation term. Let P_k be the output projected point set and $\widehat{P}_k$ be the input point set of a grid cell. The vector $\widehat{V}_k$ is defined as a vector of four vertices enclosing the grid cell that $\widehat{P}_k$ projects to, while the vector V_k denotes the same four vertices in the output grid. Each projected point from $\widehat{P}_k$ is typically not coincident with a vertex from $\widehat{V}_k$. Liu et al. applied each constraint with a bilinear interpolation of the four corners of the enclosing grid cell. The vector w_k contains four bilinear interpolation coefficients that can be computed by finding the grid cell that $\widehat{P}_k$ projects to, and $\widehat{P}_k = w_k^T \widehat{V}_k$ is a bilinear interpolation operation. The data term E_d is defined is estimated by Eq. 7.1, where w_k and P_k are known and V_k includes four unknowns.

$$E_d = \sum_k \left\| w_k^T V_k - P_k \right\|^2 \tag{7.1}$$

A similarity transformation term measures the deviation of each grid cell from a similarity transformation weighted by the salience of a grid cell. Each grid cell is split into two triangles and each vertex is represented in a local coordinate system formed by the vector between the other two vertices with the 90° rotation of that vector [20]. Thus, a vertex V_1 can be defined using the vertices V_2 and V_3 in Eq. 7.2, where x and y are the known coordinates within the local coordinate system.

$$V_1 = V_2 + x(V_3 - V_2) + y \begin{bmatrix} 0 & 1 \\ -1 & 0 \end{bmatrix} (V_3 - V_2) \tag{7.2}$$

However, a vertex V_1 may not coincide with the location calculated from the vertices V_2 and V_3. The distance between V_1 and its desired location under a similarity transformation is assessed using an energy term E_s in Eq. 7.3, where w_s is a weight. This parameter considers a salience of its enclosing grid cell.

$$E_s(V_1) = w_s \left\| V_1 - \left(V_2 + x(V_3 - V_2) + y \begin{bmatrix} 0 & 1 \\ -1 & 0 \end{bmatrix} (V_3 - V_2) \right) \right\|^2 \qquad (7.3)$$

The sum of the both energy terms E_d and E_s is a linear least-squares optimization problem in the set of unknown grid vertices $V_{i,j}$. The final energy equation is provided by Eq. 7.4, where α is the relative weight of the data term and the smoothness term.

$$E = E_d + \alpha E_s \qquad (7.4)$$

The second solution proposed by Liu et al. [13] fits a full-frame warp to the sparse displacements, such as a homography. Good results could be provided if the depth variation in the scene is not large or the desired camera path is close to the original camera path. In the general case, a homography is too constrained model and the undesired distortion and temporal wobbling can appear. One of the ways is to apply a rotation to the input camera with the following translation of the output and input cameras. In this case, a projective transformation is defined as an infinite homography $\mathbf{H} = \mathbf{K}\widetilde{\mathbf{D}}(\mathbf{KD})^{-1}$, where $\mathbf{K}$ is the shared intrinsic camera matrix, $\mathbf{D}$ and $\widetilde{\mathbf{D}}$ are the camera orientation matrices of the input and output cameras [33]. If a translation of the output camera respect to the input is large, then a computation of the best-fit homography in a least-squares sense to the set of sparse displacements is recommended.

The computational time of Liu's algorithm highly depends on the optimization process, in which the matrices sizes can be too huge to be implemented through the least squares method. Lee et al. [34] proposed the region-based CPW technique to minimize computational time for the optimization process. First, the 3D motions of feature points are estimated. Second, the Regions Of Interest (ROI) based on the estimated 3D motion information are formed. Third, the ROI-based pre-warping and content-preserving warping to the original frame are applied. As a result, the computational complexity of the proposed algorithm is noticeably decreased providing almost equivalent stabilization performance in comparison to the state-of-the-art Liu's method.

The original camera path can be estimated by various techniques, such as feature points tracking, linear motion estimation in the form of 2D transformations, or the SFM application. However, a smooth camera path is preferable [35]. Grundmann et al. [36] optimized the camera path $PH(t)$ that is composed of the following path segments:

- A constant path, representing a static camera, $DPH(t) = 0$, D is the differential operator.
- A path of constant velocity, representing a panning, $D^2PH(t) = 0$.
- A path of constant acceleration, representing the transition between the static and panning cameras, $D^3PH(t) = 0$.

These authors proposed to obtain the optimal path by a superposition of distinct constant, linear and parabolic segments and formulated it as a constrained L1 minimization problem. During retargeting, a new feature warp transform was declared using a specific map (saliency map) or convex region (face or another object) within the crop window. The feature warp transform based on the bounds of a set of salient points in frame s_i^t was estimated by Eq. 7.5, where af_t is a parameterization vector of the affine transformation, bn_x and bn_y are the bounds, ε_x and ε_y are the errors, ε_x, $\varepsilon_y \geq 0$.

$$\begin{pmatrix} 1 & 0 & s_i^x & s_i^y & 0 & 0 \\ 0 & 1 & 0 & 0 & s_i^x & s_i^y \end{pmatrix} af_t - \begin{pmatrix} bn_x \\ bn_y \end{pmatrix} \geq \begin{pmatrix} -\varepsilon_x \\ -\varepsilon_y \end{pmatrix} \qquad (7.5)$$

Such L1 optimization based approach needs in multiple simultaneous constraints and works in real-time but meets with the challengers, appearing in videos like low count of feature, excessive blur during extremely fast motions, or lack of the rigid objects in the scene.

Thereafter, Grundmann et al. [37] proposed to express the rolling shutter distortions parametrically as homography mixtures, which are used to unwarp the distortions present in the original frame. The image domain was partitioned in 10 scanline blocks, resulting in 10 unknown homographies. Then these homographies were smoothly interpolated using Gaussian weights in order to avoid the discontinuities across the scanline blocks. In general, a homography matrix $\mathbf{h}$ represented as Eq. 7.6

$$\mathbf{h} = \begin{pmatrix} h_1 & h_2 & h_3 \\ h_4 & h_5 & h_6 \\ h_7 & h_8 & 1 \end{pmatrix} \qquad (7.6)$$

includes the elements (h_7, h_8) of perspective transform and (h_1, h_5) of scale transform that can be regarded as a constant part, while the elements (h_3, h_6) of translation and (h_4, h_2) of a skew are a varying part.

Grundmann et al. [37] proposed two mixture models $\mathbf{H}_k$ and $\widehat{\mathbf{H}}_k$ with $(6 + 2k)$ and $(4 + 4k)$ degrees of freedom, respectively. They are expressed by Eq. 7.7, where $\mathbf{A}$ is a frame global 2×2 affine matrix, $\mathbf{w}^T = (w_1, w_2)^T$ is a frame constant perspective part, $\mathbf{t}_k$ is a block varying translation, a and d are frame global scale parameters, c_k and b_k are block varying rotation.

$$\mathbf{H}_k = \begin{pmatrix} \mathbf{A} & \mathbf{t}_k \\ \mathbf{w}^T & 1 \end{pmatrix} \quad \widehat{\mathbf{H}}_k = \begin{pmatrix} A & b_k & t_k^x \\ c_k & d & t_k^y \\ w_1 & w_2 & 1 \end{pmatrix} \tag{7.7}$$

As the authors clarified, these reduced models had the benefit of faster estimation and greater stability due to fewer degrees of freedom.

The depth of the scenes, especially the indoor scenes, can be estimated using the Kinect camera that was made by Liu et al. [38] in the stabilization goals. First, the 3D camera motion from the neighboring color and depth images is estimated. The availability of depth information does not cause a necessity in the long feature trajectories for 3D reconstruction. Second, the 3D camera trajectory is smoothed in order to reduce both high frequency jitters and low frequency shakes. Third, the frames are reconstructed combing information from color and depth images. The color and depth images were used to generate the projection and the motion field, respectively. The depth images are usually incomplete and include many missing pixels. The "content-preserving" warping was applied for the color images only. A fusion of a color image and a motion field provides good reconstruction result limited by the rolling shutter effects of both the color camera and the depth camera simultaneously.

7.3.3 Overview of Inpainting Methods for Boundary Completion

Since the missing regions almost always occur at the boundaries of frames, the inpainting for the stabilized videos is more an extrapolation task. The first methods for video inpainting were based on the standard mosaicing techniques with the following stitching the parts of neighboring frames in the reconstructed frame [26]. This approach is not suitable for the non-planar scenes and scenes with the moving objects that may appear at the boundary of the frames. Wexler et al. [39] solved this task by filling the holes near boundaries using the spatio-temporal volume patches from different frames of the same video. Such nonparametric implementation provided a good result, but it has high computational cost and it requires a long video sequence of the reconstructed scene. Jia et al. [40] used a video representation into two layers: a moving object layer and a static background layer. The limitation of this method is a necessity to know at least a single period of objects' periodic motion in order to find a good matching of the color patterns.

Matsushita et al. [35] proposed the state-of-the-art method that propagates 2D motion vectors of the optic flow to guide the inpainting. The main idea of such motion inpainting is to propagate a local motion instead of color/intensity into the missing regions that may be non-planar or dynamic. The basic proposition of such approach is a similarity of the local motion in the missing frame areas to the local motion in the adjacent frame areas. The motion inpainting technique proposed by Matsushita et al. includes the following steps:

- The mosaicing with consistency constraint. It is possible if the missing area is the static and planar part of an image. The validity of the mosaic may be estimated by evaluation of the consistency of the multiple mosaics, which cover the same pixels.
- The local motion computation. The local motion is estimated by computing the optical flow, using only the common coverage areas between the neighboring frames. The local motion field is built applying a pyramidal version of Lucas-Kanade optical flow computation [41]. From this step, each neighboring frame is ranked in a processing priority based on its alignment error. The nearest frames with the smallest alignment errors have the highest processing priority.
- The motion inpainting. The propagation based on the local motion begins from the boundary of the missing frame area. This procedure uses the motion values of neighboring known pixels and the boundary gradually gets ahead into the missing area until it will be completely filled. Under assumption that the local motion variation is small, the local motion $\mathbf{M}(\mathbf{p}_i, \mathbf{q}_i)$ in missing pixel $\mathbf{p}_i$ of frame t relative to the neighboring known pixel $\mathbf{q}_i$ can be expressed by the first order approximation of Taylor series expansion, Eq. 7.8, where $[u\ v]^{\mathrm{T}}$ is a vector that represents a displacement from pixel $\mathbf{q}_i$ to pixel $\mathbf{p}_i$, M_x and M_y are the projections of the motion vector on axis OX and OY, respectively.

$$\mathbf{M}(\mathbf{p}_i, \mathbf{q}_i) \approx \mathbf{M}(\mathbf{q}_i) + \begin{bmatrix} \frac{\partial M_x(\mathbf{q}_i)}{\partial x} & \frac{\partial M_x(\mathbf{q}_i)}{\partial y} \\ \frac{\partial M_y(\mathbf{q}_i)}{\partial x} & \frac{\partial M_y(\mathbf{q}_i)}{\partial y} \end{bmatrix} \begin{bmatrix} u \\ v \end{bmatrix}$$
$$= \mathbf{M}(\mathbf{q}_i) + \nabla \mathbf{M}(\mathbf{q}_i)(\mathbf{p}_i - \mathbf{q}_i) \tag{7.8}$$

The motion vector for pixel $\mathbf{p}_i$ is generated by a weighted average of the motion vectors of the pixels in neighborhood $\Omega(\mathbf{p}_i)$, Eq. 7.9, where the weighting factor w $(\mathbf{p}_i, \mathbf{q}_i)$ controls the contribution of the motion of pixel $\mathbf{q}_i \in \Omega(\mathbf{p}_i)$ to pixel $\mathbf{p}_i$.

$$\mathbf{M}(\mathbf{p}_i) = \frac{\sum_{\mathbf{q}_i \in \Omega(\mathbf{p}_i)} w(\mathbf{p}_i, \mathbf{q}_i)(\mathbf{M}(\mathbf{q}_i) + \nabla \mathbf{M}(\mathbf{q}_i)(\mathbf{p}_i - \mathbf{q}_i))}{\sum_{\mathbf{q}_i \in \Omega(\mathbf{p}_i)} w(\mathbf{p}_i, \mathbf{q}_i)} \tag{7.9}$$

The weighting factor $w(\cdot)$ reflects the geometric distance

$$g(\mathbf{p}_i, \mathbf{q}_i) = \frac{1}{\|\mathbf{p}_i - \mathbf{q}_i\|} \tag{7.10}$$

and the pseudo-similarity of colors between pixels $\mathbf{p}_i$ and $\mathbf{q}_i$ in frame t provided by Eq. 7.11, where $I_{t'}$ is a neighboring frame, ε is a small value for avoiding a division by zero.

$$c(\mathbf{p}_i, \mathbf{q}_i) = \frac{1}{\| I_{i'}(\mathbf{q}_{i'} + \mathbf{p}_i - \mathbf{q}_i) - I_{i'}(\mathbf{q}_{i'}) \| + \varepsilon} \tag{7.11}$$

The weighting factor $w(\cdot)$ is defined as a product of factors $g(\cdot)$ and $c(\cdot)$, Eq. 7.12.

$$w(\mathbf{p}_i, \mathbf{q}_i) = g(\mathbf{p}_i, \mathbf{q}_i)\, c(\mathbf{p}_i, \mathbf{q}_i) \tag{7.12}$$

The Fast Marching Method (FMM) [42] provides the texture scanning and composition in the missing frame area.

- The local adjustment with the local warping. After the loop of previous steps, most of the parts with the missing pixels are filled. The small remaining areas may be interpolated using the neighboring pixels or the diffusion methods.

As a result, the image data from the neighboring frames are locally warped to maintain the spatial and temporal continuities of the stitched images as it was suggested by Shum and Szeliski [43] in their deghosting algorithm for panoramic image construction. However, some visible artifacts occur if the blank region is large, since it is difficult to extrapolate without any underlying model. Liu and Chin [44] developed a technique based on homography field with the multi-view projective constraints that can extrapolate the large blank regions with fewer artifacts.

7.4 Scene Classification

Generally, a scene classification can be considered as a pre-processing stage, when a video sequence is divided into short-term data between the neighboring key frames (Sect. 7.4.1) and each scene between key frames ought to be classified as a scene with small and large depths (Sect. 7.4.2). In some cases, for example, the case with a smooth motion in the outdoor environment, the depth of all sub-scenes may be constant.

7.4.1 Extraction of Key Frames

In a conventional understanding, the key frames play a crucial role in the acquisition of the semantic video information. The key frames as a subset of all frames are extracted from video shots (it is considered that a video scene may include one and more video shots) and reflect the representative visual content of a given video sequence. However, in the stabilization tasks a key frame is a reference frame, relative which all subsequent frames (between two subsequent key frames) are stabilized. The conventional key frame extraction methods (regardless of the interpretation) can be categorized as shot boundary-based methods, motion analysis-based methods, clustering-based methods, visual content-based methods, compressed domain-based methods, among others. Recently, the improved methods of a key frame extraction were proposed, for example, a method based on maximum a posteriori, an aggregation method of colors and moments, the divergence-based approaches, a saliency-based model, etc.

The stabilization process does not require the content analysis but it needs a qualitative frame, for example, without blurring, with the good contrast and sufficient luminance. A shot is a set of frames captured by camera in a continuous time interval with duration of 25–30 frames. This limitation is caused by the algorithms: the feature points trajectories ought to involve not less 20 frames (in order to receive the robust trajectories without occlusions, appearance and disappearance) and a builder of 3D camera trajectory uses full information volume between two consecutive key frames (information volume may be too huge for large number of frames).

It is often assumed that a visual content changes slightly in one shot. However, this assumption is not actual in the most video sequences and a necessity of a sub-shot segmentation in a single shot exists. As a result, a non-stationary video sequence ought to be divided into shorts, each short is usually represented by the sub-shots, and the key frame/frames ought to be extracted in each sub-shot. Since a quality of a non-stationary video sequence is low, the edge and motion features are not useful for key frames extraction. In this case, the color features are preferable. Lienhart [45] claimed that 99% of all edits fall into one of the following three categories: hard cuts, fades, or dissolves.

The color content does not change rapidly within the frames in a single shot but changes significantly across different shots. Therefore, the hard cut can be successfully detected using the single peaks in the time series of the differences between color histograms of contiguous frames [45]. A hard cut is detected if the color histogram difference exceeds a certain threshold within a local frame area. The global threshold as well as the local threshold may be applied for the hard cut detection. This approach was developed by Kelm et al. [46]: the spatial correlation was considered by dividing each frame into three horizontal areas without getting sensitive to a non-significant motion of a camera and an object.

The fades and dissolves can be estimated by edge change ratio. The disappearing shot fades into a blank frame, after which the appearing shot fades in. A dissolve is defined by a temporal overlap of a few frames in the disappearing/appearing shot. For fades detection in a non-stationary video sequence, it is reasonable to analyze the luminance average μ_Y and the luminance variance σ_Y^2 in Hue, Saturation, and Value (HSV) color space. Kelm et al. [46] detected the centre of fades by thresholding the strictly monotonic decreasing first derivative of the luminance variance. The dissolve candidates are extracted by the characteristics of the first and second derivatives of the luminance variance σ_Y^2 [47]. However, due to low image quality, this approach can produce many dissolves and needs in verification, using the first and second derivatives of the candidate region and the dissolve modelled.

Two criteria, such as a good quality and a noticeable visual content, help to extract a key frame in the given limitation of a time interval (25–30 frames). The motion intensity is proportional to blurring effects. Therefore, the motion intensity ought to be minimized, as well as a frame noise. A camera motion also influences on a key frame choice. For example, the zooming-in attracts attention to details, while a fast panning or tilting camera indicates the non-significant scene fragments. Also some visual objects in a frame, such as a face or text, attract the attention and this frame can be chosen as a key frame.

7.4.2 Estimation of Scene Depth

There are many algorithms for extraction a depth from video sequences, containing the static and dynamic scenes. Usually, such techniques generate the depth maps from the arbitrary 2D videos (static/rotating camera, change in focal length, dynamic scenes) or even using the single images. The main idea is that the scenes with similar semantics are expected to have the similar depth values in the regions with similar appearance. Karsch et al. [48] proposed an approach that has three stages. Initially, the database with Red Green Blue Depth (RGBD) images ought to be collected. First, the images that are "similar" to the input image in RGB space are selected from this database. Second, a warping procedure (e.g. SIFT Flow [49]) is applied to the candidate images and depths in order to align them with the input frames. Third, an optimization procedure is used to interpolate and smooth the warped candidate depth values that results in the inferred depth. This approach is promising and provides good results for the stabilized video sequences. In the case of a non-stationary video sequence, this algorithm cannot estimate the jutted temporal volume.

Many features, describing a single frame, such as texture variations, gradients, defocus, color/haze, etc., contain useful and important depth information. The reasonable approach for a single still image was proposed by Saxena et al. [50]. The main idea is to model the relationship between the depth of a patch in a multi-scale representation and the depths of its neighboring patches, sometimes in different parts of the image, in a spatial representation. A hierarchical multi-scale Markov Random Field (MRF) models the relationship between the depth of a patch and the depths of its neighboring patches in three scales. The interactions between the depths of patches with neighbors and non-neighbors patches are simulated based on a spatial analysis, using the convolutional filters for texture energies and gradients and the Laws' masks for a local averaging, the edge and spot detection. However, the supervised learning to estimate the depth maps, as well as a 3D scanner application to collect the training data, are the necessary procedures. In this research, the approach of Saxena et al. was applied with a non-significant modification, connecting with a luminance enhancement of key frames.

Good review on 2D to 3D image and video conversion methods, including the detection of the depth from defocus, perspective geometry, models, visual saliency, motion, etc., was provided by Patil and Charles [51].

The assessment of a scene depth is required for understanding, what types of the geometrical distortions exist. Also it is necessary to estimate a number of the background motion layers in the scene with a large depth. Usually, such motion layers have the own parametrical models and this information ought to be considered in the local 3D warping during stabilization (Sect. 7.5.5). It is evident that the scenes with a small depth can be approximated by a dominant background motion layer and layers of the foreground moving objects.

7.5 3D Warping During Stabilization

The crucial issues of any image warping are the reference images and mapping functions. The stabilization task brings in a special relation to the choice of key frames and types of warping functions that may be applied to a full-frame and to the local fragments in a frame. Often the mapping function ought to capture the geometric primitives of the underlying scene, such as the straight lines and planar planes. The sources for a warping are the optical flow fields and the disparity fields based on the SFM model or a stereo disparity. In the case of the static scene, these two mappings are nearly equivalent. In the case of the dynamic scene, the multi-layered motion fields have a significant impact on the type of the warping function, which usually cannot be represented as a distinct function. The warping of pixel intensity values function during stabilization has the general form in a view of Eq. 7.13, where x_{t-1} and y_{t-1} are the pixel coordinates of a key frame in a time instant $t-1$, x_t and y_t are the pixel coordinates of the warped frame in a time instant t, $\delta(x_{t-1}, y_{t-1}, mf(x_t, y_t))$ is a generalized disparity value that depends from a motion filed $mf(x_t, y_t)$ in pixel (x_{t-1}, y_{t-1}), $f_w(\cdot)$ is a mapping function, $I(\cdot)$ is an intensity value.

$$I(x_t, y_t) = f_w(I(x_{t-1}, y_{t-1}), \delta(x_{t-1}, y_{t-1}, mf(x_t, y_t))) \qquad (7.13)$$

A view of a motion field ought to be estimated by the additional methods.

The outlines of the SFM algorithm and its application for a non-stationary video sequence are represented in Sect. 7.5.1. The building of a camera trajectory is discussed in Sect. 7.5.2. Sections 7.5.3–7.5.5 are devoted to different types of a warping, such as the warping in scenes with small depth, warping using geometrical primitives, and local 3D warping in scenes with large depth, respectively.

7.5.1 Extraction of Structure from Motion

The 3D stabilization techniques can achieve the high quality stabilized 3D camera paths through the SFM or stereoscopic applications. The goal of the SFM is to restore both the positions of 3D points in a scene (structure) and the unknown positions of a camera capturing a scene (motion). Many applications of the SFM, such as the photogrammetric tasks, the reconstruction of virtual reality models from video sequences, the camera motion detection, among others, estimate the "sparse" scene 3D points in order to construct the "dense" 3D models of a scene. The 3D stabilization methods are not the exclusion.

When two consecutive images are captured (a camera undergoes pure rotation and/or observes the same plane in a scene), the geometrical position can be modelled by homography. The measurements of the corresponding projections of 3D points can be used to form a system of equations according to the geometrical constraints of the appropriate model. Then, this system is solved by minimizing the projection error of the correspondences and the parameters of a model are obtained. Generally for given n projected points in m images u_{ij}, $i \in \{1...m\}$, $j \in \{1...n\}$, both the projection matrices $\mathbf{PR}_1...\mathbf{PR}_m$ and the consistent structures $\mathbf{S}_1...\mathbf{S}_n$ ought to be determined. In the case of video sequences, the original images are a set of the consecutive frames that may be distorted by the non-stabilized artifacts.

During image projection, two types of the coordinates—homogeneous and Euclidean (Cartesian)—are usually used. In the homogeneous coordinates, a point in N-dimensional space is expressed by a vector defined up to scale with $N + 1$ elements. If the $(N + 1)$th element is non-zero, the homogeneous coordinates may be represented in the Euclidean coordinates by dividing the first N elements by the $(N + 1)$th element, otherwise the homogeneous coordinates describe a point at infinity. Let a homogeneous 3D point has a form $\widetilde{\mathbf{S}} \approx \begin{bmatrix} \tilde{X} & \tilde{Y} & \tilde{Z} & \tilde{W} \end{bmatrix}^{\mathrm{T}}$, where a sign "$\approx$" means equality up to scale. If $\tilde{W}$ is non-zero, then $\widetilde{\mathbf{S}}$ is related to its Euclidean equivalent $\mathbf{S} \approx [X \ Y \ Z]^{\mathrm{T}}$ by Eq. 7.14.

$$\mathbf{S} = \begin{bmatrix} \tilde{X}/\tilde{W} & \tilde{Y}/\tilde{W} & \tilde{Z}/\tilde{W} \end{bmatrix}^{\mathrm{T}} \quad \widetilde{\mathbf{S}} \approx [X \ Y \ Z \ 1]^{\mathrm{T}} \qquad (7.14)$$

Similarly, a homogeneous 2D point $\tilde{\mathbf{S}} \approx \begin{bmatrix} \tilde{X} & \tilde{Y} & \tilde{W} \end{bmatrix}^{\mathrm{T}}$ is related to its Euclidean equivalent $\mathbf{S} \approx [X \ Y]^{\mathrm{T}}$:

$$\mathbf{S} = \begin{bmatrix} \tilde{X}/\tilde{W} & \tilde{Y}/\tilde{W} \end{bmatrix}^{\mathrm{T}} \quad \tilde{\mathbf{S}} \approx [X \ Y \ 1]^{\mathrm{T}}. \qquad (7.15)$$

A pinhole projection provides a good approximation to the behavior of most real cameras. In some cases, it can be improved by the non-linear effects, such as radial distortion. The relationship between 3D point and its corresponding 2D image point has three components, such as the point in the world coordinate system are related to the points in the camera coordinate system, on the camera image plane, and on the camera image plane to the pixel coordinates.

Under the assumption that the change of the camera position and orientation is small, the local surroundings of the interest elements will be similar in two nearby views. The interest elements may be the feature points described by any from famous descriptors (Maximally Stable Extremal Regions (MSER), Scale-Invariant Feature Transform (SIFT), Speeded Up Robust Features (SURF)) or the geometrical primitives like points, line segments, and planar polygons that describe a range of more complex shapes (cuboids and prisms). Usually the RANdom Sample Consensus (RANSAC) algorithm is used for correspondence of the feature points. A disadvantage of such narrow-baseline matching is that a depth computation is quite sensitive to a noise for closely spaced viewpoints. However, a tracking of the feature correspondences permits to recover the structure and motion parameters accurately.

So called the batch methods that compute the camera pose and scene geometry using all image measurements simultaneously are reasonable for video sequences. In this case, the reconstruction errors can be distributed meaningfully across all measurements. This means that the gross errors can be avoided. One family of the batch SFM algorithms is called the factorization methods. Fast and robust linear methods based on the direct Singular Value Decomposition (SVD) factorization of the point measurements had been developed for the linear (affine) camera models, e.g. orthographic, weak perspective, and para-perspective. However, these methods cannot be applied to the real-world scenes because the camera lenses have too wide-angle to be approximated as linear. Thereafter, some iterative "factorization-like" algorithms for perspective cameras were developed with a weak hope to obtain the optimal solution. The common limitation of all these algorithms is that they cannot cope with missing data: all 3D points ought to be visible in each view. Often the SFM is considered as an initial estimator of the projection matrices $\mathbf{PR}_i$ and 3D points $\mathbf{S}_j$. Usually it is necessary to refine this estimator, using an iterative non-linear optimization to minimize an appropriate cost function, i.e. to use a bundle adjustment.

The SFM algorithm has some challenges, such as a lack of parallax, camera zooming, in-camera stabilization, and rolling shutter [16]. These limitations leaded to the appearance of the SFM modifications. Even if a video sequence is stabilized, the reliable pixel correspondence is difficult to obtain, especially for long-term trajectories due to large motions, occlusions, or ambiguities. The outlier rejection techniques can reduce these problems but they cannot consider all available measurements. An approach based on a priori knowledge of camera parameters has very restricted practical application. Dellaert et al. [52] tried to implement the idea to use the SFM without correspondences. They proposed to estimate the structure and motion, given only the measurements, integrating over all possible assignments of the 3D features to the 2D measurements using the Expectation-Maximization (EM) algorithm. The conventional SFM problem is substituted the newly synthesized "virtual" measurements based on a probability distribution that is obtained using the actual image data and an initial guess for the structure and motion. One of the Markov Chain Monte Carlo methods is used to sample from this distribution. However, this approach was not tested accurately: all features were hand-picked determined and the results on sequences with occlusions or spurious features were not represented.

The application of the SFM algorithm for a non-stationary video sequence requires the removal of the outliers caused by the foreground moving objects and/or high noise level. According to our previous investigation, it is reasonable to build the short-term trajectories (between the key frames) based on the background feature descriptors without regions of foreground salient objects [53]. However, other methods are possible, for example, a previous segmentation of frames in order to find the largest visual part with the similar texture properties. Intuitively, such part may be considered as a more stable background with the minimum outliers' values. The limitation of such proposition is evident and concerns to the scenes with the large foreground object (with area more than 50% of a frame area). At current stage, a single layered motion field of background is used in the SFM algorithm for building of a 3D camera trajectory.

7.5.2   3D Camera Trajectory Building

In order to define 3D camera trajectory, the 2D analysis of motion fields ought to be executed. A typical wild scene includes many moving objects, which possess their own motion layered fields. The non-significant motion layered fields may be considered as the outliers that ought to be removed by any possible way, for example, by extraction of salient objects [54], application of the Least Median Squares (LMedS) method [55], or stereo estimator. The main goal is to detect the largest visual area that is represented by the dominant motion field. Usually such visual area is a background that is more robust to the local motions. Due to the

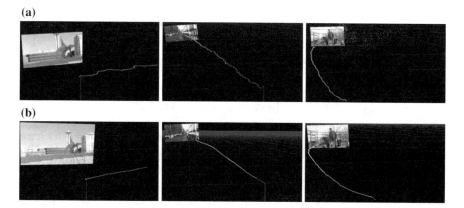

Fig. 7.2 3D camera trajectory building: **a** the original frames from Gleicher4.avi, new_gleicher. avi, and Gleicher1.avi, respectively [58], **b** the synthesis of paths based on the low-pass filter, perfect linear path, and parabolic path, respectively

theory of compactness, the background is a good representative pattern for jitter evaluation.

In current research, the well-known software tools, such as Visual SFM [56] and Voodoo Camera Tracker [57], were used for 3D camera trajectory building and 3D scene visualization, respectively. The results of processing is depicted in Fig. 7.2, when, first, a camera trajectory is built based on the SFM algorithm (Fig. 7.2a) and, second, the obtained trajectory is smoothed (Fig. 7.2b).

Two approaches can be used to smooth a camera trajectory. The first approach is a low-pass filtering, which compensates the most of high-pass shakes. The level of a smoothing is determined by a smoothing coefficient α_s. Its value is tuned in dependence of the displacement values and the motion changes in frames between the consecutive key frames (25–30 frames) from a range 0.5–0.98 defined experimentally [9]. The more is α_s value, the higher is level of smoothing. The second approach is based on a parametric smoothing, when a real camera path aligns according to one of the ideal trajectories like the linear or parabolic curves. However, if the original camera positions are strongly deformed, then too many missing fragments will appear in the target frame that leads to a loss of original information. In this case, a smoothing is implemented such that a loss would not be more than 20% of the original frame.

In our case, the background motion field was the main source for the SFM algorithm. When the novel coordinates of a camera motion are obtained, all pixel positions in a frame are recalculated globally in order to compensate the global discrepancy.

7.5.3 Warping in Scenes with Small Depth

In the case of a dynamic scene, when the visual objects can be approximated by the flat 2D surfaces, the camera induced motion can be modelled by a single global 2D parametric transformation between a pair of successive frames, which is described by Eq. 7.16, where $u(x, y)$ and $v(x, y)$ denote a relative velocity at the point (x, y), a_1 ... a_8 are the modelling parameters.

$$\begin{bmatrix} u(x,y) \\ v(x,y) \end{bmatrix} = \begin{bmatrix} a_1x + a_2y + a_5 + a_7x^2 + a_8xy \\ a_3x + a_4y + a_6 + a_7xy + a_8y^2 \end{bmatrix} \tag{7.16}$$

Equation 7.16 describe the instantaneous 2D motion of the arbitrary 3D scene undergoing the camera rotations, zooms, and small camera translations. In the case of 3D scenes with the large depth, a model "plane + parallax" is more suitable [59].

Generally, a 3D warp allows to change both a view direction and a viewpoint. Full 3D warp in scenes with a small depth is not reasonable due to a heavy computation and non-significant improvement of a visibility. Instead of this, the layering techniques for 2D warps can be applied. A scene can be separated into layers according to several criteria, such as priority rendering [60] or texture

Table 7.2 2D parametric warping techniques

2D parametric warping technique	Formulae
Translation	$x' = x + a_{00}$
	$y' = y + b_{00}$
Translation and dilation	$x' = cx + a_{00}$
	$y' = cy + b_{00}$
Translation and rotation	$x' = x \cos \theta + y \sin \theta + a_{00}$
	$y' = -x \sin \theta + y \cos \theta + b_{00}$
Procrustes	$x' = cx \cos \theta + cy \sin \theta + a_{00}$
	$y' = -cx \sin \theta + cy \cos \theta + b_{00}$
Affine	$x' = a_{10}x + a_{01}y + a_{00}$
	$y' = b_{10}x + b_{01}y + b_{00}$
Perspective	$x' = (a_{10}x + a_{01}y + a_{00})/(a_{10}x + a_{01}y + 1)$
	$y' = (b_{10}x + b_{01}y + b_{00})/(b_{10}x + b_{01}y + 1)$
Bilinear	$x' = a_{10}x + a_{01}y + a_{11}xy + a_{00}$
	$y' = b_{10}x + b_{01}y + b_{11}xy + b_{00}$
Polynomial	$x' = \sum a_{ij}x_iy_j$
	$y' = \sum b_{ij}x_iy_j$

segmentation [61]. The layered motion fields proposed in this research demonstrate the best results in the stabilization task. The main 2D transformations are given in Table 7.2, where x and y are the pixel coordinates of the reference frame, x' and y' are the pixel coordinates of the target frame, $\{a\}$, $\{b\}$ are the modelling coefficients, c is a scaling factor, θ is an angle of rotation.

It is considered that the 2D projective warps are justified if the scene is planar or the views differ purely by rotation [62]. However, for a non-stationary video sequence, the projective model cannot adequately characterize the required warp, causing misalignments or ghosting artifacts. Zaragoza et al. [63] adopted a different strategy, when a model based on the data is adjusted to improve the fit. This model called As-Projective-As-Possible (APAP) warps follows to the conception of a globally projective allowing the local deviations to account for model inadequacy. Liu and Chin [44] integrated the APAP warp with any homography-based stabilization or camera motion planning algorithm. Also, they invented a novel video inpainting algorithm based on the homography fields.

7.5.4 Warping Using Geometrical Primitives

The line segments and planes are the common features in both man-made and natural environments. The underlying geometry of the induced homographies is used in video stabilization, 3D analysis (plane and parallax), ego-motion estimation, calibration, etc. The previous extracted multi-frame context adds the stability and precision in the algorithms. The earlier algorithms required the manual identification of the corresponding geometric primitives in the images or frames. This permitted to estimate the unknown camera calibration matrix, as well as camera orientation and translation. The intrinsic and extrinsic parameters of camera were estimated by a maximum likelihood method, while the 3D point positions were obtained by a bundle adjustment. Finally, the modelling polygons were recovered using the re-calculated projection matrices and the texture of the original images.

In the case of short range motion, the line segments can be compared using the correlation approach. The main idea is to treat each segment as a list of points, to which the neighborhood correlation is applied as a measure of similarity. Only the point to point correspondence is required. If any knowledge is absent, then the corresponding points can be obtained by searching along each line segment and a winner takes all matching strategy. The epipolar geometry (if it can be used) reduces the overall search complexity due to a restriction of the line segments that are considered for a matching. For the indoor scenes, the edge pixels can be linked into line segments, especially long segments. Then the vanishing points, using statistical correlation, can be determined [64].

For the automatic detection of planar surfaces from the flow fields, a frame can be partitioned into the regions of homogeneous 2D motion based on the continuity or fitting of a parametric motion model. Sometimes a mixture of motion models is estimated enforcing by the regions' segmentation in a spatial space.

A motion of the points on a planar surface between two frames can be expressed by Eq. 7.17 [65], where $\mathbf{F}(\mathbf{p})$ is a 2×8 matrix depending on the pixel coordinates $\mathbf{p} = (x, y)$, $\mathbf{n}(\mathbf{p}) \sim N(0, \sigma)$ is a Gaussian additive noise, $\mathbf{b}$ is a 8×1 vector of the planar flow parameters.

$$\mathbf{u}(\mathbf{p}) = \mathbf{F}(\mathbf{p}) \cdot \mathbf{b} + \mathbf{n}(\mathbf{p}) \qquad (7.17)$$

The matrix $\mathbf{F}(\mathbf{p})$ is determined by Eq. 7.18

$$\mathbf{F}(\mathbf{p}) = \begin{pmatrix} 1 & x & y & 0 & 0 & 0 & x^2 & xy \\ 0 & 0 & 0 & 1 & x & y & xy & y \end{pmatrix}, \qquad (7.18)$$

while the vector $\mathbf{b}$ is factorized into a shape $\mathbf{SH}_{8 \times 6}$ part and a motion part according to Eq. 7.19, where $\mathbf{v} = (v_x, v_y, v_z)$ and $\mathbf{w} = (w_x, w_y, w_z)$ are the linear and the rotational velocities of a camera, respectively.

$$\mathbf{b} = \mathbf{SH}_{8 \times 6} \cdot \begin{pmatrix} v \\ w \end{pmatrix} \qquad (7.19)$$

A shape part can be expressed by Eq. 7.20, where $\mathbf{pl} = (pl_x, pl_y, pl_z)$ is a normal to the plane, f is a camera focal length that is assumed a constant over time.

$$\mathbf{SH} = \begin{pmatrix} fpl_z & 0 & 0 & 0 & f & 0 \\ pl_x & 0 & -pl_z & 0 & 0 & 0 \\ pl_y & 0 & 0 & 0 & 0 & -1 \\ 0 & fpl_z & 0 & -f & 0 & 0 \\ 0 & pl_x & 0 & 0 & 0 & 1 \\ 0 & pl_y & -pl_z & 0 & 0 & 0 \\ 0 & 0 & -\frac{pl_x}{f} & 0 & \frac{1}{f} & 0 \\ 0 & 0 & -\frac{pl_y}{f} & -\frac{1}{f} & 0 & 0 \end{pmatrix} \tag{7.20}$$

If the optical flow vectors $\mathbf{u}_i$ are stored in the vector $\mathbf{U}_{2N \times 1}$, N is a number of available features, the matrix $\mathbf{F}(\mathbf{p}_i)$ is transformed into the matrix $\mathbf{G}_{2N \times 8}$, and the noise $\mathbf{n}(\mathbf{p}_i)$ is converted into the vector $\mathbf{\eta}_{2n \times 1}$, then Eq. 7.17 can be rewritten in the form of Eq. 7.21.

$$\mathbf{U} = \mathbf{G} \cdot \mathbf{b} + \mathbf{\eta} \tag{7.21}$$

The maximum likelihood estimation of the planar flow parameters provided by vector $\mathbf{b}$ is given by the weighted linear least squares (Eq. 7.22), where $\mathbf{W}$ is a weight matrix.

$$\hat{\mathbf{b}} = \arg \min_{\mathbf{b}} \left((\mathbf{U} - \mathbf{Gb})^\mathsf{T} \mathbf{W} (\mathbf{U} - \mathbf{Gb}) \right) \tag{7.22}$$

The solution of Eq. 7.22 is found by solving the re-weighted system of normal equations Eq. 7.23.

$$\mathbf{G}^\mathsf{T} \mathbf{W} \mathbf{G} \hat{\mathbf{b}} = \mathbf{G}^\mathsf{T} \mathbf{W} \mathbf{U} \tag{7.23}$$

The weight matrix $\mathbf{W}$ is block diagonal, and the diagonal blocks are 2×2 matrices that represent the covariance of the estimated optical flow vectors. The covariance matrices $\mathbf{\Sigma}$ can be estimated by Eq. 7.24 [66].

$$\mathbf{\Sigma}^{-1} = \begin{pmatrix} I_{xx} & I_{xy} \\ I_{yx} & I_{yy} \end{pmatrix} \tag{7.24}$$

The introduction of re-weighting in Eq. 7.22 is particularly important since the errors are usually asymmetric due to the aperture problem. This approach was used successfully by Irani and Anandan [67] in the context of the orthographic factorization.

The idea of a scene model adaptation to the wild scene via a comparison of visual reconstruction of a modelling scene with the real input sequence began to be exploited since 1990s [68] and it is actual at present time also. During such reconstruction, the geometric primitives play a significant role.

7.5.5 Local 3D Warping in Scenes with Large Depth

The layered representation of the moving objects in video sequences appeared since 1990s [69] regarding to the 2D representations of visual objects, which motion can be described by 2D affine transformation. Afterward, a reconstruction task of a 3D model from 2D images or frames, using the information cues, such as a structure-from-motion, a structure-from-shading, or stereo information, was formulated and is solved during the present time. In the stabilization task, a definition of layers of a background motion (one or several) provides the local 3D warping more accurate. The 3D camera trajectory may be so complicated that a scene deformation cannot be described, for example, by a single projection model. It is more reasonable to recalculate the coordinates in the reconstructed frame, using the local models of the layered motion.

Most layered methods assume a parametric motion in each layer [69–71]. Consider 3D coordinate system positioned to the optical center of a camera. Suppose that a camera moves in 3D environment with a translation motion $\mathbf{TM} = (U, V, W)$ and a rotational motion $\mathbf{RM} = (\alpha, \beta, \gamma)$ but without scaling. Under perspective projection, the 2D velocity (u, v) of an image point $q(x, y)$ refers to the 3D velocity of the projected 3D point $Q(X, Y, Z)$ provided by Eq. 7.25, where f is a focal length of a camera [72].

$$
\begin{aligned}
u &= \frac{(-Uf + xW)}{Z} + \alpha\frac{xy}{f} - \beta\left(\frac{x^2}{f} + f\right) + \gamma y \\
v &= \frac{(-Vf + yW)}{Z} + \alpha\left(\frac{y^2}{f} + f\right) - \beta\frac{xy}{f} - \gamma x
\end{aligned}
\tag{7.25}
$$

Equation 7.25 describe 2D motion field of a point in 3D space projected on the image plane. The motion filed can be non-identical to the optical flow field, which reflects the apparent motion of the brightness patterns, resulting from the relative motion between an imaging system and its environment. If the projection of the optical flow on the direction of the intensity gradient, i.e. a normal flow, is estimated, then the well-known optical flow constraint equation is derived in a view of Eq. 7.26, where I_x, I_y, and I_t are the spatial and temporal partial derivatives of the intensity function, respectively, sign "·" denotes a dot product.

$$
(I_x, I_y) \cdot (u, v) = -I_t
\tag{7.26}
$$

Traditionally, the Least Squares (LS) method is used for estimation of the model parameters. However, the LS estimator becomes highly unreliable in the presence of outlines. The LMedS method proposed by Rousseuw and Leroy [73] provides the robust more estimators because this model fits the majority of the observations. Using the LMedS method, the observations can be classified into model inliers and model outliers. All normal flows are computed from a pair of the successive frames. Such computations form a linear model, when a depth Z and the 3D motion parameters are constant for all points. In terms of the LMedS estimation, the outliers of the linear model will be points, for which Z coordinates deviates from a dominant depth, points, whose 3D motion is different from the dominant motion, or points, where noise is introduced in the computation of a normal flow. The first and the third propositions can be neglected due to such points are very few if a video sequence was pre-processed well. Therefore, the model inliers correspond to the points with a dominant 3D motion in a current layer.

The parameters of the dominant 3D motions can be computed using the reliable feature points correspondences under assumption of affine (three pairs of non-collinear corresponding points) or perspective (four pairs of non-collinear corresponding points) transformation models in each motion layer in the stabilized video sequence.

Let the original non-stationary video sequence be divided into a set of key frames. Consider frames I_1, I_2, ..., I_T between two successive key frames. If the camera is purely rotating about a point, an image motion can be modelled by a homography transformation. Gleicher and Liu [74] proposed to warp each frame I_t by the update transform $\mathbf{B}_t$ providing by Eq. 7.27, where $\mathbf{R}_t$ and $\mathbf{C}_t$ are the homography matrices that warp frame I_t to the first frame I_1 following a new camera path and without following a camera path, respectively.

$$\mathbf{B}_t = \mathbf{R}_t^{-1}\mathbf{C}_t \qquad (7.27)$$

If $\mathbf{R}_t = \mathbf{C}_t$, then there is no smoothing and the update transform is an identity mapping.

The conventional approach is to perform the transform $\mathbf{C}_t$ recursively by Eq. 7.28, where $\mathbf{H}_{t,t-1}$ is a homography matrix that maps from frame I_t to frame I_{t-1}.

$$\mathbf{C}_t = \mathbf{C}_{t-1}\mathbf{H}_{t,t-1} \qquad (7.28)$$

The homography matrices $\mathbf{H}_{t,t-1}$ are computed for all $t = 2$, ..., T in the right order, using the standard estimation approach based on the detection and matching the local features in frames I_t and I_{t-1}.

A set $O(\cdot)$ of the smoothed path $\{P_t\}_{t=1}^{T}$ can be obtained by minimizing Eq. 7.29 proposed by Liu et al. [28], where Ω_t is an index in the neighborhood of frame I_t, in our experiments each frame was linked to the nearest 20–25 frames between two successive key frames.

$$O(\{\mathbf{R}_t\}) = \sum_t \|\mathbf{R}_t - \mathbf{C}_t\|^2 + \lambda \sum_{n \in \Omega_t} \|\mathbf{R}_t - \mathbf{R}_n\|^2 \to \min \qquad (7.29)$$

The first term of Eq. 7.29 stimulates $\mathbf{R}_t$ to be close to $\mathbf{C}_t$, while the second term smoothes the trajectory. The parameter λ controls the strength of the smoothing [74].

The homography-based motion modelling and smoothing is sufficient to capture and smooth the camera trajectory, even under a high degree of shakes and jitters during a hand-held shooting. As Liu and Chin claimed [44], the crucial problem lies in the update transform Eq. 7.27. Instead of a frame global homography, they proposed to map each pixel q^* in frame I_t to the output frame via the local homography (Eq. 7.30).

$$\mathbf{B}_t^* = \mathbf{R}_t^{-1} \mathbf{C}_t^* \tag{7.30}$$

Herewith, a matrix $\mathbf{C}_t^*$ is a homography that warps pixel q^* to the key frame I_1. However, this proposition is not true generality because each pixel maps a small part of a single visual object and has the own motion model in the framework of rigid or non-rigid objects. It is more reasonable to represent in Eq. 7.28 the layered motion fields (lm_i), $i = 1, \ldots, K$, as the local homographies. Let us rewrite Eqs. 7.27–7.28 in a view of Eqs. 7.31–7.32.

$$\mathbf{B}_t^*(lm_i) = \mathbf{R}_t^{-1}(lm_i) \, \mathbf{C}_t^*(lm_i) \tag{7.31}$$

$$\mathbf{C}_t^*(lm_i) = \mathbf{C}_{t-1}^*(lm_i) \, \mathbf{H}_{t,t-1}^*(lm_i) \tag{7.32}$$

Such division into the layers of motion has a connected task, viz. an interpolation between the layers of motion in the possible blank places if the shapes of the layered motion fields are changed. The compactness theory of visual objects' representation provides a hope that the layers of motion describe robustly and adequately the most types of the natural motions. In this case, we can speak about the Structure-From-Layered-Motion (SFLM) with the simplified algorithmic complexity due to the restricted number of the reliable feature points' trajectories into each SFLM. Each SFLM ought to be weighted due to its contribution in a video visibility. The feature points' trajectories in the SFLM provide a spatial channel effect in a 3D space and a spatial chaining effect in a 2D space on a set of homographies $\mathbf{H}_{t,t-1}^*(lm_i)$ between frames I_t and I_{t-1}. A set of homographies defines a spatially varying warp.

Equation 7.32 can be simplified by the frame-global homography chain $\mathbf{C}_{t-1}$ (i.e. the camera trajectory obtained from a smoothing step) provided by Eq. 7.33.

$$\mathbf{C}_t^*(lm_i) = \mathbf{C}_{t-1}^*(lm_i) \, \mathbf{H}_{t,t-1}^*(lm_i) \approx \mathbf{C}_{t-1} \mathbf{H}_{t,t-1}^*(lm_i) \tag{7.33}$$

However, the assumption about parametric motion in each layer is a rough approximation for natural scenes. As a result, some improvements of the SFM algorithm were proposed. Thus, Weiss [75] developed an approach, where the motion of each group is described by a smooth dense flow field. This assumption leaded to the nonparametric mixture estimation based on the EM algorithm, where the stability of the estimation process was ensured by a prior distribution on the class of flow fields. Sun et al. [76] proposed a discrete layered model based on a sequence of ordered MRFs. They introduced the overall energy function, in which every layer was associated with an affine motion field. The binary visibility masks provided a segmentation of the scene into the motion layers that is reasoning to the objects, tracking under occlusions.

In spite of the layered representation is very reasonable, the problem of appearance/disappearance of multiple objects in multiple layers exists. Jojic and Frey [70, 77] introduced the term "flexible sprites", which can deform from frame to frame. Their algorithm is based on the probabilistic representation of "flexible sprites", unsupervised learning, and the EM algorithm. The occluding objects are modelled using the layers of sprites and the sprite masks. Sun et al. [78] suggested a probabilistic graphical model explicitly captures occlusions/non-occlusions, a depth ordering of the layers, and a temporal consistency of the layer segmentation. Additionally, the optical flow in each layer was modelled by a combination of a parametric model and a smooth deviation based on the MRFs. The resulting model allows some roughness in the layers.

7.6 Warping During Video Inpainting

The conventional approach for video inpainting supposes a stitching the reliable parts of neighboring frames to the reconstructed frame in order to fill the blank regions on boundaries. For this goal, several techniques were developed, among which the inpainting based on the local motion [35], homography warps [26], and homography fields [44] are the most interesting from a viewpoint of a higher accuracy. Especially, the non-planar scenes produce the artifacts of such stitching. Our approach directs on the minimization of the stitching procedures and preparing of pseudo-panoramic key frame during a stabilization stage. When the stabilized video frames with the decreased sizes will be obtained and aligned respect to the original sizes, they can be reconstructed using the pseudo-panoramic key frame as a first approximation. Then an inpainting based on the MLMFs is applied in order to improve the details of the missing regions. A pseudo-panoramic key frame is an extension of the original key frame by the parts from the warped following frames with the random sizes. In the case of the non-planar frames, several pseudo-panoramic key frames can be constructed but their number is very restricted and helps to find the coinciding regions very fast. A schema of this process is depicted in Fig. 7.3.

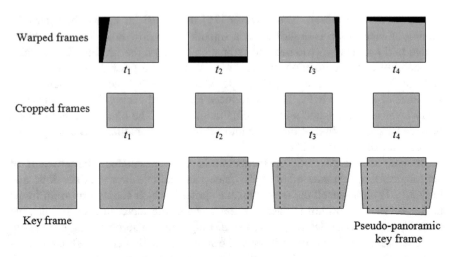

Fig. 7.3 Scheme of pseudo-panoramic key frame receiving

The use of such pseudo-panoramic key frame/frames provides the background information in the missing boundary areas. The following clarification is based on the SFLMs of foreground objects. The inpainting is realized according to the approach of Matsushita et al. [35] (Eqs. 7.8–7.12) but applied for each significant SFLM. For the complicated scenes, the local homography fields [44] can be computed for the necessary SFLMs.

During a filling process, some discontinuities may appear. This problem can be solved by the simplest smoothing based on the pixels' intensities, extrapolation, or smoothing based on the Poisson equation. The Poisson equation in the spatial domain has a view of simultaneous linear equations Eq. 7.34 [79], where Ω is a missing area, p and q are the pixels in the missing and known areas, respectively, $|N_p|$ is a number of neighboring pixels N_p, f_p and f_q are the correct pixel values, div_{pq} is a divergence of pixels p and q, $\partial\Omega$ is a region, surrounding the missing area Ω in the known areas, f_q^* is a known value of pixel q in a region $\partial\Omega$.

$$\text{for all } p \in \Omega \quad |N_p| f_p - \sum_{q \in N_p \cap \Omega} f_q = \sum_{q \in N_p \cap \partial\Omega} f_q^* + \sum_{q \in N_p} \mathrm{div}_{pq} \qquad (7.34)$$

Equations 7.34 form a classical, sparse, symmetric, positive-definite system. Because of the arbitrary shape of region $\partial\Omega$, the iterative solvers ought to be attracted, for example, either Gauss-Seidel iteration. If a missing area Ω contains pixels on the border of a frame, $|N_p| < 4$, then there are no boundary terms in the right hand side of Eqs. 7.34, and 7.35 is obtained.

$$|N_p|f_p - \sum_{q \in N_p} f_q = \sum_{q \in N_p} \mathrm{div}_{pq} \tag{7.35}$$

Note that the Poison equation can be applied in the temporal domain, when the neighboring pixels are attracted from the adjacent frames.

In additional to the stitching problem, the blurring effects become noticeable in some frames. Matsushita et al. [80] proposed a deblurring based on the assumption that the "relative blurriness" of each frame can be evaluated and fragments from the neighboring sharper frames will be copied in the blurry frames. This method is often applied in many researches, including the current one. Objectively, the total gradient value of a blurry image is smaller than that of a sharper frame at the same regions. The blurriness of frame bl_{I_t} is defined by Eq. 7.36, where $\mathbf{p}_i$ is a pixel in frame I_t, g_x and g_y are the gradients along the axis OX and OY, respectively.

$$bl_{I_t} = \sum_{i \in I_t} \left((g_x(\mathbf{p}_i))^2 + (g_y(\mathbf{p}_i))^2 \right) \tag{7.36}$$

A recovering of a blurry pixel $\mathbf{p}_i$ in frame I_t, using a non-blurry pixel $\mathbf{p}_{i'}$ in frame $I_{t'}$, is executed by Eq. 7.37, where $\tilde{\mathbf{p}}_i$ is the transformed value after a deblurring procedure, N is the neighboring frames relative to the current frame, $\mathbf{T}_{t,t'}$ is a warping transform, $w_{t,t'}$ is a weight factor.

$$\tilde{I}_t(\mathbf{p}_t) = \frac{I_t(\mathbf{p}_t) + \sum_{t' \in N} w_{t,t'}(\mathbf{p}_t) I_{t'}\left(\mathbf{T}_{t,t'}\mathbf{p}_t\right)}{1 + \sum_{t' \in N} w_{t,t'}(\mathbf{p}_t)} \tag{7.37}$$

The weight factor is a composition of a relative blurriness $b_t/b_{t'}$ and an alignment error $e_{t,t'}$ and is expressed by Eq. 7.38

$$w_{t,t'}(\mathbf{p}_t) = \begin{cases} 0 & \text{if } b_t/b_{t'} < 1 \\ \frac{b_t}{b_{t'}} \frac{\alpha_{al}}{e_{t,t'}(\mathbf{p}_t) + \alpha_{al}} & \text{otherwise} \end{cases}, \tag{7.38}$$

where $\alpha_{al} \in [0, \infty]$ controls a sensitivity on the alignment error and the pixel-wise alignment error $e_{t,t'}$ is estimated according to Eq. 7.39.

$$e_{t,t'}(\mathbf{p}_t) = \left| I_{t'}\left(\mathbf{T}_{t,t'}\mathbf{p}_t\right) - I_t(\mathbf{p}_t) \right| \tag{7.39}$$

The weighting factor permits to choose only those frames, which are sharper than the current frame.

Table 7.3 Descriptions of some test video sequences

Title	Representative frame	Resolution, pixels	Number of frames	Type of motion	Foreground objects	Scene structure
volley_10100153.mp4 [81]		270 × 480	480	Several moving objects, fixed camera	Several small objects	Scene with small depth, several moving layers
diving_1104.mp4 [81]		270 × 480	180	Single moving object, fast camera displacement	Single moving object	Scene with small depth, several moving layers, complicated scene structure
running_546.mp4 [81]		270 × 480	210	Single moving object, fast camera displacement	Single moving object	Scene with large depth, several moving layers, changing of scene structure
Gleicher2.avi [58]		640 × 360	374	Unstable camera, movement in the direction of the shooting	Single moving object	Scene with large depth, several moving layers, complicated scene structure
New_gleicher.avi [58]		480 × 270	275	Unstable camera, movement in the direction of the shooting	Single moving object	Scene with large depth, several moving layers, complicated scene structure

(continued)

Table 7.3 (continued)

Title	Representative frame	Resolution, pixels	Number of frames	Type of motion	Foreground objects	Scene structure
Gleicher4. avi [58]		640 × 360	412	Unstable camera, slow movement parallel to the shooting	No moving objects	Scene with large depth, several moving layers
sam_1.avi [58]		630 × 360	330	Unstable camera, irregular object movement	Single moving object	Scene with small depth, changing of scene structure

Fig. 7.4 Sparse structure
reconstruction of Gleicher2.
avi [58]

7.7 Experimental Results

The dataset Sports Videos in the Wild [81] and the additional non-stationary video
sequences [58] were used in the experiments. The dataset Sports Videos in the Wild
includes more than 200 video sequences categorized into 30 types of sports.
Herewith, 80 video sequences are annotated regarding to the moving objects. All
test video sequences are non-stationary with cluttered background and involve one
or several foreground moving objects. Short descriptions of some test video
sequences are given in Table 7.3.

The 3D camera trajectory building and 3D scene visualization were executed by
use of the well-known software tools Visual SFM [56] and Voodoo Camera Tracker
[57], respectively. The Kanade-Lucas Tracker (KLT) based on the SURF
descriptors is implemented in the mentioned above software tools that provides the
estimation and tracking of the motion vectors. The foreground moving objects were
not considered during a camera path building because they bring the instability in
the global motion field and usually do not have stable feature points in the

(a)   (b)

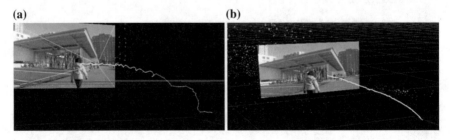

Fig. 7.5 Compensation of 3D camera trajectory for Gleicher2.avi [58]: **a** building of 3D camera
path, **b** smoothing of camera path using a parabolic curve

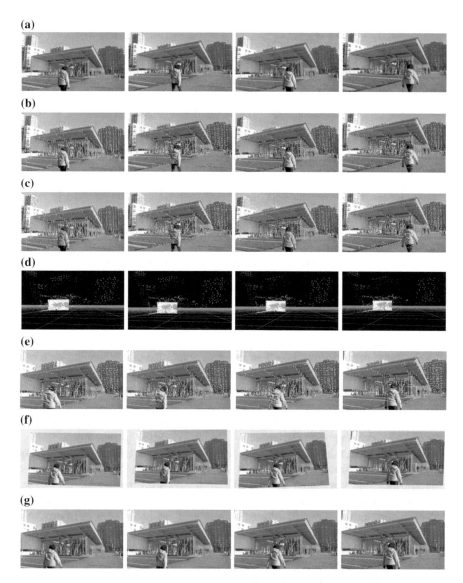

Fig. 7.6 Video stabilization stages on the example of video sequence Gleicher2.avi [58]: **a** original frames 140, 160, 180, and 200, **b** feature points tracking, **c** clustering of motion vectors, **d** building of 3D camera path, **e** smoothing of camera path, **f** cropped frames, **g** warping and inpainting of frames

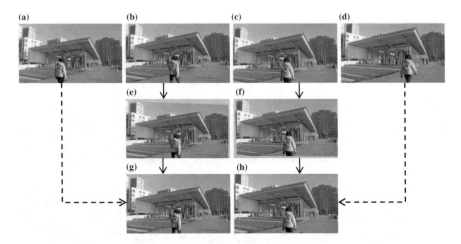

Fig. 7.7 Frame inpainting between two consecutive key frames in video sequence gleicher1.avi video [58], where **a**, **b**, **c**, and **d** are the non-stabilized frames 140, 160, 180, and 200, respectively, **e** and **f** are the stabilized and cropped frames 110 and 120, **g** and **h** are the stabilized frames 110 and 120 after inpainting

sequential frames. Note that it is required to have not less than 20 sequential frames with the same set of feature points in order to obtain the stable moving path of a camera. The examples of the 3D sparse structure reconstruction using [56] and the compensation of the 3D camera path by a parabolic curve using [57] are depicted in Figs. 7.4 and 7.5, respectively.

The stages of the applied stabilization technique based on the SFLMs on the example of video sequence Gleicher2.avi [58] are depicted in Fig. 7.6.

The frame inpainting in a stationary video sequence is executed between two consecutive key frames. The results of the proposed technique, using a pseudo-panoramic key frame and the MLMF analysis on the boundaries of each frame, are depicted in Fig. 7.7. This approach shows the best results in scenes that do not include fast motions of a lot of objects near the boundaries of frames.

The plots of displacements for three video sequences from Fig. 7.2 with the path based on the low-pass filter, perfect linear path, and parabolic path are depicted in Fig. 7.8. For all three cases, the displacements without trajectories of the foreground objects achieve the best smoothing results.

Experimental results demonstrate that the application of the proposed approaches increases the quality of digital video stabilization preserving the sizes of the original frames.

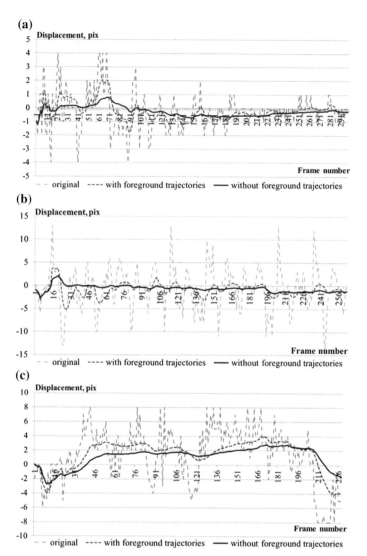

Fig. 7.8 Evaluation of displacements (Gleicher4.avi, new_gleicher.avi, and Gleicher1.avi, respectively [58]): **a** path based on the low-pass filter, **b** perfect linear path, **c** parabolic path

7.8   Conclusions

In this chapter, many techniques of 2D and 3D warping in video stabilization task were discussed. All modern approaches require several loops of processing even between the consecutive key frames that concerns this issue to the video editing activity more than the on-line application. Our strategy is directed on the reasonable combination of the 2D and 3D analysis based on the multi-layered motion fields that reflect the scene content. The distinctive features of scenes with small and large depths were considered in the warping during stabilization and inpainting. Our experiments confirm the quality of the stabilized test video sequences in terms of the content alignment and the frame size preserving.

Acknowledgements The reported study was funded by the Russian Fund for Basic Researches according to the research project № 16-07-00121 A.

References

1. Barr, A.H.: Global and local deformations of solid primitives. Comput. Graph. **18**(3), 21–30 (1984)
2. Bechmann, D.: Space deformation models survey. Comput. Graph. **18**(4), 571–586 (1994)
3. Anandan, P.: A computational framework and an algorithm for the measurement of visual motion. Int. J. Comput. Vis. **2**, 283–310 (1989)
4. Black, M.J., Anandan, P.: The robust estimation of multiple motions: parametric and piecewise smooth flow fields. Comput. Vis. Image Underst. **63**(1), 75–104 (1996)
5. Brox, T., Bruhn, A., Papenberg, N., Weickert, J.: High accuracy optical flow estimation based on a theory for warping. In: Pajdla, T., Matas, J. (eds.) Proceedings of the 8th European Conference on Computer Vision. Springer LNCS 3024, vol. 4, pp. 25–36. Springer, Berlin, Heidelberg (2004)
6. Papenberg, N., Bruhn, A., Brox, T., Didas, S., Weickert, J.: Highly accurate optic flow computation with theoretically justified warping. Int. J. Comput. Vis. **67**(2), 141–158 (2006)
7. Memin, E., Perez, P.: Hierarchical estimation and segmentation of dense motion fields. Int. J. Comput. Vis. **46**(2), 129–155 (2002)
8. Alvarez, L., Weickert, J., Sanchez, J.: Reliable estimation of dense optical flow fields with large displacements. Int. J. Comput. Vis. **39**(1), 41–56 (2000)
9. Favorskaya, M., Jain, L.C., Buryachenko, V.: Digital video stabilization in static and dynamic scenes. In: Favorskaya, M.N,. Jain. L.C. (eds.) Computer Vision in Control Systems-1, ISRL, vol. 73, pp. 261–309. Springer International Publishing, Switzerland (2015)
10. Favorskaya, M., Buryachenko, V.: Fuzzy-based digital video stabilization in static scenes. In: Tsihrintzis, G.A., Virvou, M., Jain, L.C., Howlett, R.J., Watanabe, T. (eds.) Intelligent Interactive Multimedia Systems and Services in Practice, SIST, vol. 36, pp. 63–83. Springer International Publishing, Switzerland (2015)
11. Battiato, S., Gallo, G., Puglisi, G., Scellato, S.: SIFT features tracking for video stabilization. In: 14th IEEE International Conference on Image Analysis and Processing (ICIAP'2007), pp. 825–830 (2007)
12. Cai, J., Walker, R.: Robust video stabilisation algorithm using feature point selection and delta optical flow. IET Comput. Vis. **3**(4), 176–188 (2009)
13. Liu, F., Gleicher, M., Jin, H., Agarwala, A.: Content-preserving warps for 3D video stabilization. ACM Trans. Graph. **28**(3), 44.1–44.9 (2009)
14. Wang, J.M., Chou, H.P., Chen, S.W., Fuh, C.S.: Video stabilization for a hand-held camera based on 3D motion model. In: IEEE International Conference on Image Processing (ICIP'2009), pp 3477–3480 (2009)
15. Zhang, G., Hua, W., Qin, X., Shao, Y., Bao, H.: Video stabilization based on a 3D perspective camera model. Vis. Comput.: Int. J. Comput. Graph. **25**(11), 997–1008 (2009)
16. Liu, F., Gleicher, M., Wang, J., Lin, H., Agarwala, A.: Subspace video stabilization. ACM Trans. Graph. **30**(1), 4.1–4.10 (2011)
17. Milliron, T., Jensen, R.J., Barzel, R., Finkelstein, A.: A framework for geometric warps and deformations. ACM Trans. Graph. **21**(1), 20–51 (2002)

18. Alexa, M., Cohen-Or, D., Levin, D.: As-rigid-as-possible shape interpolation. In: 27th Annual Conference on Computer Graphics and Interactive Techniques (SIGGRAPH'2000), pp. 157–164 (2000)
19. Liu, L., Zhang, L., Xu, Y., Gotsman, C., Steven, S.J.: A local/global approach to mesh parameterization. In: Symposium on Geometry Processing (SGP'2008), pp. 1495–1504 (2008)
20. Igarashi, T., Moscovich, T., Hughes, J.F.: As-rigid-as-possible shape manipulation. ACM Trans. Graph. **24**(3), 1134–1141 (2005)
21. Schaefer, S., McPhail, T., Warren, J.: Image deformation using moving least squares. ACM Trans. Graph. **25**(3), 533–540 (2006)
22. Zhou, K., Huang, J., Snyder, J., Liu, X., Bao, H., Guo, B., Shum, H.Y.: Large mesh deformation using the volumetric graph Laplacian. ACM Trans. Graph. **24**(3), 496–503 (2005)
23. Sheffer, A., Kraevoy, V.: Pyramid coordinates for morphing and deformation. In: 2nd International Symposium on 3D Data Processing, Visualization, and Transmission (3DPVT'2004), pp. 68–75 (2004)
24. Weng, Y., Xu, W., Wu, Y., Zhou, K., Guo, B.: 2D shape deformation using nonlinear least squares optimization. Vis. Comput. **22**(9), 653–660 (2006)
25. Gal, R., Sorkine, O., Cohen-Or, D.: Feature-aware texturing. In: 17th Eurographics Conference on Rendering Techniques (EGSR'2006), pp. 297–303 (2006)
26. Litvin, A., Konrad, J., Karl, W.C.: Probabilistic video stabilization using Kalman filtering and mosaicking. In: IS&T/SPIE Symposium on Electronic Imaging, Image and Video Communications, Santa Clara, CA, USA, pp. 663–674 (2003)
27. Wang, Y.S., Liu, F., Hsu, P.S., Lee, T.Y.: Spatially and temporally optimized video stabilization. IEEE Trans. Vis. Comput. Graph. **19**(8), 1354–1361 (2013)
28. Liu, S., Yuan, L., Tan, P., Suny, J.: Bundled camera paths for video stabilization. ACM Trans. Graph. **32**(4), 78.1–78.10 (2013)
29. Liu, S., Yuan, L., Tan, P., Sun, T.: SteadyFlow: spatially smooth optical flow for video stabilization. In: IEEE Conference on Computer Vision and Pattern Recognition (CVPR'2014), pp. 4209–4216 (2014)
30. Smith, B.M., Zhang, L., Jin, H., Agarwala, A.: Light field video stabilization. In: IEEE International Conference on Computer Vision (ICCV'2009), pp. 341–348 (2009)
31. Huynh, L., Choi, J., Medioni, G.: Aerial implicit 3D video stabilization using epipolar geometry constraint. In: 22nd International Conference on Pattern Recognition (ICPR'2014), pp. 3487–3492 (2014)
32. Goldstein, A., Fattal, R.: Video stabilization using epipolar geometry. ACM Trans. Graph. **31**(5), 126.1–126.10 (2012)
33. Hartley, R.I., Zisserman, A.: Multiple View Geometry in Computer Vision. Cambridge University Press (2000)
34. Lee, D.B., Choi, I.H., Song, B.C., Lee, T.H.: ROI-based video stabilization algorithm for hand-held cameras. In: IEEE International Conference on Multimedia and Expo Workshops (ICMEW'2012), pp. 314–318 (2012)
35. Matsushita, Y., Ofek, E., Ge, W., Tang, X., Shum, H.Y.: Full-frame video stabilization with motion inpainting. IEEE Trans. Pattern Anal. Mach. Intell. **28**(7), 1150–1163 (2006)
36. Grundmann, M., Kwatra, V., Essa, I.: Auto-directed video stabilization with robust L1 optimal camera paths. In: IEEE Conference on Computer Vision and Pattern Recognition (CVPR'2011), pp. 225–232 (2011)
37. Grundmann, M., Kwatra, V., Castro, D., Essa, I.: Calibration-free rolling shutter removal. In: IEEE International Conference on Computational Photography (ICCP'2012), pp. 4.1–4.8 (2012)
38. Liu, S., Wang, Y., Yuan, L., Bu, J., Tan, P., Sun, J.: Video stabilization with a depth camera. IEEE Conf Computer Vision Pattern Recogn (CVPR'2012), pp. 89–95
39. Wexler, Y., Shechtman, E., Irani, M.: Space-time video completion. In: IEEE Conference on Computer Vision and Pattern Recognition (CVPR'2004), vol. 1, pp. 120–127 (2004)

40. Jia, J., Wu, T., Tai, Y., Tang, C.: Video repairing: inference of foreground and background under severe occlusion. In: IEEE Conference on Computer Vision and Pattern Recognition (CVPR'2004), vol. 1, pp. 364–371 (2004)
41. Bouguet, J.Y.: Pyramidal Implementation of the Lucas Kanade Feature Tracker: Description of the Algorithm. Intel Corporation, Microprocessor Research Labs, OpenCV Documents (2000)
42. Telea, A.: An image inpainting technique based on the fast marching method. J. Graph. Tools 9(1), 25–36 (2004)
43. Shum, H.Y., Szeliski, R.: Construction of panoramic mosaics with global and local alignment. Int. J. Comput. Vis. 36(2), 101–130 (2000)
44. Liu, W.X., Chin, T.J.: Smooth globally warp locally: video stabilization using homography fields. In: International Conference on Digital Image Computing: Techniques and Applications (DICTA'2015), pp. 1–8 (2015)
45. Lienhart, R.: Comparison of automatic shot boundary detection algorithms. In: SPIE Conference on Storage and Retrieval for Image and Video Databases VII, vol. 3656, pp. 290–301 (1999)
46. Kelm, P., Schmiedekem, S., Sikora, T.: Feature-based video key frame extraction for low quality video sequences. In: 10th Workshop on Image Analysis for Multimedia Interactive Services (WIAMIS'2009), pp. 25–28 (2009)
47. Won, J.U., Chung, Y.S., Kim, I.S., Choi, J.G., Park, K.H.: Correlation based video-dissolve detection. Inf. Technol.: Res. Educ. 104–107 (2003)
48. Karsch, K., Liu, C., Kang, S.B.: Depth extraction from video using non-parametric sampling. In: 12th European Conference on Computer Vision (ECCV'2012), vol. V, pp. 775–788 (2012)
49. Liu, C., Yuen, J., Torralba, A.: SIFT flow: dense correspondence across scenes and its applications. IEEE Trans. Pattern Anal. Mach. Intell. 33(5), 978–994 (2011)
50. Saxena, A., Chung, S.H., Ng, A.Y.: 3-D Depth reconstruction from a single still image. Int. J. Comput. Vis. 76(1), 53–69 (2008)
51. Patil, S., Charles, P.: Review on 2D to 3D image and video conversion methods. In: International Conference on Computing Communication Control and Automation (ICCUBEA'2015), pp. 728–732 (2015)
52. Dellaert, F., Seitz, S.M., Thorpe, C.E., Thrun, S.: Structure from motion without correspondence. In: IEEE Conference on Computer Vision and Pattern Recognition (CVPR'2000), pp. 557–564 (2000)
53. Favorskaya, M., Buryachenko, V., Tomilina, A.: Global motion estimation using saliency maps in non-stationary videos with static scenes. In: De Pietro, G., Gallo, L., Howlett, R.J., Jain, L.C. (eds.) Intelligent Interactive Multimedia Systems and Services, SIST, vol. 55, pp. 133–144. Springer International Publishing, Switzerland (2016)
54. Favorskaya, M., Buryachenko, V.: Fast salient object detection in non-stationary video sequences based on spatial saliency maps. In: De Pietro, G., Gallo, L., Howlett, R.J., Jain, L.C. (eds.) Intelligent Interactive Multimedia Systems and Services, SIST, vol. 55, pp. 121–132. Springer International Publishing, Switzerland (2016)
55. Argyros, A.A., Lourakis, M.I.A., Trahanias, P.E., Orphanoudakis, S.C.: Independent 3D motion detection through robust regression in depth layers. In: British Machine Vision Conference (BMVC'1996), pp. 9–12 (1996)
56. Visual SFM: A Visual Structure from Motion System. http://ccwu.me/vsfm/. Accessed 21 Oct 2016
57. Voodoo Camera Tracker: A tool for the integration of virtual and real scenes. http://www.viscoda.com/index.php/en/products/non-commercial/voodoo-camera-tracker. Accessed 21 Oct 2016
58. Auto-Directed Video Stabilization with Robust L1 Optimal Camera Paths video dataset. http://www.cc.gatech.edu/cpl/projects/videostabilization/. Accessed 21 Oct 2016
59. Irani, M., Anandan, P.: A unified approach to moving objects detection in 2D and 3D scenes. IEEE Trans. Pattern Anal. Mach. Intell. 6, 577–589 (1998)

60. Regan, M., Pose, R.: Priority rendering with a virtual reality address recalculation pipeline. In: Computer Graphics Annual Conference Series (SIGGRAPH'1994), pp. 155–162 (1994)
61. Aliaga, D.G.: Visualization of complex models using dynamic texture-based simplification. In: IEEE 7th Conference on Visualization (VIS'1996), pp. 101–106 (1996)
62. Alexa, M., Behr, J., Cohen-Or, D., Fleishman, S., Levin, D., Silva, C.T.: Computing and rendering point set surfaces. IEEE Trans. Vis. Comput. Graph. 9(1), 3–15 (2003)
63. Zaragoza, J., Chin, T.J., Brown, M., Suter, D.: As-projective-as-possible image stitching with moving DLT. In: IEEE Conference on Computer Vision and Pattern Recognition (CVPR'2013), pp. 2339–2346 (2013)
64. Yu, S.X., Zhang, H., Malik, J.: Inferring spatial layout from a single image via depth-ordered grouping. In: IEEE Computer Society Conference on Computer Vision and Pattern Recognition Workshops (CVPRW'2008), pp. 1–7 (2008)
65. Kanatani, K.: Geometric Computation for Machine Vision. Clarendon Press (1993)
66. Shi, J., Tomasi, C.: Good features to track. In: IEEE Conference on Computer Vision and Pattern Recognition (CVPR'1994), pp. 593–600 (1994)
67. Irani, M., Anandan, P.: Factorization with uncertainty. In: European Conference on Computer Vision (ECCV'2000), vol. 1, pp. 539–553 (2000)
68. Koch, R.: Dynamic 3D scene analysis through synthesis feedback control. IEEE Trans. Pattern Anal. Mach. Intell. 15(6), 556–568 (1993)
69. Wang, J.Y.A., Adelson, E.H.: Representing moving images with layers. IEEE Trans. Image Process. 3(5), 625–638 (1994)
70. Jojic, N., Frey, B.: Learning flexible sprites in video layers. In: IEEE International Conference on Computer Vision and Pattern Recognition (CVPR'2001), vol. I, pp. 199–206 (2001)
71. Kumar, M., Torr, P., Zisserman, A.: Learning layered motion segmentations of video. Int. J. Comput. Vis. 76(3), 301–319 (2008)
72. Longuet-Higgins, H.C., Prazdny, K.: The interpretation of a moving retinal image. Proc. R. Soc. Lond. Ser. B Biol. Sci. 208(1173), 385–397 (1980)
73. Rousseeuw, P.J., Leroy, A.M.: Robust Regression and Outlier Detection. Wiley, New York (1987)
74. Gleicher, M.L., Liu, F.: Re-cinematography: improving the camerawork of casual video. ACM Trans. Multimed. Comput. Commun. Appl. 5(1), 2:1–2:28 (2008)
75. Weiss, Y.: Smoothness in layers: motion segmentation using nonparametric mixture estimation. In: IEEE International Conference on Computer Vision and Pattern Recognition (CVPR'1997), pp. 520–526 (1997)
76. Sun, D., Sudderth, E.B., Black, M.J.: Layered segmentation and optical flow estimation over time. In: IEEE International Conference on Computer Vision and Pattern Recognition (CVPR'2012), pp. 1768–1775 (2012)
77. Frey, B.J., Jojic, N., Kannan, A.: Learning appearance and transparency manifolds of occluded objects in layers. In: IEEE Conference on Computer Vision Pattern and Recognition (CVPR'2003), vol. 1, pp. 45–52 (2003)
78. Sun, D., Sudderth, E.B., Black, M.J.: Layered image motion with explicit occlusions, temporal consistency, and depth ordering. In: Advances in Neural Information Processing Systems (NIPS'2010), vol. 23, pp. 2226–2234 (2010)
79. Pérez, P., Gangnet, M., Blake, A.: Poisson image editing. ACM Trans. Graph. 22(3), 313–318 (2003)
80. Matsushita, Y., Ofek, E., Tang, X., Shum, H.Y.: Full-frame video stabilization. In: IEEE Computer Vision and Pattern Recognition (CVPR'2005), vol. 1, pp. 50–57 (2005)
81. Sports Videos in the Wild (SVW): A Video Dataset for Sports Analysis. http://www.cse.msu.edu/~liuxm/sportsVideo/. Accessed 22 Oct 2016

Chapter 8
Image Deblurring Based on Physical Processes of Blur Impacts

Andrei Bogoslovsky, Irina Zhigulina, Eugene Bogoslovsky and Vitaly Vasilyev

Abstract Main methods of image deblurring, as well as their advantages and disadvantages, are considered in this chapter. It is revealed that these methods do not take into account the physical processes that occur during a blur impact. It is shown that instead of the continuous function of illuminating intensity it is convenient to consider its discrete-analog modification, in which the size of a discrete will be equal to the size of a photosensitive element. In the case of the stationary images, such replacement will not impact on the formed video signal in any way. On the contrary, when objects move relative to the fixed sensor, the processes of blur may appear. Due to these reasons, the models of various blur types, such as the linear, non-linear, and vibrational models, were built. Note that these models are subdivided on the models with the small and large blur according to the ratio of movement displacement of an object and its length. The constructed models are the systems of the simple non-uniform algebraic equations with non-singular matrixes that allows to build the deblurring algorithms. The proposed algorithms were tested using natural images obtained from the imaging device layout.

Keywords Image deblurring · Blind deconvolution · Small blur
Large blur · Vibrational blur · System of equations · Image restoration
Matrix methods

A. Bogoslovsky (✉) · I. Zhigulina · E. Bogoslovsky · V. Vasilyev
Air Force Military Training and Scientific Center "Air Force Academy",
54a Starykh Bolshevikov, Voronezh 394064, Russian Federation
e-mail: p-digim@mail.ru

I. Zhigulina
e-mail: irazhigulina@gmail.com

E. Bogoslovsky
e-mail: qro76@yandex.ru

V. Vasilyev
e-mail: warner88@mail.ru

© Springer International Publishing AG 2018   217
M.N. Favorskaya and L.C. Jain (eds.), *Computer Vision in Control Systems-3*,
Intelligent Systems Reference Library 135, https://doi.org/10.1007/978-3-319-67516-9_8

8.1 Introduction

Quite often in the wild video sequences, the specific distortions are caused by movement. These distortions appear, when the sizes of dynamic objects in an image are larger than the sizes of the same objects in a static position. At the same time, owing to loss of high-frequency structure in video signals corresponding to them such objects are identified more difficult. These distortions are called "blur" due to the inconsistency of the scanning vision system with the parameters of mutual motion of the object and the optical system. The blur distortion significantly impacts the quality of the images, for example, when the playback capabilities of some parts of objects are limited (see Fig. 8.1).

The changes in the image happening at blur depend on the contrast and size of its details, from which it is possible to distinguish conditionally the small low-contrast details, small details of average and high contrast, and large details [1, 2]. Figure 8.2 shows the undistorted and blurred images. Note that small low contrast objects marked in Fig. 8.2a with white framework do not display in the blurred image.

Even if the objects remain in the blurred image, their shape, size, and contrast relative to the background are significantly distorted that leads to substantial

Fig. 8.1 Frame from video sequence containing blurred parts

(a)   **(b)**

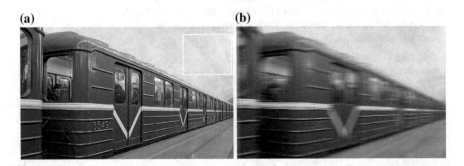

Fig. 8.2 Examples of images: **a** undistorted image, **b** blurred image

difficulties and, sometimes, to inability of their identification. In addition, the resolution of the optical system itself may bring a blur in an image. Therefore, the deblurring is necessary for reliable definition of parameters of the distorted dynamic objects.

The chapter is organized as follows. In Sect. 8.2, the overview of blur and deblurring methods, as well as the problem statement of image blur, are represented. The model of the linear blur formation and deblurring method are detailed in Sect. 8.3. Section 8.4 presents a model of the non-linear blur formation. Conclusions and future development remarks are given in Sect. 8.5.

8.2 Overview of Blur and Deblurring Methods

There are several approaches for blur description. Basically, all investigations are dedicated to the study of the uniform direct movement (receiver of image, object, or their mutual movement) that is known as a linear blurring [1–17]. It is very suitable to use the grayscale images for study the nature of blur and image restoration techniques. In the case of color images, all required steps for each of color RGB channels can be repeated [4, 8].

Distorted images can be described using Eq. 8.1 [18, 19], where $f(x, y)$ is an original image, $h(x, y)$ is a motion blur kernel, $n(x, y)$ is an additive noise (observation noise), $g(x, y)$ is a distorted (blurred) image, sign «*» is a convolution process.

$$g(x, y) = h(x, y) * f(x, y) + n(x, y) \qquad (8.1)$$

The task of image deblurring is to search such function $f'(x, y)$ that is the best estimation of the latent image. Function $h(x, y)$ has some different titles. It is known as the distorting function [1], Point Spread Function (PSF) [4], transfer function [5], pulse response function [7], blur kernel [9], or blur operator [10]. This function describes a functional according to law that is smearing each pixel of the image. Afterwards, we should call it a Distortion Function (DF).

There are various image deblurring techniques in image processing. In the case of the prior knowledge about the DF, we should use such techniques as inverse filtering, Wiener filtering, Kalman recursive filtering, minimum Mean Square Error (MSE) method, and various forced iterative methods. Those techniques are also known as the non-blind deconvolution methods [1, 2, 4, 7–10, 20–32].

Using the convolution theorem in Eq. 8.1, we can obtain Eq. 8.2, where u and v are the spatial frequencies.

$$G(u, v) = H(u, v)F(u, v) + N(u, v) \qquad (8.2)$$

The easiest way to reconstruct the blurred image is to apply the inverse filtering. In this case, we can obtain the estimation of origin image Fourier transform $\widehat{F}(u, v)$

by dividing a distorted image in Fourier domain on the frequency analogue of DF. Then the estimation itself can be received in the form of Eq. 8.3 [1, 4].

$$\widehat{F}(u,v) = F(u,v) + \frac{N(u,v)}{H(u,v)} \tag{8.3}$$

As it is followed from Eq. 8.3, the reconstructed image is a sum of origin image and observation noise passed by the inverse filter. The absence of noise allows to reconstruct the origin image accurately using the inverse filtering.

The inverse filter provides an undesirable restoration because of the edge effects. These effects look like the powerful oscillation interference, which fully disguised the restored image. Such effects arise even in absence of an observation noise. There are some methods that we can use for noise reducing. All of them limit the bandwidth of an inverse filter. It is necessary to place the corrective module consecutive to inverse filter. Moreover, the pulse characteristic of this unit should approach zero beyond a certain given cut-off frequency. In this case, the cut-off frequency is out of a compromise between the noise reduction and clearness of restorable image. However, these methods do not solve the problems of the edge effects and presence of zeros in pulse characteristic of formation system within the operating frequency band. Therefore, despite the obvious simplicity of the inverse filtration it can be used successfully to restore a limited range of images, whose background level around the edges is constant. Besides, the inverse filtering possesses very low noise stability [1, 2, 4].

Wiener filtering allows to consider the properties of the DF and noise characteristics. This type of filtering is based on the image and the DF as random processes. Wiener filter minimizes the MSE between the estimated random process and desired process. Three conditions must be executed for guaranteeing the MSE [1]:

- The noise and undistorted image do not correlate with each other.
- The undistorted image or noise must have zero average value.
- The estimation linearly depends on the distorted image.

The minimum of the MSE of the difference between the estimation and undistorted image attained with a type of estimation (in the frequency domain) is given by Eq. 8.4, where $S_\eta(u,v)$ and $S_f(u,v)$ are the energy spectrums of the noise and undistorted image, respectively [1, 4].

$$\widehat{F}(u,v) = \left(\frac{1}{H(u,v)} \frac{|H(u,v)|^2}{|H(u,v)|^2 + S_\eta(u,v)/S_f(u,v)} \right) G(u,v) \tag{8.4}$$

Inverse Fourier transform of Eq. 8.4 gives the restored image. Usually we have not any information about the spectrum of the undistorted image. In this case, we can use Eq. 8.5 as the approximation of $\widehat{F}(u, v)$. In Eq. 8.5, parameter K is some constant with no influence on the decision.

$$\widehat{F}(u, v) = \left(\frac{1}{H(u, v)} \frac{|H(u, v)|^2}{|H(u, v)|^2 + K} \right) G(u, v) \tag{8.5}$$

Analyzing the above, it can be noted that Wiener filter passes to the inverse filter in the case of noise absence. Pulse characteristics of Wiener filter and inverse filter are practically identical at low frequencies (where the signal-noise ratio is large as a rule). Due to the use of information about spectral characteristics of the image and noise, Wiener filter has relatively high noise immunity and it does not have a singularity caused by zeros of pulse characteristic of the forming system. However, existence of edge effects that are shown in the form of the oscillating hindrance masking the restored image remains the main lack of Wiener filter [1, 4, 5]. Blurred images restored by use the inverse filter and Wiener filter usually contain an oscillating hindrance of great intensity, which arises because both of these filters are synthesized without the fact that observed images have a finite size (occurrence of edge effect) [4]. Thus, in a truncated image there is a loss of information contained in the original image near the borders. Therefore, the correction of linear distortions of a truncated image leads to the false details in the form of the ripples or bands, whose intensity is especially great with the cylindrical shape of the DF and uniform blur.

Wiener filter with the estimation of the original image in the form of Eq. 8.4 can be generalized by Eq. 8.6, where α and β are positive real constants [1].

$$\widehat{F}(u, v) = \left(\frac{H^*(u, v)}{|H(u, v)|^2} \right)^{\alpha} \left[H^*(u, v) \middle/ \left(|H(u, v)|^2 + \beta \left(\frac{S_{\eta}(u, v)}{S_f(u, v)} \right) \right) \right]^{1-\alpha} G(u, v) \tag{8.6}$$

Filter issuing the assessment in the form of Eq. 8.6 is called the geometric mean [1]. When $\alpha = 1$, this filter reduces to the inverse filter. When $\alpha = 0$, it reduces to the so-called parametric Wiener filter, which in turn becomes an ordinary Wiener filter, when $\beta = 1$. Thus, within Eq. 8.6 the whole family of the filters used for the blurred images restoration is described.

It is not possible to solve the equation of Wiener-Hopf for the signals and images observed in a limited interval. Therefore, there is no optimal space-invariant filters sensitive to edge effects. For compensation of edge effects, various heuristic algorithms are used. If the central part of an image (and its sizes much more than the sizes of DF frame) enters the area of interest, then one can apply a multiplication of the observed image to function of a window $w(i_1, i_2)$, which smoothly decreases to zero at the edges of a frame and zero everywhere outside it. After that, an image is restored by Wiener filter.

The 1D window function for vertical or horizontal blur is used. This function multiplies columns or rows of the observed image, respectively. Many 1D window functions $w(i)$ can be used for image restoration, for example the Bartlett window, Kaiser window, or Blackman window [4]. Good results are given by a function of the window, a shape of which is defined by two independent parameters γ and β [4]. This function is defined by Eq. 8.7, where parameter γ influences the window sizes, β is a speed of recession of edges of a window to zero.

$$w(i) = 0,5\left[\text{th}\left(\left(i+\frac{\gamma}{2}\right)\Big/\beta\right) - \text{th}\left(\left(i-\frac{\gamma}{2}\right)\Big/\beta\right)\right] \qquad (8.7)$$

The brightness level at the edges of the image multiplied by the window goes to zero together with a reduction of edge effects narrow borders of the restored image. Besides, the optimum parameters of windows depend on parameters of the distorting system and are defined by the applied field that complicates the practical application of the recovery algorithms [4].

It is possible to consider the restricted sizes of the observed image at the stage of Wiener filter synthesis, which uses information about the spectral and correlation characteristics of an image. Getting images of limited-size is equivalent to multiplication of infinite images on a window of single brightness, which is equal to the size of a frame.

Note that the discussed recovery methods are linear. They are widespread due to the rather simple methods of the synthesis and analysis of linear systems. Also they have high computational efficiency. However, these methods are not optimal and cannot always provide an effective compensation of distortion. The linear processing is only an approximation to the optimal processing because the statistical characteristics of the vast majority of the images are non-Gaussian [4]. In addition, the linear methods do not take into account a priori information about the recoverable images. Therefore, interest to the non-linear methods of image processing arises. Synthesis of optimal non-linear algorithms is usually much more complicated than linear. However, there are linear recovery methods that are quite simply can be converted into non-linear taking into account a priori information about the images and disturbances. Striking examples of such methods are the iterative methods (the methods of successive approximations) [1, 4]. Iterative methods are methods that choose some initial approximate solution, calculate the next (more accurate) approximation using the previous solutions [4].

Consider one of the ways of constructing iterative procedures based on the decomposition in the number of inverse filter frequency response. The spectrum of estimation of original image using inverse filtration (excluding the noise component) is determined by Eq. 8.8:

$$\widehat{F}^{(2)}(u,v) = G(u,v) + (1 - H(u,v))\widehat{F}^{(1)}(u,v), \qquad (8.8)$$

and so on until Eq. 8.9:

$$\widehat{F}^{(n)}(u,v) = G(u,v) + (1 - H(u,v))\widehat{F}^{(n-1)}(u,v). \tag{8.9}$$

The implementation of Eq. 8.8 can be interpreted as a procedure of serial finding of amendments to the distorted image. If we will find the exact solution relative to the original image because of successive approximations to the nth step, then an assessment will not change any more on the subsequent steps [4].

The considered iteration scheme is the linear and has no advantages in comparison with the linear algorithms. However, this method allows to solve the edge problems effectively but with an excessive amplification of noise during image restoration. The iterative process can always be stopped if the noise and oscillating hindrance on the image amplify sharply. The stop of an iterative process means a truncation of a row (Eq. 8.8) that leads to restriction of an intensification coefficient outside some boundary frequency. With increasing length of a row, the boundary frequency and intensification coefficient of the filter are enlarged.

Iterative algorithms can be easily converted into non-linear algorithms by introducing the non-linear constraints for the reconstructed image (Lucy-Richardson method) [4, 8]. These constraints are formulated on the basis of a priori information about the form or structure of objects in the original image. Such properties of an image, as the range of brightness change, minimum power of a signal, restricted space, and spectral extent, can be referred to the priori data. Brightness range leads to a significant improvement in the quality of recovery [4].

The given methods and algorithms can be used for the restoration of blurred images only using a knowledge of the DF. However, in practice most processing tasks are facing the problem of lack of information about it. Moreover, in certain cases a receiving analytical expression of the DF can be just impossible or accompanied by danger to life of the operator. In these cases, the assessment of the DF in the algorithms described above or methods of a blind deconvolution can be used.

There are three main ways to assess the DF for further use in the restoration of images: the visual analysis, experiment, and mathematical modeling [1]. For example, suppose that we have a distorted image but information about the DF is missing. It is possible to identify information about the DF directly from the available image. It is advisable to consider a small piece of this image that contains a simple structure and background [1, 9]. To reduce the influence of background, we need to select an image area that contains the useful signal of large amplitude. Using the available values of brightness of the object and background, one can build an approximately undistorted image of the same size and with the same features as the original distorted image. Let us designate the considered part of the image as $s(x,y)$ and the constructed part of the image representing (in fact, an approximation for a part of the undistorted image in this area) as $\hat{f}_s(x,y)$. Analyzing the selected area with a high level of the desired signal, it can be assumed that the effect of noise is negligible, and then we can get Eq. 8.10 as the DF version in the frequency domain.

$$H_s(u, v) = \frac{S(u, v)}{\hat{F}_s(u, v)} \tag{8.10}$$

Based on the properties of $H_s(u, v)$, one can be made conclusions about the DF $H(u, v)$ properties. At the same time, the approximation of $H_s(u, v)$ is used further in algorithms of recovery of the blurred images with the known DF. The DF type can be obtained from experiments if there is an inventory similar to that, which was used when obtaining the distorted image. The function $H(u, v)$ assessment received as a result, also as well as in the previous case, is used for recovery of the blurred images. Particular interest represents receiving the DF assessment based on the modeling, since in many cases it is possible to consider the main physical principles of operation of vision systems and the external environment causing a blurring. Thus, for example a turbulence of the atmosphere, which makes interfering impact in the aerial photographs, can be considered in the form of Eq. 8.11 [1, 5].

$$H(u, v) = \exp\left(-k\left(u^2 + v^2\right)^{\frac{5}{6}}\right) \tag{8.11}$$

In Eq. 8.11 the constant k describes the turbulent properties of the atmosphere.

For simulation of the blurring that occurs during a movement of the image receiver and/or object, a mathematical model of the DF described by Eq. 8.12, where T is an exposure time, $x_0(t)$ defines a law of motion in the x direction and $y_0(t)$ defines a law of motion in the y direction, is often used [1, 5].

$$H(u, v) = \int_0^T \exp[-j2\pi(u\,x_0(t) + v\,y_0(t))]dt \tag{8.12}$$

If the laws $x_0(t)$ and $y_0(t)$ are known, then the DF for each concrete type of blur can be defined immediately from Eq. 8.12. Furthermore, as it is evident from Eq. 8.12, the H finding problem can be factorized, i.e. separately considered in the direction of motion x and y. Thus, if in the x direction the uniform rectilinear motion with a speed of $x_0(t) = at/T$ (for time T image shifted to the distance a) is observed, then we will obtain Eq. 8.13 [1, 4, 5].

$$H(u) = \frac{T}{\pi u a} \sin(\pi u a) \exp(-j\pi u a) \tag{8.13}$$

It is visible that the DF (Eq. 8.13) becomes zero in $u = n/a$ points. If there is a simultaneous movement as in the x direction and the y direction (at a speed of $y_0(t) = bt/T$), then the function H takes a form of Eq. 8.14 [1].

$$H(u, v) = \frac{T}{\pi(ua + vb)} \sin(\pi(ua + vb)) \exp(-j\pi(ua + vb)) \qquad (8.14)$$

Often in theoretical studies found the model of the DF determined by the Eq. 8.15 [1, 4, 6, 7], where Δ is a value of the uniform rectilinear image blur in the direction of the axis OX.

$$h(x - \xi) = \begin{cases} 1/\Delta & -\Delta \leq x - \xi \leq 0 \\ 0 & \text{otherwise} \end{cases} \qquad (8.15)$$

The given Eq. 8.1 and disabling the consideration of noise (i.e. greatly simplifying the task) lead to the solution of Fredholm I kind equation of the convolution type, which can be written as Eq. 8.16.

$$g(x) = \int_{-\infty}^{\infty} h(x - \xi) f(\xi) d\xi \qquad (8.16)$$

Equation 8.16 is usually solved by the method of Fourier transform. In accordance with it, the kernel of the DF in the frequency domain has the form of Eq. 8.17 [6].

$$H(\omega) = \frac{\sin(\omega\Delta)}{\omega\Delta} + j\frac{\cos(\omega\Delta) - 1}{\omega\Delta} \qquad (8.17)$$

Use the DF model in the form of Eq. 8.15 or Eq. 8.17 does not lead to the improvement of the quality in restoring blurred images due to the strong instability of the solution. Besides, during creation of such model a noise does not considered. It also can impact on the quality of the original image.

In order to produce acceptable results, it is necessary to obtain a stable solution (Eq. 8.16). For this purpose it is widely used the Tikhonov regularization method [6] or, in other words, a filtering technique for minimizing the smoothing functional with communication [1]. The solution of Eq. 8.16 by the Tikhonov regularization method is given by the Eq. 8.18, where, α is the regularization parameter ($\alpha > 0$) [5, 6].

$$f_\alpha(\xi) = \frac{1}{2\pi} \int_{-\infty}^{\infty} F_\alpha(\omega) \exp(-j\omega\xi) d\omega \qquad (8.18)$$

Several ways may be offered for a choice of value α, [14, 15]. However, the most efficient way of selection α is considered in [33, 34]. A selection gets out of set $\alpha = \alpha_1, \alpha_2, \ldots, \alpha_n$, at which the visibility of the restored image and stability of the decision (Eq. 8.18) are reached the best. The operator, proceeding from physiological reasons, carries out a selection of parameter α. That is for everyone α

restored image is output, and the operator chooses the best of them (the procedure is similar to control of contrast of the TV image). The regularization method gives acceptable results of image processing but has a significant computational complexity [6]. This can be seen particularly well if Eq. 8.16 goes to a vector-matrix form [1]. This means a necessity to work with the huge arrays of numbers presented in the form of matrices and vectors (at the same time the analyzed matrixes are very sensitive to noise).

It is possible to reduce a computing capacity of the Tikhonov regularization algorithm approximately twice (if a decision of Eq. 8.16 does not use Fourier transform and Hartley's method [6, 35]). In its application, the processing of real functions g, h, and f will be carried out in the field of real numbers, unlike Fourier transform that displays the physical functions in the complex area [6].

A few more methods of regularization are known (for example, Shepp-Logan intuitive regularization method, or Arsenin method of local regularization [36, 37]). Thus, Arsenin method yields the results of restoration of the blurred images even the better than Tikhonov method. However, all of the methods of regularization have the inherent disadvantage of high computational complexity.

Methods of blind deconvolution have received great attention in recent studies [20–30, 32]. In practice, the DF is often unknown, and the amount of information about the natural image is very small. Therefore, the natural image $f(x, y)$ should be recovered directly from function $g(x, y)$ using partial information about the blurring process and a natural image. Such evaluation approach, in which the distortion model is assumed to be linear, is called the blind deconvolution. At blind recovery of images, according to characteristics of the distorted image, the initial image and DF have to be found.

The main reasons for use of the blind deconvolution methods are the inability to obtain preliminary information about the scene and physical danger [2]. The purpose of blind deconvolution is to display the functions h and f from a single input g (see Fig. 8.1). Additionally, it is assumed that the blurring kernel h is a non-negative and its value is small compared to the size of an image. There are two basic approaches of blind deconvolution:

1. Definition of the DF separately from the restored image means a use of this information later by applying one of the known classical methods of restoration (non-blind deconvolution). Estimation of the DF and image restoring are separate procedures. The algorithms applied to implement such methods are computationally simple.
2. Inclusion of the identification procedure of the DF in the restoring algorithm. This approach involves the simultaneous evaluation of the distorting function and recovered image that leads to more sophisticated computational algorithms.

The blind deconvolution of an individual object is an incorrect task, since a number of unknowns exceeds a number of known data. Early approaches to the blurring kernel imposed the limitations and parameterized form of the kernel [2]. Recently, several methods for processing of more general cases of blur, using the image of an individual object, were offered. Although these methods can provide excellent recovery results, they are very computationally capacious [5]. Most of the methods of blind restoration work iteratively, alternately optimizing the blur kernel and the latent image. During the processing, the blur kernel is calculated based on the detected variant of hidden (desired) image and the original blurred image.

The kernel is then used for the calculation of new latent image by applying "non-blind" deconvolution to the initial blurred image. The obtained latent image is used to calculate the next iteration of the kernel. Intensive calculations of the existing methods are based on application of the difficult algorithms that find the kernel and hidden image. For the calculation of kernel optimization, including large matrices and vector, and management of "non-blind" deconvolution with a priori non-linear probability distributions, the methods of the improved optimization are required [7]. Introduction the consecutive optimization of functions f and h, as well as their prediction at each step allows to avoid a use of the computationally inefficient priori probability distributions for non-blind deconvolution, which are needed to determine the latent image. Small artifacts of deconvolution can be rejected at the step of predictions of the following iteration without delay of calculation of the kernel. The combination of non-blind deconvolution and predictions provides an efficient way for calculation of the hidden image, which is used for kernel definition.

The main purpose of the iterative alternating optimization is the gradual improvement of the blurring kernel h. The last result of the recovery obtained using the last operation of non-blind deconvolution, which is performed with the final kernel value h and blurred image g. The intermediate representation of the original image obtained during the iterations does not influence directly on the result of the recovery. They affect the outcome only indirectly helping to improve the kernel h.

The success of the iterative methods is based on two important properties of their way to calculate a clear picture: the restoration of sharp edges and noise reduction in homogeneous areas. These properties allow to compute the kernel accurately. Although it is assumed that a blur is constant in the whole image, more accurately blur kernel may be obtained in areas with sharp edges. For example, it is impossible to determine the blur kernel in areas of constant intensity. Since natural image usually contains regions with sharp edges, the blur kernel can be effectively found on the edges, restored in the calculation of the latent image. The noise reduction in homogeneous areas is also important, as these areas usually occupy a much larger part of the natural image than the expressed edges. If the noise in homogeneous areas does not reduce, it will significantly impact the data fitting component, reducing the accuracy of the kernel computation on sharp edges.

Thus, we can say that the methods of blind deconvolution can effectively solve the problem of image deblurring but have the significant disadvantage of high computational complexity.

8.3 Model of Linear Blur Formation

Without loss of generality, consider the 1D case. The blur that occurs, when the object moves with a constant magnitude and direction velocity, we will call "linear". We will distinguish between the "small" and "large" blur [38]. For a small blur (Fig. 8.3a), the image of the object projected on the photosensitive line at the end of the exposure time $t = T$ does not go beyond the boundaries of its own image at the beginning of exposure ($t = 0$). If the object length is $m\Delta x$ (i.e. it is projected on m photosensitive cells) and distance of the moved object is $n\Delta x$ for exposure time, then $n \leq m$ at a small blur and $n > m$ at a large blur (Fig. 8.3b).

Further, the discrete-analog representation of the illumination is discussed in Sect. 8.3.1, while mathematical models of small and large linear blur are developed in Sects. 8.3.2–8.3.3, respectively.

8.3.1 Discrete-Analog Representation of Illumination

In order to understand the image deformation mechanism, it is necessary to analyze the energy distribution of an entrance luminous flux on discrete space coordinates during a formation of video sequence frame.

Formation the counts of video signal occurs during the accumulation process T of projecting an image of a natural scene on the photosensitive surface, which is always discrete (Charge-Coupled Device (CCD), retina, etc.) on the spatial coordinates x and y. Assume that the elements that form the counts of a video signal have a rectangular shape with sides Δx and Δy. The process of forming the video signal counts by the expression that is up to a constant size ratio is conveniently written in the form of Eq. 8.19, where $E(x, y, t) = \frac{\partial \Phi(x,y,t)}{\partial S}$ is the illuminating intensity of the photosensitive surface, $\Phi(x, y, t)$ is the emission energy flow, ∂S is the element of the photosensitive surface, $i, j \in Z$.

$$f_{i,j} = \int_0^T \int_{(j-1)\Delta y}^{j\Delta y} \int_{(i-1)\Delta x}^{i\Delta x} E(x, y, t)\,dx\,dy\,dt \qquad (8.19)$$

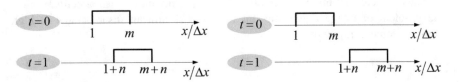

Fig. 8.3 Types of blur: **a** "small" blur, **b** "large" blur

If the spatial position of objects' projections on the photosensitive surface is invariable for exposure time. It means that it is possible to consider illuminating intensity during this time depending only on spatial coordinates. However, the actual scene may contain dynamic objects or the optical properties of the environment may change over time (for example, the turbulent flows in the atmosphere, water, etc.). Then under the certain conditions, a situation can occur, when the same parts of the images are consistently projected onto a few light-sensitive elements for exposure time T. As a result, the formed images of objects become blur.

Consider the moving objects on a stationary background. For exposure time, the image of a dynamic object moved along a photosensitive surface is larger than Δx. Assume that within a photosensitive line the illuminating intensity does not depend on coordinate y. We will neglect the distance between the photosensitive cells. If there are no moving objects, the output line of the photosensitive image forms a line in a set of counts of the video signal $\{f_i\}$. This line can be determined by the Eq. 8.20.

$$f_i = E_i \Delta_x \Delta_y T \qquad (8.20)$$

In Eq. 8.20, parameter E_i is defined by Eq. 8.21.

$$E_i = \frac{1}{\Delta_x} \int\limits_{(i-1)\Delta x}^{i\Delta x} E(x)dx \qquad (8.21)$$

Thus, each count of the video signal of line is proportional to the luminance averaged over the coordinate x. It is equivalent to as though instead of the continuous illuminating intensity $E(x)$ the photosensitive line was affected by the discrete-analog illuminating intensity $\{E_i\}$ presented in Fig. 8.4.

The illuminating intensity of a photosensitive surface $E^b(x,t)$ corresponding to a dynamic object will depend on spatial and temporary coordinates, i.e. a movement along a photosensitive line with some speed of $v(t)$, generally changeable. Illumination $E^a(x)$ corresponding to the stationary background during the exposure time remains unchanged if the photosensitive element begins to be projected an image of the moving object.

When constructing a model of the blur formation, it is assumed that the photosensitive line is affected by its discrete-analog modification $\{E_i(t)\}$ but not illuminating intensity $E(x,t)$. Then, the video signal corresponding to a stationary background remains unchanged.

Fig. 8.4 The discrete-analog illuminating intensity

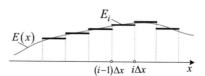

8.3.2 Small Linear Blur

Discrete-analog representation of the illumination at the beginning and end of the exposure with a small blur is depicted in Fig. 8.5. Thin lines represent the illumination corresponding to a stationary background and bold lines show the moving object.

Consider a formation of count f_{m+n+1}. At the end of the accumulation time T the rightmost value of E_m^b will correspond to the cell with the number $(m+n)$ and, therefore, for the entire frame of the cell $(m+n+1)$ is projected luminous flux corresponding to the stationary background (Fig. 8.5). Therefore, we can get Eq. 8.22, where a_{m+n+1} is a background count in a pixel of video signal with the number $(m+n+1)$ [38].

$$f_{m+n+1} = \Delta_y \int_{(m+n)\Delta x}^{(m+n+1)\Delta x} \int_0^T E_{m+n+1}^a \, dt \, dx = E_{m+n+1}^a \Delta x \Delta y T = a_{m+n+1} \quad (8.22)$$

In forming the count f_{m+n} for a period from zero to $T - \frac{T}{n} = \frac{n-1}{n}T$ on the light-sensitive cell, a stationary background E_{m+n}^a will be displayed and then the corresponding part E_m^b of the image of a moving object will also formed for the period of time $\left[\frac{n-1}{n}T; T\right]$, i.e. an image of the object begins to "run into" the image of the background with a speed $\frac{n\Delta x}{T}$ and obstruct it (Fig. 8.6). As a result, we get Eq. 8.23, where $\sigma(\cdot)$ is Heaviside function.

$$f_{m+n} = \Delta y \int_{(m+n-1)\Delta x}^{(m+n)\Delta x} \left(\int_0^{\frac{n-1}{n}T} E_{m+n}^a \, dt + \int_{\frac{n-1}{n}T}^T \left[\begin{array}{l} E_{m+n}^a \sigma\left(x - m\Delta x - \frac{n\Delta x}{T}t\right) + \\ + E_m^b \left(1 - \sigma\left(x - m\Delta x - \frac{n\Delta x}{T}t\right)\right) \end{array} \right] dt \right) dx$$

$$(8.23)$$

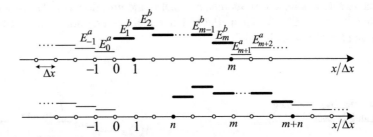

Fig. 8.5 Discrete-analog representation of illumination at the beginning and end of the exposure with a small blur

Fig. 8.6 Projections of elements E^a_{m+n} and E^b_m during the formation of the count f_{m+n}

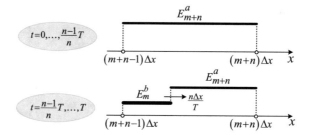

Fig. 8.7 Integral domain for Heaviside function

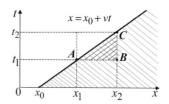

Calculation of Eq. 8.23 involves finding the integrals of the form $\int_{x_1}^{x_2} \int_{t_1}^{t_2} \sigma(x - x_0 - vt)dxdt = \frac{1}{2}(x_2 - x_1)(t_2 - t_1)$. In Fig. 8.7, the area, in which Heaviside function $\sigma(x - x_0 - vt)$ is not equal to zero, is shaded. At the same time, value of integral is equal to the area of a triangle ABC.

The final expression for the count f_{m+n} takes the form Eq. 8.24, where $a_{m+n} = E^a_{m+n}\Delta x \Delta y T$ and $b_m = E^b_m \Delta x \Delta y T$ [38].

$$f_{m+n} = \frac{2n-1}{2n} a_{m+n} + \frac{1}{2n} b_m \qquad (8.24)$$

The process of formation the count f_{m+n-1} is shown schematically in Fig. 8.8. During the time from 0 to $T - \frac{2T}{n} = \frac{n-2}{n}T$ on the corresponding photosensitive cell will be displayed a stationary background (i.e. we have E^a_{m+n-1}). Then for the period of time $\left[\frac{n-2}{n}T; \frac{n-1}{n}T\right]$, the background will be blocked by the "bump" of the object portion, which corresponds to E^b_m. At time $t = \frac{n-1}{n}T$, the projection of this area will completely cover the photosensitive cell with number $(m+n-1)$. Finally, at $t \in \left[\frac{n-1}{n}T; T\right], E^b_m$ is gradually being replaced by E^b_{m-1}.

Thus, we have Eq. 8.25.

$$f_{m+n-1} = \frac{2n-3}{2n} a_{m+n-1} + \frac{1}{n} b_m + \frac{1}{2n} b_{m-1} \qquad (8.25)$$

By similar reasoning, we can obtain Eq. 8.26, where $\{f_1, \ldots, f_{m+n}\}$ are the counts of video signal of blurred image, $\{b_1, \ldots, b_m\}$ are the counts of undistorted

Fig. 8.8 The process of
formation the count f_{m+n-1}

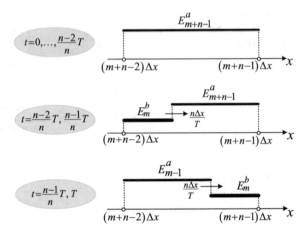

image of the object, $\{a_1, \ldots, a_n, a_{m+1}, \ldots, a_{m+n}\}$ are the counts of stationary
background.

$$
\begin{cases}
f_{m+n} = \frac{1}{2n}b_m + \frac{2n-1}{2n}a_{m+n} \\
f_{m+n-1} = \frac{1}{2n}b_{m-1} + \frac{1}{n}b_m + \frac{2n-3}{2n}a_{m+n-1} \\
\cdots\cdots\cdots\cdots\cdots\cdots\cdots\cdots\cdots\cdots\cdots\cdots\cdots\cdots\cdots \\
f_{m+1} = \frac{1}{2n}b_{m-n+1} + \frac{1}{n}b_{m-n+2} + \cdots + \frac{1}{n}b_m + \frac{1}{2n}a_{m+1} \\
f_m = \frac{1}{2n}b_m + \frac{1}{n}b_{m-1} + \ldots + \frac{1}{n}b_{m-n+1} + \frac{1}{2n}b_{m-n} \\
f_{m-1} = \frac{1}{2n}b_{m-1} + \frac{1}{n}b_{m-2} + \cdots + \frac{1}{n}b_{m-n} + \frac{1}{2n}b_{m-n-1} \\
\cdots\cdots\cdots\cdots\cdots\cdots\cdots\cdots\cdots\cdots\cdots\cdots\cdots\cdots\cdots \\
f_{n+1} = \frac{1}{2n}b_{n+1} + \frac{1}{n}b_n + \cdots + \frac{1}{n}b_2 + \frac{1}{2n}b_1 \\
f_n = \frac{1}{2n}b_n + \frac{1}{n}b_{n-1} + \cdots + \frac{1}{n}b_1 + \frac{1}{2n}a_n \\
f_{n-1} = \frac{1}{2n}b_{n-1} + \frac{1}{n}b_{n-2} + \cdots + \frac{1}{n}b_1 + \frac{3}{2n}a_{n-1} \\
\cdots\cdots\cdots\cdots\cdots\cdots\cdots\cdots\cdots\cdots\cdots\cdots\cdots\cdots\cdots \\
f_1 = \frac{1}{2n}b_1 + \frac{2n-1}{2n}a_1
\end{cases} \tag{8.26}
$$

Equation 8.26 is a system of $(m+n)$ linear algebraic equations with $(m+n)$
unknowns $b_1, \ldots, b_m, a_1, \ldots, a_n$, describing the process of forming counts of the
video signal distorted by a small linear blur. Values $f_1, \ldots, f_{m+n}$ are determined by
the current frame and values $a_{m+1}, \ldots, a_{m+n}$ are determined by the previous frame.

The counts of motionless background $a_1, \ldots a_n$ shaded by the object in the
previous frame are located from the Eq. 8.25 according to the previously found
values of $b_1, \ldots b_m$. For finding the counts $b_1, \ldots b_m$, it is necessary to use only m
first equations from Eq. 8.26 which are convenient for rewriting as Eq. 8.27 or in
matrix form as Eq. 8.28, where $\mathbf{A}$, $\mathbf{B}$ and $\mathbf{F}$ are defined by Eq. 8.29.

$$\begin{cases} b_1 + 2b_2 + \cdots + 2b_n + b_{n+1} = 2nf_{n+1} \\ b_2 + 2b_3 + \cdots + 2b_{n+1} + b_{n+2} = 2nf_{n+2} \\ \dots\dots\dots\dots\dots\dots\dots\dots\dots\dots\dots \\ b_{m-n-1} + 2b_{m-n} + \cdots + 2b_{m-2} + b_{m-1} = 2nf_{m-1} \\ b_{m-n} + 2b_{m-n+1} + \cdots + 2b_{m-1} + b_m = 2nf_m \\ b_{m-n+1} + 2b_{m-n+2} + \cdots + 2b_{m-1} + 2b_m = 2nf_{m+1} + a_{m+1} \\ \dots\dots\dots\dots\dots\dots\dots\dots\dots\dots\dots \\ b_{m-1} + 2b_m = 2nf_{m+n-1} + (2n-3)a_{m+n-1} \\ b_m = 2nf_{m+n} - (2n-1)a_{m+n} \end{cases} \qquad (8.27)$$

$$\mathbf{AB} = \mathbf{F} \qquad (8.28)$$

$$\mathbf{A} = \begin{Vmatrix} \overbrace{1\ 2\ \dots\ 2\ 1}^{n+1} & \overbrace{}^{m-n-1} & \\ \quad 1\ 2\ \dots\ 2\ 1 & \mathbf{0} & \\ \qquad 1\ 2\ \cdots\ 2\ 1 & & \\ \dots\ \dots\ \dots & & \\ \qquad 1\ 2 \quad\ \dots\ 2\ 1 & & \\ \qquad\quad 1\ 2\ \dots \quad\ 2 & & \\ \mathbf{0} \qquad 1\ 2\ \cdots\ 2 & & \\ \qquad\qquad \dots\ \dots & & \\ \qquad\qquad\quad 1\ 2 & & \\ \qquad\qquad\qquad 1 & & \end{Vmatrix} \quad \mathbf{B} = \begin{Vmatrix} b_1 \\ b_2 \\ \vdots \\ b_m \end{Vmatrix} \quad \mathbf{F} = \begin{Vmatrix} 2nf_{n+1} \\ 2nf_{n+2} \\ \vdots \\ 2nf_m \\ 2nf_{m+1} - a_{m+1} \\ 2nf_{m+2} - 3a_{m+2} \\ \vdots \\ 2nf_{m+n} - (2n-1)a_{m+n} \end{Vmatrix} \qquad (8.29)$$

When solving Eq. 8.28, the counts of video signal of the blurred object image in the current frame and counts of video signal of the motionless background from the previous frame are used. As a result, the counts of video signal of the undistorted image of an object can be found.

For finding the counts, which corresponds to the site of background shaded by an object in the previous frame, it is necessary to solve the Eq. 8.30.

$$\begin{cases} a_n = 2nf_n - 2b_1 - 2b_2 - \cdots - 2b_{n-1} - b_n \\ a_{n-1} = \frac{2n}{3}f_{n-1} - \frac{2}{3}b_1 - \cdots - \frac{2}{3}b_{n-2} - \frac{1}{3}b_{n-1} \\ \dots\dots\dots\dots\dots\dots\dots\dots\dots\dots\dots\dots \\ a_1 = \frac{2n}{2n-1}f_1 - \frac{1}{2n-1}b_1 \end{cases} \qquad (8.30)$$

Note 1. If the object moves as it is shown in Fig. 8.9, then in systems (Eqs. 8.26–8.28, 8.30) it is necessary to replace $(m+n-i)$ on $(1-n+i)$ for all $i = \overline{0, m+n-1}$.

Fig. 8.9 Moving of the object

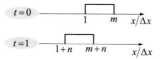

Fig. 8.10 Constant
brightness on the uniform
background and a scheme of
object movement (*gray
rectangle*)

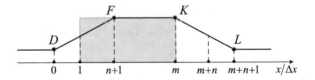

Let us investigate the above-described processes of forming the counts of the
signal of the blurred image in the simple case of the constant brightness of an object
on a uniform background. The scheme of the object movement (gray rectangle) and
video signal formed at the same time with the front *DE* and a cut *KL* is shown in
Fig. 8.10 (the black continuous line).

The analysis of Fig. 8.10 shows that the most essential changes happen on fronts
and cuts of impulses of video signal. Besides, finding of undistorted counts of video
signal of an object in Eq. 8.27 or the counts corresponding to the shaded background in Eq. 8.30 is based on counts of a cut $f_m, \ldots, f_{m+n}$ or the front $f_1, \ldots, f_n$.
Therefore, it is necessary to consider the processes of formation of fronts and cuts of
video pulses in details.

If we consider that all points of the front or cut are formed in the same manner,
i.e. without features, then it is possible to receive, for example Eq. 8.31, which
describes the cut equation—a straight line *KL* (the dashed line in Fig. 8.11).

$$f^*(x) = \frac{b-a}{n+1}(m-x) + b, \quad x \in [m; m+n+1] \tag{8.31}$$

However, as follows from Eq. 8.26, the formation of counts m and $(m+n+1)$
has the features: if the difference between any two adjacent samples has a form of
Eq. 8.32, then for counts f_m and f_{m+n+1} we can get Eq. 8.33.

$$f_{m+n-i} - f_{m+n-i+1} = \frac{b-a}{n} \tag{8.32}$$

$$f_m - f_{m+1} = f_{m+n} - f_{m+n+1} = \frac{b-a}{2n} \tag{8.33}$$

Fig. 8.11 Forming the cut of
video pulses

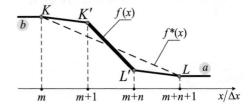

Thus, it turns out that the cut of the pulse is a broken line (solid line in Fig. 8.11). Section $K'L'$ then get by the Eq. 8.34.

$$f(x) = -\frac{b-a}{n}x + \frac{(2m+1)(b-a)+2nb}{2n}, \quad x \in [m+1; m+n] \quad (8.34)$$

If Eq. 8.34 does not consider, then the undistorted samples $b_1, \ldots, b_m$ can be detected incorrectly. We show this by the blurred image restoration using Eq. 8.31 and Eq. 8.34 in a frame with a dynamic rectangular object of constant brightness on a uniform background.

Video signal of a line passing through the blurred image of an object is shown in Fig. 8.12. In Fig. 8.13a–b, the video signals of a line restored by system (Eq. 8.27) are given according to Eqs. 8.31–8.34, respectively. It is evident that if the changes in the form of the front and cut of the distorted pulse do not take into account (Eq. 8.34), the counts of video signal are not restored, although the size of the object is determined.

Consider an example of compensation of a small linear blur on the actual video sequence, two adjacent frames of which are given in Fig. 8.14. Video signals of similar lines that pass through an image of the dark central object for two adjacent frames are shown in Fig. 8.15. The size of shift of the white rectangle for one frame is $n = 46$ and its size in line is $m = 337$.

The restoration result of the first frame is shown in Fig. 8.16. It is visible that indistinct borders of the blurred white rectangle and a dark figure in its center are restored clearly. At the same time, the restored borders of the central figure became broken lines.

The fragments of the blurred image (shown in Fig. 8.14a) and a fragment of the restored image (shown in Fig. 8.16) increased twice are presented in Fig. 8.17a–b. Figure 8.17c–f shows the deblurring results by known methods implemented in the MatLAB software package.

Comparison of Fig. 8.17b, c–f shows that the neglect of forming processes in the restoration of blurred images may give rise to "rings" type defects. In addition,

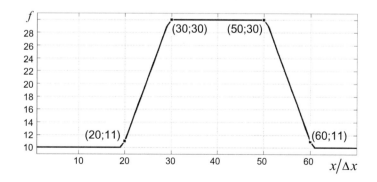

Fig. 8.12 Video signal of a line passing through the blurred image of an object

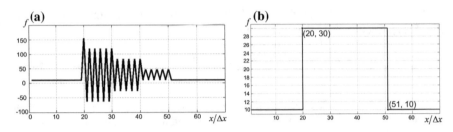

Fig. 8.13 Video signals of a line restored (Eq. 8.27): **a** according to Eq. 8.31, **b** according to Eq. 8.34

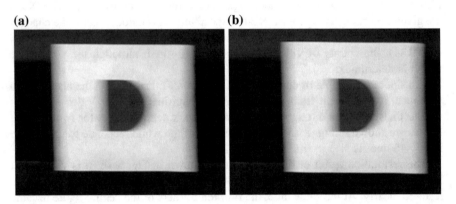

Fig. 8.14 Two adjacent frames of natural video sequence with small linear blur

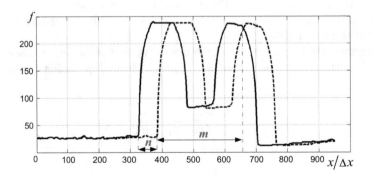

Fig. 8.15 Video signals of similar lines for two adjacent frames

when using the proposed approach there is no need of a priori knowledge about the DF and do not require additional algorithms to estimate the noise and calculate the number of iterations.

Fig. 8.16 The restoration result of the first frame in Fig. 8.15a

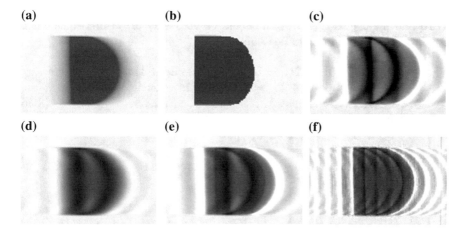

Fig. 8.17 Restoration process: **a** fragments of blurred image (shown in Fig. 8.14a) increased twice, **b** fragment of the restored image (shown in Fig. 8.16) increased twice, **c** deblurring results by Wiener filtering, **d** deblurring results by regularization method, **e** deblurring results by Lucy-Richardson algorithm, **f** deblurring results by blind deconvolution method

8.3.3   Large Linear Blur

Large blur occurs, when the object moves by a distance exceeding its length, i.e. $n > m$ for exposure time of the frame (Fig. 8.3b). As well as in the case of small blur based on discrete-analog representation of illuminating intensity, it is possible to obtain the system, describing a formation of the counts of video signal of the image distorted by the large blur. This system is represented by Eq. 8.35.

$$\begin{cases} f_{m+n} = \frac{1}{2n}b_m + \frac{2n-1}{2n}a_{m+n} \\ f_{m+n-1} = \frac{1}{2n}b_{m-1} + \frac{1}{n}b_m + \frac{2n-3}{2n}a_{m+n-1} \\ \cdots\cdots\cdots\cdots\cdots\cdots\cdots\cdots\cdots\cdots \\ f_{n+1} = \frac{1}{2n}b_1 + \frac{1}{n}b_2 + \cdots + \frac{1}{n}b_m + \frac{2n-2m+1}{2n}a_{n+1} \\ f_n = \frac{1}{n}b_1 + \frac{1}{n}b_2 + \cdots + \frac{1}{n}b_m + \frac{2n-2m}{2n}a_n \\ \cdots\cdots\cdots\cdots\cdots\cdots\cdots\cdots\cdots\cdots\cdots \\ f_{m+1} = \frac{1}{n}b_1 + \frac{1}{n}b_2 + \cdots + \frac{1}{n}b_m + \frac{2n-2m}{2n}a_{m+1} \\ f_m = \frac{1}{n}b_1 + \frac{1}{n}b_2 + \cdots + \frac{1}{n}b_{m-1} + \frac{1}{2n}b_m + \frac{2n-2m+1}{2n}a_m \\ f_{m-1} = \frac{1}{n}b_1 + \frac{1}{n}b_2 + \cdots + \frac{1}{n}b_{m-2} + \frac{1}{2n}b_{m-1} + \frac{2n-2m+3}{2n}a_{m-1} \\ \cdots\cdots\cdots\cdots\cdots\cdots\cdots\cdots\cdots\cdots\cdots \\ f_1 = \frac{1}{2n}b_1 + \frac{2n-1}{2n}a_1 \end{cases} \qquad (8.35)$$

The system (Eq. 8.35) contains $(m+n)$ equations with $2m$ unknowns $b_1, \ldots, b_m, a_1, \ldots, a_m$. Values $f_1, \ldots, f_{m+n}$ are determined by the current frame, sizes $a_{m+1}, \ldots, a_{m+n}$ are defined by previous values. Since $n > m$, Eq. 8.33 is an over determined system.

To find the counts $b_1, \ldots, b_m$, it should be used only the m first equations of Eq. 8.35. These equations are Eq. 8.36.

$$\begin{cases} b_1 + 2b_2 + \cdots + 2b_m = 2nf_{n+1} - (2n - 2m + 1)a_{n+1} \\ b_2 + 2b_3 + \cdots \in + 2b_m = 2nf_{n+2} - (2n - 2m + 3)a_{n+2} \\ \cdots\cdots\cdots\cdots\cdots\cdots\cdots\cdots\cdots\cdots\cdots\cdots\cdots\cdots \\ b_{m-1} + 2b_m = 2nf_{m+n-1} - (2n - 3)a_{m+n-1} \\ b_m = 2nf_{m+n} - (2n - 1)a_{m+n} \end{cases} \qquad (8.36)$$

It is convenient to rewrite Eq. 8.36 in a matrix form as Eq. 8.37, where $\mathbf{C}$, $\mathbf{B}$ and $\mathbf{F}$ defined as Eq. 8.38.

$$\mathbf{CB} = \mathbf{F} \qquad (8.37)$$

$$\mathbf{C} = \begin{Vmatrix} 1 & 2 & \cdots & \cdots & 2 \\ & 1 & 2 & \cdots & 2 \\ & & \cdots & \cdots & \cdots \\ \mathbf{0} & & & 1 & 2 \\ & & & & 1 \end{Vmatrix} \quad \mathbf{B} = \begin{Vmatrix} b_1 \\ b_2 \\ \vdots \\ b_m \end{Vmatrix} \quad \mathbf{F} = \begin{Vmatrix} 2nf_{n+1} - (2n - 2m + 1)a_{n+1} \\ 2nf_{n+2} - (2n - 2m + 3)a_{n+2} \\ \vdots \\ 2nf_{m+n-1} - (2n - 3)a_{m+n-1} \\ 2nf_{m+n} - (2n - 1)a_{m+n} \end{Vmatrix} \qquad (8.38)$$

To find the counts corresponding to the background shaded by the object in the previous frame, it is necessary to solve the system of equations, all of which have already been resolved relatively unknown. This system is defined as Eq. 8.39.

$$\begin{cases} a_m = \frac{1}{2n-2m+1}(2nf_m - 2b_1 - 2b_2 - \cdots - 2b_{m-1} - b_m) \\ a_{m-1} = \frac{1}{2n-2m+3}(2nf_{m-1} - 2b_1 - 2b_2 - \cdots - 2b_{m-2} - b_{m-1}) \\ \cdots\cdots\cdots\cdots\cdots\cdots\cdots\cdots\cdots\cdots\cdots\cdots\cdots\cdots\cdots\cdots \\ a_1 = \frac{1}{2n-1}(2nf_1 - b_1) \end{cases} \qquad (8.39)$$

Finally, from Eq. 8.35 we can obtain Eq. 8.40, which may serve to clarify the values $m, n, b_1, \ldots, b_m$.

$$b_1 + b_2 + \cdots + b_m = nf_{m+i} - (n - m)a_{m+i}, \quad i = 1, 2, \ldots, n - m \qquad (8.40)$$

As well as in the case of small blur, for systems (Eqs. 8.35–8.37, 8.39–8.40) the note 1 is fair.

We investigated the processes of formation of the counts of video signal distorted by the large blur using an object of constant brightness in the uniform background as the simplest case (Fig. 8.10a). The scheme of movement of an object (a gray rectangle) and a video signal (a black solid line) formed at the same time are shown in Fig. 8.18.

In this case, the maximum level of video signal of the blurred image of an object b' is less than the level of video signal of an undistorted object b. We show this by the example of count f_{m+1}. From Eq. 8.35, it follows that $f_{m+1} = \frac{m}{n}b + \frac{2n-2m}{2n}a = \frac{m}{n}(b - a) + a = b'$. Since $b' - b = \frac{(m-n)(b-a)}{n} < 0$, then $b' < b$. Therefore, the contrast of an image of the object distorted by the large blur will be less contrast of an image of the object without distortion or blurred with the small blur $b' - a = \frac{m}{n}(b - a) < (b - a)$.

As for the small blur, it is necessary to consider the formation of the front and cut of video signal. In particular, for the site of a cut $K'L'$, the Eq. 8.41 is fairly.

$$f(x) = -\frac{b' - a}{m}x + \frac{(2m + 2n + 1)b'}{2m}, \quad x \in [n + 1; m + n] \qquad (8.41)$$

Let us consider a possibility of recovery of the actual images distorted by the large blur on system (Eq. 8.36) taking into account the Eq. 8.41. Figure 8.19a shows two motionless images. Using the experimental setup that ensures a uniform motion of objects at a predetermined speed, the video sequences have been obtained, each of which contains the images of blurred objects. The frames of these video sequences distorted by large blur are shown in Fig. 8.19b, while the recovery results are represented in Fig. 8.19c.

The analysis of Fig. 8.19b shows that a presence of large blur causes the strong deformation of an object form and change of its contrast. Images of objects and

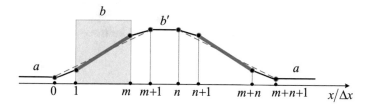

Fig. 8.18 Scheme of the object movement and video signal formed at the same time (*black solid line*)

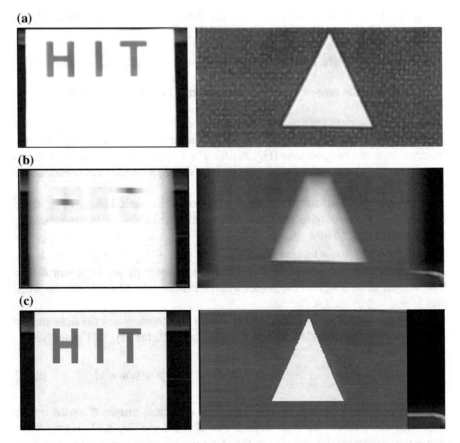

Fig. 8.19 Example of large blur: **a** two motionless images, **b** frames distorted by large blur, **c** recovery results

their contrast are restored, but as well as at a small blur, the objects have slight breaks of borders. In Fig. 8.20a the fragment of the blurred image (Fig. 8.19b) is shown, while the results of restoration of this image by known methods are presented in Fig. 8.20b–e.

The analysis of Fig. 8.20 shows that the restoration of object image distorted by a large blur using known methods does not lead to the correction of the contrast level. This lack coupled with characteristic distortion type "rings" can cause the situation, when the low-contrast objects would not be discovered or the multiple elements would be false detected.

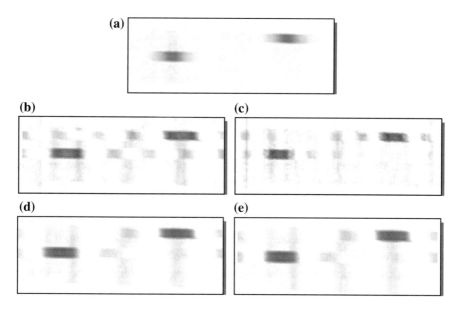

Fig. 8.20 Deblurring results: **a** fragments of the blurred image (shown in Fig. 8.19b), **b** deblurring results by Wiener filtering, **c** deblurring results by regularization method, **d** deblurring results by Lucy-Richardson's algorithm, **e** deblurring results by blind deconvolution method

8.4 Non-linear Blur

In general, a law of the object motion is non-linear. Consider a movement of an object in the positive direction of an OX axis. In order to analyze the processes of emergence of non-linear blur, it is convenient to use dimensionless coordinates $\tilde{x} = \frac{x}{\Delta x}$ and $\tilde{t} = \frac{t}{T}$.

When forming any count of the output signal $f_i, i \in \overline{1, m+n}$, it is necessary to know the contribution of the element E_p^b of the discrete-analog illuminating intensity corresponding to the moving object. For this purpose, consider the integral in the form of Eq. 8.42, where $\tilde{x}(\tilde{t})$ is a function that describes a projection of object movement on the photosensitive surface, $\left(\tilde{t}_{i-1}^p, \tilde{t}_i^p\right)$ is a time of any part of the element E_p^b on the photosensitive cell with the coordinate $\tilde{x} \in [i-1;\ i]$.

$$Y_p(i) = \iint_D \sigma(\tilde{x} - \tilde{x}(\tilde{t}))d\tilde{t}d\tilde{x} = \int_{i-1}^{i} \int_{\tilde{t}_{i-1}^p}^{\tilde{t}_i^p} \sigma(\tilde{x} - \tilde{x}(\tilde{t}))d\tilde{t}d\tilde{x} \qquad (8.42)$$

In Fig. 8.21, the shaded area shows the region, where the kernel $\sigma(\tilde{x} - x(\tilde{t}))$ of integral $Y_p(i)$ is equal to unity.

Fig. 8.21 The allocated
region of integration D

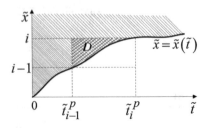

Assume that function f_i corresponds to a pixel of the blurred image of an object, which does not belong to its edges. Then, at time $\tilde{t} \in \left(\tilde{t}_{i-1}^{\,p}, \tilde{t}_i^{\,p} \right)$ the greater part of element E_p^b is projected on the ith light-sensitive cell and at time $\tilde{t} = \tilde{t}_i^{\,p}$ the element E_p^b is the fully projected. Then, when $\tilde{t} \in \left(\tilde{t}_i^{\,p}, \tilde{t}_{i+1}^{\,p} \right)$, the element E_p^b will "move" with the ith cell to $(j+1)$ th one, which is shown schematically in Fig. 8.22.

Thus, a contribution of the element E_p^b in a formation of the count f_i of the video signal will be proportional to the sum of the integrals, which can be written as Eq. 8.43.

$$\int_{i-1}^{i} \int_{\tilde{t}_{i-1}^{\,p}}^{\tilde{t}_i^{\,p}} \sigma(\tilde{x} - \tilde{x}(\tilde{t}))d\tilde{t}d\tilde{x} + \int_{i}^{i+1} \int_{\tilde{t}_i^{\,p}}^{\tilde{t}_{i+1}^{\,p}} [1 - \sigma(\tilde{x} - \tilde{x}(\tilde{t}))]d\tilde{t}d\tilde{x}$$
$$= \tilde{t}_{i+1}^{\,p} - \tilde{t}_i^{\,p} + Y_p(i) - Y_p(i+1) \tag{8.43}$$

Using Eq. 8.43, it is possible to obtain Eq. 8.44 for parts $f_i(p)$ of any count f_i of output video signal caused by counts b_p of video signal of a moving object.

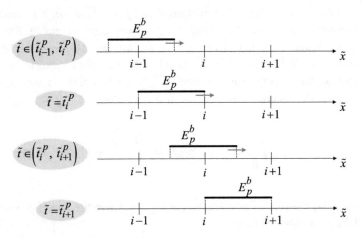

Fig. 8.22 The movement of the element E_p^b

$$f_i(p) = b_p\big(Y_p(i) - Y_p(i+1) + \tilde{t}^p_{i+1} - \tilde{t}^p_i\big) \tag{8.44}$$

When forming the count, a function f_i must also consider the impact of the remaining counts $\ldots, b_{p-2}, b_{p-1}, b_{p+1}, b_{p+2}, \ldots$ corresponding to moving object and the count a_p corresponding to the motionless background. For them, we will get Eq. 8.45.

$$f_{m+n} = b_m Y_m(m+n) + \tilde{t}^m_{m+n-1} a_{m+n}; \quad f_1 = b_1 Y_1(1) + \big(1 - \tilde{t}^1_1\big) a_1 \tag{8.45}$$

Generally, if a moving object is moved as a whole, the illumination on the photosensitive surface moves in the same way. In this case, we have Eq. 8.46.

$$\tilde{t}^p_i - \tilde{t}^p_{i-1} = \tilde{t}^{p-1}_{i-1} - \tilde{t}^{p-1}_{i-2}, \quad \forall p \in \{b_p\}, \quad \forall i = \overline{2, m+n} \tag{8.46}$$

Taking into account Eq. 8.46, the upper indexes that have normalized time counts $\tilde{t}^p_i$ and subscripts of functions $Y_p(i)$ will not burn.

It is possible to get a set of equations for finding the counts of video signal of the output image of an object distorted by a non-linear blur using ratios (Eqs. 8.44–8.45). This set is represented by Eq. 8.47.

$$\begin{cases} f_{m+n} = b_m Y(m+n) + \tilde{t}_{m+n-1} a_{m+n} \\ f_{m+n-1} = b_{m-1} Y(m+n) + b_m[Y(m+n-1) - Y(m+n) + \tilde{t}_{m+n} - \tilde{t}_{m+n-1}] + \tilde{t}_{m+n-2} a_{m+n-1} \\ f_{m+n-2} = b_{m-1} Y(m+n) + b_{m-1}[Y(m+n-1) - Y(m+n) + \tilde{t}_{m+n} - \tilde{t}_{m+n-1}] + \\ \quad + b_m[Y(m+n-2) - Y(m+n-1) + \tilde{t}_{m+n-1} - \tilde{t}_{m+n-2}] + \tilde{t}_{m+n-3} a_{m+n-2} \\ \cdots\cdots\cdots\cdots\cdots\cdots\cdots\cdots\cdots\cdots\cdots\cdots\cdots\cdots\cdots\cdots\cdots\cdots\cdots\cdots \end{cases}$$

$$\tag{8.47}$$

The system (Eq. 8.47) can be written in matrix form. For example, for a large blur it is submitted in the form $\mathbf{C} \cdot \mathbf{B} = \mathbf{F}$ with matrices $\mathbf{C}, \mathbf{B}$ and $\mathbf{F}$, which defined by Eq. 8.48.

$$\mathbf{C} = \begin{Vmatrix} 1 & c_{1,2} & \cdots & \cdots & c_{1,m} \\ & 1 & c_{2,3} & \cdots & c_{2,m} \\ & & \ddots & \cdots & \\ & \mathbf{0} & & 1 & c_{m-1,m} \\ & & & & 1 \end{Vmatrix} \quad \mathbf{B} = \begin{Vmatrix} b_1 \\ b_2 \\ \vdots \\ b_m \end{Vmatrix} \quad \mathbf{F} = \frac{1}{Y(m+n)} \begin{Vmatrix} f_{n+1} - \tilde{t}_n a_{n+1} \\ f_{n+2} - \tilde{t}_{n+1} a_{n+2} \\ \vdots \\ f_{m+n-1} - \tilde{t}_{m+n-2} a_{m+n-1} \\ f_{m+n} - \tilde{t}_{m+n-1} a_{m+n} \end{Vmatrix}$$

$$\tag{8.48}$$

The elements $c_{i,j}$ are defined by Eq. 8.49.

$$c_{i,j} = \frac{Y(m+n+i-j) - Y(m+n+i-j+1) + \tilde{t}_{m+n+i-j+1} - \tilde{t}_{m+n+i-j}}{Y(m+n)}, \quad j > i$$

$$\tag{8.49}$$

For the linear blur, all values $Y(i) = \frac{1}{2n}$ and $\tilde{t}_i - \tilde{t}_{i-1} = \frac{1}{n}$. Thus, the matrix of the system (Eq. 8.47) coincides with the matrix $\mathbf{C}$ of the system (Eq. 8.37).

Also note that in this case the note 1 is fair.

In order to restore the counts distorted by blur, it is necessary to know a law of motion $\tilde{x} = \tilde{x}(\tilde{t})$ describing a projection of an object movement on the photosensitive surface. From this, the time moments $\tilde{t}_i$ instants and then the sizes $Y(i) = i(\tilde{t}_{i-1} - \tilde{t}_i) + \int_{\tilde{t}_{i-1}}^{\tilde{t}_i} \tilde{x}(\tilde{t})d\tilde{t}$ are determined by them.

It is convenient to look for the law of motion in the form of Makloren's decomposition (Eq. 8.50).

$$\tilde{x}(\tilde{t}) = \tilde{x}_1\tilde{t} + \tilde{x}_2\tilde{t}^2 + \tilde{x}_3\tilde{t}^3 + \cdots + \tilde{x}_z\tilde{t}^z + \cdots \tag{8.50}$$

If the sum of the first z items of $\tilde{x}(\tilde{t}) = \tilde{x}_1\tilde{t} + \tilde{x}_2\tilde{t}^2 + \tilde{x}_3\tilde{t}^3 + \cdots + \tilde{x}_z\tilde{t}^z + \cdots$ is limited, then for determination of coefficients of a polynomial (Eq. 8.42) it is necessary to have $(z + 1)$ consecutive frames of the video.

Thus, assuming a law of motion $\tilde{x}(\tilde{t}) = \tilde{x}_1\tilde{t} + \tilde{x}_2\tilde{t}^2 + \tilde{x}_3\tilde{t}^3$, we can obtain the system of equations in the form of Eq. 8.51, where n_1, n_2, n_3 are the movements of an object in the first, second, and third frames, respectively.

$$\begin{cases} n_1 = \tilde{x}_1 + \tilde{x}_2 + \tilde{x}_3 \\ n_2 = 2\tilde{x}_1 + 4\tilde{x}_2 + 8\tilde{x}_3 \\ n_3 = 3\tilde{x}_1 + 9\tilde{x}_2 + 27\tilde{x}_3 \end{cases} \Leftrightarrow \begin{cases} \tilde{x}_1 = \frac{1}{6}(18n_1 - 9n_2 + 2n_3) \\ \tilde{x}_2 = \frac{1}{2}(-5n_1 + 4n_2 - n_3) \\ \tilde{x}_3 = \frac{1}{6}(3n_1 - 3n_2 + n_3) \end{cases} \tag{8.51}$$

In this way, it is possible to get the criteria for various laws of motion, for example:

- The condition of uniformly accelerated motion (Eq. 8.52).

$$n_3 = 3(n_2 - n_1) \tag{8.52}$$

- The condition of uniform motion (Eq. 8.53).

$$n_2 = 2n_1 \tag{8.53}$$

Consider a case of uniformly accelerated movement of an object, i.e., when a law of motion is set by a function, which can be written as Eq. 8.54, where the ratio $\tilde{x}_1$ can be interpreted as the initial velocity, $\tilde{x}_2$ can be interpreted as a half the acceleration.

$$\tilde{x}(\tilde{t}) = \tilde{x}_1\tilde{t} + \tilde{x}_2\tilde{t}^2 \tag{8.54}$$

Then, from Eq. 8.54 for the time moments we will obtain Eq. 8.55.

$$\tilde{t}_i = \frac{\sqrt{\tilde{x}_1^2 + 4i\tilde{x}_2} - \tilde{x}_1}{2\tilde{x}_2} \tag{8.55}$$

Equation 8.55 (based on the system (Eq. 8.51) and condition (Eq. 8.52)) can be written as Eq. 8.56.

$$\tilde{t}_i = \frac{\sqrt{\left(\frac{4n_1-n_2}{2}\right)^2 + 2i(n_2 - 2n_1)} - \frac{4n_1-n_2}{2}}{n_2 - 2n_1}$$
$$= \frac{\sqrt{(4n_1 - n_2)^2 + 8i(n_2 - 2n_1)} - (4n_1 - n_2)}{2(n_2 - 2n_1)} \tag{8.56}$$

Also for integral we can get Eq. 8.57.

$$Y(i) = \frac{\tilde{x}_1^2 + 4\tilde{x}_2 - \sqrt{(\tilde{x}_1^2 + 4i\tilde{x}_2) \cdot (\tilde{x}_1^2 + 4(i-1)\tilde{x}_2)}}{6\tilde{x}_2\left(\sqrt{\tilde{x}_1^2 + 4i\tilde{x}_2} + \sqrt{\tilde{x}_1^2 + 4(i-1)\tilde{x}_2}\right)} \tag{8.57}$$

Using Eqs. 8.55, 8.57, and 8.49, the matrix $\mathbf{C}$ and column-vector $\mathbf{F}$ can be found and then determined the counts of undistorted video signal corresponding to the moving object.

Another example is a non-linear vibrational blur that can occur, when the unstable position of the camera, the reciprocating motion of the object, or atmospheric turbulence take place.

Consider the simplest case of vibrational blur, which can be modeled by two consecutive linear blurs directed in opposite directions and characterized by the kernels $\sigma(\tilde{x} - n_1\tilde{t})$ and $\sigma(\tilde{x} + n_2\tilde{t})$. Figure 8.23 schematically shows the projection of the moving object on the photosensitive surface at the beginning of the frame $(\tilde{t} = 0)$ in an intermediate stop position and the end of frame $(\tilde{t} = 1)$. In this case, $n_1 > m, n_2 > m - 1$.

It is possible to obtain five systems of linear equations for unknown counts $b_1, \ldots, b_m$ of undistorted image of a dynamic object taking into account the

Fig. 8.23 The projection of the moving object on the photosensitive surface

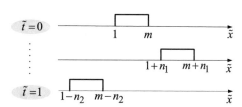

processes of forming the counts of video signal of blurred image. Each of these systems can be written in the form $\mathbf{C} \cdot \mathbf{B} = \mathbf{F}$. The matrix-column of unknowns for all systems is the same $\mathbf{B}^T = \| b_1 \quad b_2 \quad \ldots \quad b_m \|$, and the matrix of coefficients and matrix-column of free terms are determined by the Eqs. 8.58–8.62.

$$
\mathbf{C}_1 =
\begin{Vmatrix}
1 & 2 & \cdots & \cdots & \cdots & 2 \\
0 & 1 & 2 & \cdots & \cdots & 2 \\
\cdots & \cdots & \cdots & \cdots & \cdots & \cdots \\
0 & \cdots & 0 & 1 & 2 & 2 \\
0 & \cdots & \cdots & 0 & 1 & 2 \\
0 & \cdots & \cdots & \cdots & 0 & 1
\end{Vmatrix}
$$

(8.58)

$$
\mathbf{F}_1 =
\begin{Vmatrix}
(2n_1 + n_2)f_{1+n_1} - a_{1+n_1}(2n_1 + n_2 - 2m + 1) \\
(2n_1 + n_2)f_{2+n_1} - a_{2+n_1}(2n_1 + n_2 - 2m + 3) \\
\vdots \\
(2n_1 + n_2)f_{m+n_1-2} - a_{m+n_1-2}(2n_1 + n_2 - 5) \\
(2n_1 + n_2)f_{m+n_1-1} - a_{m+n_1-1}(2n_1 + n_2 - 3) \\
(2n_1 + n_2)f_{m+n_1} - a_{m+n_1}(2n_1 + n_2 - 1)
\end{Vmatrix}
$$

$$
\mathbf{C}_2 = 2
\begin{Vmatrix}
1 & \cdots & 1 \\
1 & \cdots & 1 \\
\cdots & \cdots & \cdots \\
1 & \cdots & 1
\end{Vmatrix}
\qquad
\mathbf{F}_2 =
\begin{Vmatrix}
(2n_1 + n_2)f_{n_1} - a_{n_1}(2n_1 + n_2 - 2m) \\
(2n_1 + n_2)f_{n_1-1} - a_{n_1-1}(2n_1 + n_2 - 2m) \\
\vdots \\
(2n_1 + n_2)f_{m+1} - a_{m+1}(2n_1 + n_2 - 2m)
\end{Vmatrix}
$$

(8.59)

$$
\mathbf{C}_3 =
\begin{Vmatrix}
3 & 2 & \cdots & \cdots & \cdots & \cdots & 2 \\
4 & 3 & 2 & \cdots & \cdots & \cdots & 2 \\
4 & 4 & 3 & 2 & \cdots & \cdots & 2 \\
\cdots & \cdots & \cdots & \cdots & \cdots & \cdots & \cdots \\
4 & \cdots & \cdots & 4 & 3 & 2 & 2 \\
4 & \cdots & \cdots & \cdots & 4 & 3 & 2 \\
4 & \cdots & \cdots & \cdots & \cdots & 4 & 3
\end{Vmatrix}
$$

(8.60)

$$
\mathbf{F}_3 =
\begin{Vmatrix}
(2n_1 + n_2)f_1 - a_1[2(2n_1 + n_2) - 2m - 1] \\
(2n_1 + n_2)f_2 - a_2[2(2n_1 + n_2) - 2m - 3] \\
(2n_1 + n_2)f_3 - a_3[2(2n_1 + n_2) - 2m - 5] \\
\vdots \\
(2n_1 + n_2)f_{m-2} - a_{m-2}[2(2n_1 + n_2) - 4m + 5] \\
(2n_1 + n_2)f_{m-1} - a_{m-1}[2(2n_1 + n_2) - 4m + 3] \\
(2n_1 + n_2)f_m - a_m[2(2n_1 + n_2) - 4m + 1]
\end{Vmatrix}
$$

$$\mathbf{C}_4 = \left\| \begin{matrix} 1 & \cdots & 1 \\ 1 & \cdots & 1 \\ \cdots & \cdots & \cdots \\ 1 & \cdots & 1 \end{matrix} \right\| \quad \mathbf{F}_4 = \left\| \begin{matrix} (2n_1 + n_2)f_0 - a_0(2n_1 + n_2 - m) \\ (2n_1 + n_2)f_{-1} - a_{-1}(2n_1 + n_2 - m) \\ \vdots \\ (2n_1 + n_2)f_{m-n_2+1} - a_{m-n_2+1}(2n_1 + n_2 - m) \end{matrix} \right\|$$

$$(8.61)$$

$$\mathbf{C}_5 = \left\| \begin{matrix} 2 & 2 & \cdots & \cdots & 2 & 1 \\ 2 & \cdots & \cdots & 2 & 1 & 0 \\ \cdots & \cdots & \cdots & \cdots & \cdots & \cdots \\ 2 & 2 & 1 & 0 & \cdots & 0 \\ 2 & 1 & 0 & \cdots & \cdots & 0 \\ 1 & 0 & \cdots & \cdots & \cdots & 0 \end{matrix} \right\|$$

$$\mathbf{F}_5 = \left\| \begin{matrix} 2(2n_1 + n_2)f_{m-n_2} - a_{m-n_2}[2(2n_1 + n_2) - 2m + 1] \\ 2(2n_1 + n_2)f_{m-n_2-1} - a_{m-n_2-1}[2(2n_1 + n_2) - 2m + 3] \\ \vdots \\ 2(2n_1 + n_2)f_{3-n_2} - a_{3-n_2}[2(2n_1 + n_2) - 5] \\ 2(2n_1 + n_2)f_{2-n_2} - a_{2-n_2}[2(2n_1 + n_2) - 3] \\ 2(2n_1 + n_2)f_{1-n_2} - a_{1-n_2}[2(2n_1 + n_2) - 1] \end{matrix} \right\|$$

$$(8.62)$$

Figure 8.24 shows the video signal that distorted by vibrational blur (corresponding to Fig. 8.23). For calculation, the following values of parameters are chosen: $a = 1, b = 2, m = 10, n_1 = 15, n_2 = 13$.

By the form the distorted video signal, it is possible to judge a type of vibrational blur. For example, in this case the maximum is shifted to the right of the coordinate's origin. If the object was moved firstly to the left and secondly to the right, then the maximum of the plot would be situated to the left of the origin. Under this dependence, the parameter values m, n_1, n_2 may be determined.

Similarly, the distortions and other types of oscillatory movement of the object including a non-linear law of motion might be considered.

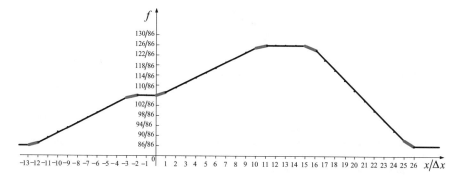

Fig. 8.24 The video signal distorted by vibrational blur

At such approach, there are difficulties in finding the values of the counts of video signal corresponding to the stationary background. In systems with matrices (Eq. 8.59) and (Eq. 8.61), it is possible to express all their background counts through any one of them. However, the remaining counts of the background included in the system (Eqs. 8.58, 8.60, and 8.62) do not determined in this manner. Therefore, in any case of vibrational blur to compensate such distortion, the mechanism of evaluating the counts of motionless background is required.

8.5 Conclusions

In this chapter, the model of linear blur considering the formation features of video signal fronts and cuts is constructed. In this model, the cases of the small and large blur are allocated. The model is described by system of the linear algebraic equations, which connect unknown values of the undistorted image of an object, values of video signal of a background in the previous frame, and values of video signal in the current frame. A comparison of the restoration results using the known deblurring methods shows that application of this model allows to restore the smaller distorted images with low computing complexity. The model of non-linear blur is constructed on the basis of the generalized description for processes of video signal distortion. Particularly, it is convenient that all equations of finite algebraic systems are written in relative coordinates. The problem of determination of the object movement laws' parameters is solved, in particular, for the most important case—the uniformly accelerated movement. Vibrational blur is also considered. It shows the need to use the individual methods for evaluation of stationary background for each type of vibrational blurring.

References

1. Gonzalez, R.C., Woods, R.E.: Digital Image Processing, 2nd edn. Prentice Hall, Upper Saddle River (2002)
2. Jahne, B.: Digital Image Processing: with CD-ROM. Springer, Berlin; Heidelberg; New York, Barcelona, Hong Kong, London, Milan, Paris, Tokyo (2002)
3. William, K., Pratt, W.K.: Digital Image Processing: PIKS Inside, 3rd edn. Wiley, New York (2001)
4. Gruzman, I.S., Kirichuk, V.S., Kosykh, V.P., Peretyagin, G.I., Spector, A.A.: Digital image processing in information systems: Tutorial. NSTU (2000) (in Russian)
5. Vasilenko, G.I., Taratorin, A.M.: Image Restoration. Radio and Communication, Moscow (in Russian) (1986)
6. Sizikov, V.S.: Sustainable methods of processing measurement results: Tutorial. "SpetsLit", Saint Petersburg (1999) (in Russian)
7. Pereslavtseva, E.E., Philippov, M.V.: Rapid method for restoration of images blurred when moving. Electronic scientific and technical edition "Science and Education", № 77-30569/340562 (2012) (in Russian)

8. Kokoshkin, A.V., Korotkov, V.A., Korotkov, K.V., Novichikhin, E.P.: Blind Restoration of Images Distorted by Blur and Defocusing, with Unknown Shape and Parameters of AF. Radio Electronics Magazine, Moscow (in Russian) (2014)
9. Yadav, S., Jain, C., Chugh, A.: Evaluation of image deblurring techniques. Int. J. Comput. Appl. **139**(12), 32–36 (2016)
10. Al-Ameen, Z., Sulong, G., Johar, M.G.M.: A comprehensive study on fast image deblurring techniques. Int. J. Adv. Sci. Technol. **44**, 1–10 (2012)
11. Wang, J., Lu, K., Wang, Q., Jia, J.: Kernel Optimization for Blind Motion Deblurring with Image Edge Prior. Mathematical Problems in Engineering, vol. 2012, Article ID 639824. Hindawi Publishing Corporation (2012)
12. Vio, R., Nagy, J., Tenorio, L., Wamsteker, W.: A simple but efficient algorithm for multiple-image deblurring. Astron. Astrophys. **416**, 403–410 (2004)
13. Singh, M.K., Tiwary, U.S., Hoon-Kim, Y.: An adaptively accelerated Lucy-Richardson method for image deblurring. EURASIP J. Adv. Signal Process. **2008**, Article ID 52 (2008)
14. Sorel, M., Flusser, J.: Space-variant restoration of images degraded by camera motion blur. IEEE Trans. Image Process. **17**(2), 105–116 (2008)
15. Malgouyres, F.: A framework for image deblurring using wavelet packet bases. Appl. Comput. Harmon. Anal. **12**(3), 309–331 (2002)
16. Daubechies, I., Teschke, G.: Variational image restoration by means of wavelets: simultaneous decomposition, deblurring, and denoising. Appl. Comput. Harmon. Anal. **19**(1), 1–16 (2005)
17. Yuzhikov, V.: Restoration of defocused and blurred images. http://habrahabr.ru/post/136853. Accessed 25 June 2017 (in Russian)
18. Yitzhaky, Y., Mor, I., Lantzman, A., Kopeika, N.S.: Direct method for restoration of motion-blurred images. J. Opt. Soc. Am. **15**(6), 1512–1519 (1998)
19. Yap, K.H., Guan, L., Perry, S.W., Wong, H.S.: Adaptive Image Processing: A Computational Intelligence Perspective, 2nd edn. CRC Press, Taylor & Francis Group (2010)
20. Levin, A., Weiss, Y., Durand, F., Freeman, W.T.: Understanding and evaluating blind deconvolution algorithms. In: IEEE Conference on Computer Vision and Pattern Recognition (CVPR'2009), pp. 1964–1971 (2009)
21. Babacan, S.D., Molina, R., Katsaggelos, A.K.: Parameter estimation in total variation blind deconvolution. In: 16th European Signal Processing Conference (EUSIPCO'2008), pp. 25–29 (2008)
22. Kundur, D., Hatzinakos, D.: Blind image deconvolution. IEEE Signal Process. Mag. **13**(3), 43–64 (1996)
23. Wu, J.M., Chen, H.C., Wu, C.C., Hsu, P.H.: Blind image deconvolution by neural recursive function approximation. World Acad. Sci. Eng. Technol. **4**(10), 1383–1390 (2010)
24. Martinello, M., Favaro, P.: Single image blind deconvolution with higher-order texture statistics. In: Cremers, D., Magnor, M., Oswald, M.R., Zelnik-Manor, L. (eds.) Video Processing and Computional Video. LNCS, vol. 7082, pp. 124–151. Springer, Berlin, Heidelberg (2011)
25. Ayers, G.R., Dainty, J.C.: Iterative blind deconvolution method and its applications. Opt. Lett. **13**(7), 547–548 (1988)
26. Holmes, T.J., Biggs, D., Abu-Tarif, A.: Blind deconvolution. In: Pawler, J.B. (ed.) Handbook of Biological Confocal Microscopy, 3rd edn, pp. 468–487. Springer Science + Business Media, LLC, New York (2006)
27. Bhuiyan, M.I., Sacchi, M.D.: Two-stage blind deconvolution. In: GeoConvention 2013: Integration, pp. 1–6 (2013)
28. Kwon, T.M.: Blind Deconvolution of Vehicle Inductance Signatures for Travel-Time Estimation. Minnesota Department of Transportation, MN/RC-2006-06 (2006)
29. Hirsch, M., Harmeling, S., Sra, S., Schölkopf, S.: Online multi-frame blind deconvolution with super-resolution and saturation correction. Astron. Astrophys. **531**, A9.1–A9.11 (2011)
30. Bell, A.J., Sejnowski, T.J.: An information-maximization approach to blind deconvolution. Neural Comput. **7**(6), 1129–1159 (1995)

31. Shan, Q., Jia, J., Agarwala, A.: High-quality motion deblurring from a single image. ACM Trans. Graphics **27**(3), Article 73.1–73.10 (2008)
32. Bando, Y., Chen, B.Y., Nishita, T.: Motion deblurring from a single image using circular sensor motion. Comput. Graph. Forum **30**(7), 1869–1878 (2011)
33. Sizikov, V.S., Rossiyskaya, M.V., Kozachenko, A.V.: Blurred Images Processing by the Methods of Differentiation, Hartley Transform and Tikhonov Regularization. Priborostroenie, Moscow (in Russian) (1999)
34. Sizikov, V.S.: The truncation—blurring—rotation technique for reconstructing distorted images. J. Opt. Technol. **78**(5), 298–304 (2011)
35. Breiswell, R.: Hartley Transformation. Mir, Moscow (in Russian) (1990)
36. Tikhonov, A.N., Arsenin, V.Y., Timonov, A.A.: Mathematical Problems of Computer Tomography. Science, Moscow (1987) (in Russian)
37. Sizikov, V.S.: Analysis of local regularization method and formulation of the suboptimal filtration method for solution of equation I kind. Journal of Computer Mathematics and Mathematical Physics, Moscow (in Russian) (1999)
38. Bogoslovskiy, A.V., Zhigulina, I.V., Bogoslovskiy, E.A., Ponomarev, A.V., Vasilyev, V.V.: Linear Blur. Radiotec, Moscow (in Russian) (2015)

Chapter 9
Core Algorithm for Structural Verification of Keypoint Matches

Roman O. Malashin

Abstract Outlier elimination is a crucial stage in keypoints-based methods, especially in extreme conditions. In this chapter, a fast and robust "Core" Structural Verification Algorithm (CSVA) for a variety of applications and feature extraction methods is developed. The proposed algorithm pipeline involves many-to-one matches' exclusion, the improved Hough clustering of keypoint matches, and cluster verification procedure. The Hough clustering is improved through an accurate incorporation of translation parameters of similarity transform and "partially ignoring" the boundary impact using two displaced accumulators. The cluster verification procedure involves the use of modified RANSAC. It is also shown that the use of the nearest neighbour ratio may eliminate too many inliers, when two images are matched (especially in extreme conditions), and the preferable method is a simple many-to-one matches exclusion. The theory and experiment prove the propriety of the suggested parameters, algorithms, and modifications. The developed cluster analysis algorithms are robust and computationally efficient at the same time. These algorithms use some specific information (rigidity of objects in a scene), consume low volume memory and only 3 ms in average on a standard Intel i7 processor for verification of 1,000 matches (i.e. magnitudes less than the time needed to generate those matches). The CSVA has been successfully applied to practical tasks with minor adaptation, such as the matching of 3D indoor scenes, retrieval of images of 3D scenes based on the concept of Bag of Words (BoWs), and matching of aerial and cosmic photographs with strong appearance changes caused by season, day-time, and viewpoint variation. Eliminating a huge number of outliers using geometrical constraints allowed to reach the reliability and accuracy in all solutions.

R.O. Malashin (✉)
National Research University of Information Technology, Mechanics and Optics,
49 Kronverksky Pr., St. Petersburg 197101, Russian Federation
e-mail: malashinroman@mail.ru

R.O. Malashin
Pavlov Institute of Physiology Russian Academy of Sciences,
Makarova emb. 6, St. Petersburg 199034, Russian Federation

© Springer International Publishing AG 2018 251
M.N. Favorskaya and L.C. Jain (eds.), *Computer Vision in Control Systems-3*,
Intelligent Systems Reference Library 135, https://doi.org/10.1007/978-3-319-67516-9_9

Keywords Keypoint-based methods · Scene geometry · Structural analysis
Outlier elimination · Hough clustering · RANSAC · Bag of words
Nearest neighbour ratio · Aerospace images · 3D images

9.1 Introduction

The keypoint-based (feature-based) algorithms like Scale-Invariant Feature
Transform (SIFT), Speeded Up Robust Features (SURF), and Binary Robust
Invariant Scalable Keypoints (BRISK) have many applications in computer vision,
for example, the panorama stitching, object detection, robot navigation, among
others. Both the advantage and drawback of the approach is that the keypoint
descriptors are matched independently. Small patches are usually geometrically
more stable than the entire image but they do not contain a lot of information.
Therefore, micro changes of the patch can lead to a mismatch. Many incorrect
matches arise from the occluded regions and unstable keypoints (for example, in
images of tree branches or waves). At the same time even in adverse conditions, one
can expect that some of the patches in both images are unique and, hence, matched
correctly. Therefore, the robustness of outlier elimination algorithm is critical for
the successful results.

The algorithm suggested in this chapter uses the spatial model that approximates
the surfaces of arbitrary 3D scenes with a set of simple transformations. The
approach allows successful registration with less than 5% of inliers. The basic idea
is taken from the SIFT algorithm that recognizes objects through the cluster analysis
of parameters of the matched keypoints. The procedure was modified to be more
general and more efficient. The proposed algorithm called as the CSVA significantly outperforms the methods used by developers in most cases.

The CSVA was mainly introduced and tested for an image matching task (i.e.
image against image, image against a set of images), as it is the main area of local
features application (panorama stitching, unmanned aerial vehicle position rectifi-
cation, etc.). Therefore, all the conclusions are mainly stated for image matching
tasks though mostly remain actual for other cases as well.

The chapter is organized as follows. A short overview of keypoint-based
methods is given in Sect. 9.2. Mathematic notation used in the chapter is mentioned
in Sect. 9.3. Section 9.4 describes the datasets that were used for testing and
comparison. The algorithms of primary elimination of local mismatches are discussed in Sect. 9.5. The clustering and cluster verification algorithms are represented in Sect. 9.6. Section 9.7 describes the algorithm to estimate a confidence of
the verified cluster hypothesis. The practical applications of the proposed structural
verification pipeline are discussed in Sect. 9.8. Conclusions are given in Sect. 9.9.

9.2 Overview of Keypoint-Based Methods

Approach of keypoint-based matching usually involves the following stages: feature detection, feature description, descriptors matching, and structural verification. The goal of detection is to select such image patches (keypoints) that are unique and stable, in other words, they preserve the particularities in different lighting and viewpoint conditions. Then the selected patches are described by a scale and orientation invariant feature vector (descriptor). The nearest neighbour search in feature space gives a set of keypoint-to-keypoint matches. The last stage involves the geometrical constraints on mutual relations between the keypoints to eliminate the outliers. Figure 9.1 illustrates the stages of the keypoint-based approach.

A brief overview of the methods applied at every step is below.

Detection. It is common to select corners [2–4] and blob centres [5, 6] as keypoints. Usually the position (x, y) in the image, orientation and size are recovered by detectors, though there are algorithms that give some extra parameters [7] as well not every algorithm is truly scale invariant; size of the feature is not always supported by detectors (for example by naive Features from Accelerated Segment Test (FAST) detector [8]).

There are target specific detectors that use heuristics. For example, Hauagge and Snavely [9] applied the hypothesis of the artificial objects symmetry. In [10], the Kalman filter was used to track the most stable keypoints in a video sequence. The repeatability is one of the main objectives considered while developing a detection algorithm. This is an ability to detect with high probability the same keypoints (that refer to the same elements of the objects) in different images. The density is another required quality: there should be as many keypoints as possible to cover images evenly.

Description. Description algorithms must be robust and distinctive to natural alterability of an image patch (illumination and view point changes). The most popular algorithms can be classified into two categories: floating-point or gradient

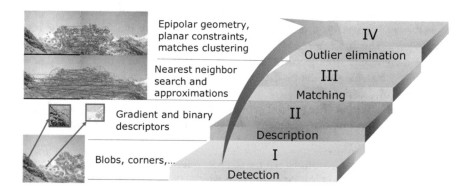

Fig. 9.1 Stages of keypoint-based methods (The image of a mountain view used in the figure is borrowed from [1].)

(using gradient of an image) and binary (operating intensities directly). The most well-known descriptors of the first category are the SIFT [11] and the SURF [12] that are created by computing histograms of gradient orientation in subregions of an image patch. The histogram is normalized according to dominant gradient direction. The SURF uses integral image [13], and, hence, it is more computationally effective. However, many authors consider the SIFT to be more robust to the affine transformations.

Contour structural elements can be matched to each other directly according to their type (line, corner, circle, etc.) [14] that makes the matches less ambiguous. An image patch around these elements can be used to calculate the descriptors [15].

Binary descriptors, including Binary Robust Independent Elementary Features (BRIEF) [16], Oriented fast and Rotated BRIEF (ORB) [17], BRISK [18], and Fast REtinA Keypoints (FREAK) [19] are from the second category. A binary descriptor is a binary vector; each element corresponds to the result of an application of a relational operator to two pixels in the keypoint neighbourhood (1 means greater or equal, 0 means less). The main advantage is low resource consumption due to the reasons mentioned below:

1. There is no convolution with differential kernels.[1]
2. Similarity measure is Hamming weight (which can be calculated effectively by most of processor architectures).
3. Binary vectors are extremely compact.[2]

There are also descriptions that do not directly fall in both aforementioned categories. They are, for example, texture features [20, 22, 23, 24], descriptors that consider spatial-temporal [25, 26] or depth information [27, 28]. Some algorithms use the non-rectangular image patches. For example, Gaussian derivatives calculated inside the oval region compose the descriptor [29].

Typically, the descriptors are extremely sensitive to the estimated direction of an image patch, thus the features with inaccurate orientations are unmatchable. The Fourier-Mellin transform can help to reduce the problem [30]. However, the log-polar transform that is the core of the method leads to heavy computations and in practice the independency on the orientation is not achieved.

Matching. Generally, a descriptor matching is done by the nearest neighbour search [31]. The brute force solution is inefficient but some approximates like nearest neighbours for gradient [32] and binary features [33] are applied. Accelerated matching of the FREAK rejects most of the candidates by inspecting only a part of descriptors.

Outlier elimination. The general geometrical constraint to an image pair of static 3D scenes is the epipolar geometry. An alternative solution is to approximate

---

[1]Though in more recent the BRISK and the FREAK, noise is suppressed by convolution with Gaussian of a variable size (size increases in peripheral areas of the patch), it is applied only to part of the pixels according to pairwise comparison template.

[2]In [20], Principal Component Analysis (PCA) is applied to build more compact gradient descriptors.

shapes of all objects in a scene with one plane. There are more general models that use some heuristics to fit several planes in a set of matches [34]. Another approach exploits the principle of minimum description length [35]. Keypoint coordinates detected in the second image can be effectively encoded as the set of transformation parameters and the set of deviations if there is a good transformation hypothesis given. However, this method does not provide any effective way to construct the hypothesis.

Another aspect of outlier elimination is a method used to fit a model into a set of matches. Well known approaches are RANdom SAmple Consensus (RANSAC) [36] and the least squares. The least squares solution fits model parameters by minimizing the sum of squared deviations of keypoints locations. It is extremely sensitive to outliers.[3] The RANSAC (and its modifications like Progressive SAmple Consensus PROSAC [37]) is more robust. It iteratively selects a small amount of matches, relying on which it constructs the hypotheses on the model parameters. The best hypothesis for the RANSAC is the one that supports the largest number of matches. Unfortunately, even the RANSAC is not able to find correct epipolar geometry, when the number of outliers is more than 70–75%.

The Hough clustering of matches in the similarity transform space proposed in the SIFT [11] is robust because the size and orientation of keypoints are considered (that is not true for the most of other methods). However, it was suggested for a specific object detection task (actually, it has modifications for 3D object recognition [38] and scene recognition [1]) and, therefore, it does not attract a lot of interest compared to the SIFT detection and description part.[4] The work [39] showed that the approach was likely to be improved and generalized for aerospace images [40] and even non-static 3D scenes [41]. The chapter gives detailed description of this method.

9.3 Notation

The appearance of the SIFT [42] was one of the factors that highly increased an interest to the keypoint-based methods because of its efficiency. One of the reasons is that the SIFT has the detailed and robust implementation of every step mentioned above. The SIFT is applied in many tasks, such as the object detection, 3D modelling [43], panorama stitching [1, 44, 45], gesture recognition [46, 47], faces recognition [48, 49], fingerprint detection [50, 51], and object recognition [52].

Most of the algorithms that appeared after SIFT targeted to improve separate steps of the SIFT (often detection or description) in terms of robustness,

---

[3]There is the iterative least squares: after the transformation parameters are found, the outliers can be eliminated and parameters can be recalculated. Still if the initial guess is too bad, then the iterative least squares converge to wrong solution.

[4]For example, there is no open source implementation.

computational efficiency that sometimes leads to wrong conclusions because the improving commonly used metrics of separate steps (for example, repeatability for detectors) does not necessarily lead to the overall improvement of the result. The chapter considers the outlier elimination stage, which is the outcome of the keypoint-based method, and, therefore, it provides a useful analysis on the previous stages.

Assume a matching two images ε_1 and ε_2 (in general case, it can be an image vs. a dataset of images). The order of the images is usually important and we refer to ε_1 as a query image and ε_2 as a test image. Final algorithm ξ inputs two images and outputs a set of n local correspondences $\xi(\varepsilon_1, \varepsilon_2) \rightarrow \{(a_i, b_i), i = 1, \ldots, n\}$, where a_i and b_i are n matched patches from ε_1 and ε_2 respectively.

Define a detection algorithm as a function α that associates the image ε with a set of keypoints, $Q = \{q_1, q_2, \ldots\}$, where q_i is an ith set of keypoint parameters (Eq. 9.1).

$$\alpha(\varepsilon) \rightarrow Q \tag{9.1}$$

We consider that the recovered parameters of a keypoint are the location, size, and orientation.[5] Description algorithm calculates a descriptor in an image patch for every keypoint and can be denoted as function β that associates the set of keypoints Q and image ε with a set of features F ("described keypoints"). It can be defined by Eq. 9.2, where $f \in F = \{p_f, S_f, \theta_f, d_f\}$ is a feature, $p_f = (x_f, y_f)$ is a keypoint location in pixels, S_f is a feature size in pixels, θ_f is its orientation in radians, d_f is an invariant descriptor vector of $q = \{p_f, S_f, \theta_f\}$.

$$\beta(\varepsilon, Q) \rightarrow F \tag{9.2}$$

Sometimes it is convenient to operate with the "extraction" algorithm δ as "superposition" of α and β (Eq. 9.3).

$$\delta(\varepsilon) \rightarrow F \quad \delta(\varepsilon) = \beta(\alpha(\varepsilon), \varepsilon) \tag{9.3}$$

The local match for two images is a pair (u, v), where $u \in F_1$ and $v \in F_2$. Then the set of all possible keypoint-to-keypoint matches between F_1 and F_2 is denoted as $\Omega = F_1 \times F_2$. For sake of simplicity in all the following notations, we assume $u \in F_1$, $v \in F_2$, $(u, v) \in \Lambda$ by default, where Λ is a subset of Ω.

We can define the descriptor matching algorithm γ as a function that maps two sets of keypoints into some subset of Ω (Eq. 9.4).

$$\gamma(F_1, F_2) \rightarrow 2^\Omega \tag{9.4}$$

---

[5]Developed algorithms prefer detectors that can recover comparatively fair size of the features. Most of commonly used detectors can do this (SIFT, SURF, BRISK, FREAK).

For the feature sets F_1 and F_2 from concrete images, the matching algorithm γ extracts a subset M_γ from Ω (Eq. 9.5).

$$\gamma(F_1, F_2) = M_\gamma \subseteq \Omega \qquad (9.5)$$

An outlier elimination algorithm ϕ extracts a subset M_{final} from M_γ (Eq. 9.6).

$$\phi(M_\gamma) = M_{final} \subseteq M_\gamma \qquad (9.6)$$

Further, we usually omit F_1 and F_2 as input parameters assuming them for matching algorithms by default (where it does not lead to ambiguous understanding). We also assume that parameter γ implements the nearest neighbour algorithm and use a notation M instead of M_γ. If $d(u, v) = d(v, u)$ is the distance between d_u and d_v in multidimensional feature space according to the chosen metric, then the nearest neighbour algorithm generates a set of matches M according to Eq. 9.7.

$$M = \left\{ (u, v) | v = \underset{w \in F_2}{\arg\min}\, d(u, w) \right\} \qquad (9.7)$$

Function ξ is a superposition of the extraction, matching and outlier elimination (Eq. 9.8). As soon as γ is always the same, we can think of it as a part of ϕ (Eq. 9.9).

$$\xi(\varepsilon_1, \varepsilon_2) = \phi(\gamma(\delta(\varepsilon_1), \delta(\varepsilon_2))) \qquad (9.8)$$

$$\xi(\varepsilon_1, \varepsilon_2) = \phi(\delta(\varepsilon_1), \delta(\varepsilon_2)) \qquad (9.9)$$

The CSVA inputs the nearest neighbour solution for keypoint descriptors (set M) and then takes three steps (Fig. 9.2):

Step 1. Primary outlier elimination. We denote a produced set of matches after the step as a set M_{prime}.
Step 2. The Hough clustering (set of clusters M_{cl}).
Step 3. Verification of the largest cluster(s) relying on mutual relation of the keypoints of different matches (set M_{final}). Estimating a confidence of the verified clusters.

Generally, any set of matches Λ can be decomposed according to Eq. 9.10, where $G(\Lambda)$ is a set of good (correct) matches (inliers), $B(\Lambda)$ is a set of bad (incorrect) matches (outliers).

$$\Lambda = G(\Lambda) \cup B(\Lambda) \qquad (9.10)$$

Maximizing $G(M_{final})$ and minimizing $B(M_{final})$ is the goal of ξ (and ϕ as well). However, it is hard to determine an exact objective function without any generality loss in practice. The function does not only depend on the quantity or ratio of

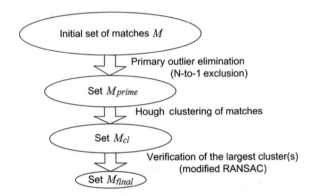

Fig. 9.2 Outline of the structural verification algorithm

inliers, but on the parameters of the matches themselves as well. For example, if we use M_{final} to fit a homography using the least squares, then several outliers with small deviation give a better solution than a set with one but bad mistake. The only fair request for ξ is sufficiency to solve the task on as many image pairs as possible. Hence, all the conclusions made in the chapter are verified by experiments with train datasets containing image pairs $\{(\varepsilon_1, \varepsilon_2)\}$. Some tasks need good density of matches, others need precise accuracy, therefore, the cardinalities of $G(\Lambda)$ and $B(\Lambda)$ across the datasets have been also considered.

For example, let us consider an objective function of a detector via a ratio of inliers and outliers. It appears that the repeatability (a commonly used metric for detectors comparison) can be misleading. Detector determines the properties of the set Ω, thus the repeatability $0 \leq R \leq 1$ can be expressed by Eq. 9.11.

$$R = |G(\Omega)|^2 \big/ |\Omega| \tag{9.11}$$

Parameter Ω is the input for the descriptor objective function L_β (Eq. 9.13) that has to consider a ratio of correct matches after the nearest neighbour search (Eq. 9.12).

$$K = |G(M)| / |M| \tag{9.12}$$

$$L_\beta = K = f(\beta, \Omega) \to \max \tag{9.13}$$

Greater R does not strictly lead to greater K because the uniqueness of the keypoints is not considered, therefore, commonly used objective function $L_\alpha = R \to \max$ does not maximize L_β. Moreover, a detector can have a good repeatability in terms of keypoint position but a poor repeatability, when a keypoint size is considered. In all experiments, the default settings of detectors (implemented in OpenCV 3.1 library) were used.

9.4 Training Datasets

Algorithms proposed in the chapter are verified by the experiments with three datasets containing the manually labelled image pairs. Table 9.1 summarises general information about them.

Dataset db1 has 20 challenging image pairs of 3D scenes (mostly under strong viewpoint changes). The pairs were matched by different combinations of the SURF and the SIFT detectors and descriptors. For every pair from the dataset, set M was decomposed into the set of correct matches $G(M)$ and set of incorrect matches $B(M)$ manually inspecting every match. The corresponding keypoints of correct matches are situated in the image patches of the same object elements (with the acceptance of small inaccuracies). The overall number of labelled matches is around 24,000 and only 4% of them are correct. Figure 9.3 illustrates two pairs from dataset db1.

Dataset db2 consists of 120 pairs of aerial and cosmic image pairs under the strong viewpoint, season, day-time, and anthropogenous (man-made) changes. Several manually matched points were used to calculate a ground truth homography H for every image pair. In dataset db2, $G(M)$ is composed of matches that satisfy condition given by Eq. 9.14,[6] where "$\approx$" denotes "agrees within considered error bounds" (significant deviations are accepted to cover mountain landscapes, non-orthogonal views, detectors inaccuracies), θ_H and S_H are the orientation and scale parameters obtained from H.[7]

$$G(M, H) = \{(u, v) | \|Hp_u - p_v\| \approx 0 \quad (\theta_v - \theta_u) \approx \theta_H \quad S_v / S_u \approx S_H\} \quad (9.14)$$

Dataset db3 has more than 300 pairs of images of static 3D scenes (mostly indoor, like the ones in Fig. 9.3). Few manually labelled corresponding points were used to calculate etalon fundamental matrices E by the least squares. In this case, the outliers in the set M are matches (u, v), where keypoint location p_v lies far away from the epipolar line corresponding to p_u (Eq. 9.15).

$$G(M, E) = \{(u, v) | d(p_v, Ep_u) \approx 0\} \quad (9.15)$$

Note that dataset db1 is more accurate than dataset db2 (because in the first case every match is labelled manually), and dataset db2 is more accurate than dataset db3 (because a homography is more rigid than epipolar geometry).

---

[6]It is assumed that p_v and p_u are presented in homogenous coordinates.

[7]These parameters can be recovered by decomposition of H ignoring projective components of the homography matrix [52]. Inefficient but more accurate solution is calculating the needed parameters for every keypoint individually by applying H to keypoint location and several points nearby.

Table 9.1 Summary of image matching datasets

Dataset	Content	Number of image pairs	Ground truth information
db1	3D scenes	20	Labels for each match
db2	Aerospace	120	Homography H for each pair
db3	3D scenes	352	Fundamental matrix E for each pair

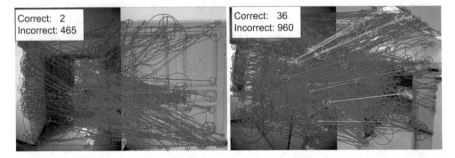

Fig. 9.3 Manually labelled matches M of two image pairs from dataset db1 matched by the SURF. Correct matches are shown as *green lines*, while the incorrect ones are shown as *red lines*

9.5 Algorithms of Primary Outlier Elimination

The ratio of correct to incorrect matches is a key factor for the successful application of geometrical constraints. There are approaches to increase this ratio without involving structural information. Three methods, such as the nearest neighbour ratio test, crosscheck, and exclusion of ambiguous n-to-1 matches, are considered in Sects. 9.5.1–9.5.3, respectively. The intermediate comparison and conclusion are represented in Sect. 9.5.4.

9.5.1 Nearest Neighbour Ratio Test

The most popular technique to eliminate outliers is the use of the nearest neighbour ratio as a signal of mismatch [11]. In this case, for each descriptor of the query image, two closest descriptors from the test image are found. If two distances are similar, then a probability of the match to be correct decreases as the match is not truly unique. Let $d(u, F, k)$ be a distance from d_u to kth closest descriptor from set F.[8] Then, for any match $(u, v) \in M$ the closest/next closest ratio can be defined as $nnr(u, F)$ according to Eq. 9.16.

$$nnr\,(u, F) = d(u, F, 1)\,/\,d(u, F, 2) \qquad (9.16)$$

---

[8]$d(d_u, d_v) = d(u, F, 1)$.

Then, a set of matches that pass through the nearest neighbour test is M_{NNR} defined by Eq. 9.17, where $0 < \lambda < 1$ is a predefined threshold.

$$M_{NNR}(\lambda) = \{(u, v) \in M \,|\, nnr(u, F_2) < \lambda\} \qquad (9.17)$$

There is no obvious way to choose λ. In [11], the experiments were conducted by simulating the image transformations. They resulted in Probability Density Function (PDF) shown in Fig. 9.4a and the best nearest neighbour ratio around 0.7–0.8.

Experiments with datasets db1 and db2 resulted in different PDFs (Fig. 9.4b–c). The thresholding ratio of distances to two closest neighbours eliminates a major part of inliers. This holds for different combinations of the SIFT, the SURF, the BRISK detectors and descriptors. Datasets db1 and db2 contain natural images making the results more reliable than in work [11]. At least, this is true for image against image matching (work [11] considers image against dataset), when the image pairs are challenging.

9.5.2 Crosscheck

According to the crosscheck, only those matches (u, v) remain if descriptor d_v is the closest to d_u among all descriptors in set F_2 and descriptor d_u is the closest to d_v among all descriptors in F_1 at the same time. More formally the matches that pass the crosscheck test compose a set M_{CC} according to Eq. 9.18, where sign "T" denotes that pair elements are reordered: $(u, v) \rightarrow (v, u)$, γ is the nearest neighbour algorithm (Eq. 9.7).

$$M_{CC} = \gamma(F_1, F_2) \cap \gamma(F_2, F_1)^{\mathrm{T}} \qquad (9.18)$$

Despite the correct/incorrect ratio increase, the crosscheck makes the results invariant to images order.[9]

9.5.3 Exclusion of Ambiguous Matches

The approach proposed in [40] is a simple one-to-many exclusion: the matches that link several keypoints from the first image with the same keypoint from the second image are eliminated. One match corresponding to the shortest distance in multi-dimensional feature space remains from each ambiguous match. Then the produced set of matches M_{ex} can be formalized by Eq. 9.19.

---

[9]Note that $\gamma(S_1, S_2) \neq \gamma(S_2, S_1)^{\mathrm{T}}$.

Fig. 9.4 Probability density function of correct and incorrect matches for: **a** artificially generated images [11], **b** dataset db1 (natural and manually labelled pairs of images of 3D scenes), **c** dataset db2 (natural and manually labelled aerial and cosmic images)

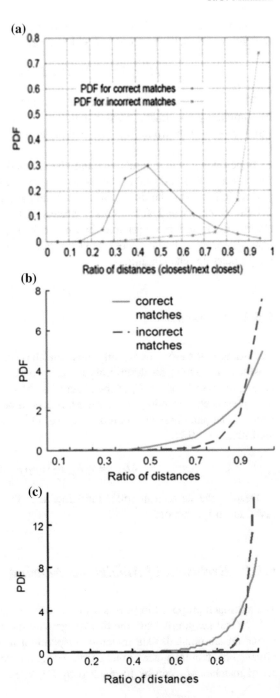

Fig. 9.5 Average number of false clusters of matches (with 3 and more votes) before and after exclusion of N-to-1 matches. The plot was generated by matching hundreds of images having non-overlapping content

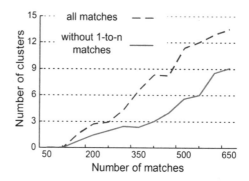

Table 9.2 Comparison of primary elimination methods on db1

Matches set	Precision	Recall	F_1-score
M	0.039	1	0.076
$M_{NNR}(0.91)$	0.064	0.619	0.117
M_{ex}	0.068	0.607	0.112
M_{CC}	0.109	0.489	0.178
$M_{NNR}(0.77)$	0.161	0.24	0.192
$M_{CC} \cap M_{NNR}(0.81)$	0.211	0.201	0.202
$M_{ex} \cap M_{NNR}(0.82)$	0.181	0.25	0.21

$$M_{ex} = \left\{ (u,v) \in M \mid u = \underset{w \in F_1, (w,v) \in M}{\arg\min}\ d(w,v) \right\} \qquad (9.19)$$

The ambiguous many-to-one matches have a higher probability to be incorrect. Moreover, they have a higher probability to form incorrect (false) clusters of matches with relevant parameters (Fig. 9.5) in similarity transform space [40]. The algorithms of cluster analysis are described in Sect. 9.7.

9.5.4 Comparison and Conclusion

Table 9.2 contains the precision, recall, and F_1-score estimations that all the methods got for the most accurate dataset db1.

Precision and recall for $M_{NNR}(0.91)$ and M_{ex} are similar. This shows that many-to-one exclusion has the similar characteristics with the closest/next closest ratio test.[10] All other nearest neighbour ratios are chosen to provide the best F_1-

---

[10]This does not hold for the experiments with dataset db2 (the reasons are not clear). Nevertheless, for all datasets M_{ex} is advantageous to $M_{NNR}(0.91)$ because this set has fewer clusters that are difficult to verify with the CSVA.

score. Parameter $\Lambda = 0.77$ is the best for M and is close to the one chosen by PDF analysis in [11].

As seen from Table 9.2, the combining different methods can help to improve the precision and F_1-score. For example, $M_{ex} \cap M_{NNR}(0.81)$ provides better precision and recall than $M_{NNR}(0.77)$ at the same time. The crosscheck removes too much inliers from M. In this case, the most challenging pairs remain unrecognized, the crosscheck is not used in the CSVA and, in general, $M_{prime} = M_{ex}$.

Because it is hard to define the constraints on a set of matches after the preliminary elimination, we carried out the experiments on image matching task for datasets db2 and db3. A small improvement is achieved in dataset db2 by removing matches with bad closest/next closest ratio from M_{ex}, i.e. $M_{prime} = M_{ex} \cap M_{NNR}(0.98)$ provides the best results for aerospace image matching. In some tasks, it is impossible to use the nearest neighbour ratio or the many-to-one exclusion (see Sect. 9.8.2) than $M_{prime} = M$.

9.6 Geometrical Verification of Matches

The CSVA is based on clustering of feature matches in similarity transform parameter space and consists of the Hough clustering (Sect. 9.6.1) and cluster verification (Sect. 9.6.2).

9.6.1 Hough Clustering of Keypoint Matches

The main idea of clustering is to eliminate matches that do not form enough large clusters in the space of similarity transform parameters. Assume that a match (u, v) that forms two sets of parameters $\{p_u, S_u, \theta_u\}$ and $\{p_v, S_v, \theta_v\}$, respectively (descriptors do not used for structural verification). Hence, a match (u, v) determines the unique translation, rotation, and scaling of the coordinate system, or in other words the similarity transform (Eq. 9.20).

Note that the chosen spatial model is only a rough approximation of natural non-flat transformations. Therefore, large deviations must be accepted. The similarity transform maps an arbitrary point (x, y) to a new point (x', y') via Eq. 9.20, where S is a scaling factor, θ is an angle of rotation, t_x, t_y are translation parameters along the image axes.

$$\begin{bmatrix} x' \\ y' \end{bmatrix} = \begin{bmatrix} S\cos\theta & -S\sin\theta \\ S\sin\theta & S\cos\theta \end{bmatrix} \begin{bmatrix} x \\ y \end{bmatrix} + \begin{bmatrix} t_x \\ t_y \end{bmatrix} \tag{9.20}$$

Equation 9.20 can be rewritten as Eq. 9.21, where $\mathbf{T}$ is 2×2 matrix, $\mathbf{p} = (x, y)$, $\mathbf{p'} = (x', y')$, and $\mathbf{t} = (t_x, t_y)$.

$$\mathbf{p}' = \mathbf{Tp} + \mathbf{t} \tag{9.21}$$

Each match from Affine-SIFT (ASIFT) [53] algorithm[11] can describe affine distortion. Therefore, a match (u, v) from the ASIFT determines six parameters $(S, \theta, t_x, t_y, \beta, \mu)$, where β and μ are direction and factor of anisotropic scaling, respectively [52]. The drawback of the ASIFT is heavy extra computations. By that and some other reasons, the CSVA is proposed for classical 4D parameter space. If application of the ASIFT is justified (very strong affine distortions are expected), then a good alternative to clustering in affine space is applying the CSVA for each pair of simulated images.

Elements of $\mathbf{T}$ depend on S and θ, which can be recovered from u and v parameters directly.

$$S = S_v / S_u \tag{9.22}$$

$$\theta = \theta_v - \theta_u \tag{9.23}$$

If a match (u, v) is correct, we can assume $\mathbf{Tp}_u + \mathbf{t} = \mathbf{p}_v$ to get equations for translation parameters (Eq. 9.24)

$$\mathbf{t} = \mathbf{p}_v - \mathbf{Tp}_u \tag{9.24}$$

Tuples (S, θ, t_x, t_y) can be clustered directly, but there is a reason to substitute the translation parameters with "model" [11] or "reference" point [55]. Consider an arbitrary reference point with the location $\mathbf{p}_r = (x_r, y_r)$ in the query image. Then $\mathbf{p}'_r = [x'_r, y'_r]$ is the reference point location in the test image predicted by unknown similarity transform parameters (Eq. 9.25).

$$\mathbf{p}'_r = \mathbf{Tp}_r + \mathbf{t} \tag{9.25}$$

Applying Eq. 9.24, we get Eq. 9.26 for the predicted reference point position.

$$\mathbf{p}'_r = \mathbf{T}(\mathbf{p}_r - \mathbf{p}_u) + \mathbf{p}_v \tag{9.26}$$

Equation 9.26 is a matrix form equivalent of Eqs. 9.27–9.28.

$$x'_r = x_v + S(x_r - x_u)\cos\theta - S(y_r - y_u)\sin\theta \tag{9.27}$$

$$y'_r = y_v + S(x_r - x_u)\sin\theta + S(y_r - y_u)\cos\theta \tag{9.28}$$

Note that the influence of S and θ on $\mathbf{p}'_r$ is proportional to the distance between $\mathbf{p}_r$ and $\mathbf{p}_u$. Recall that keypoint parameters θ and S (especially) are not precise due

---

[11]There are some other approaches to form affine [7] or even projective-invariant [54] descriptors but they are less popular then the ASIFT.

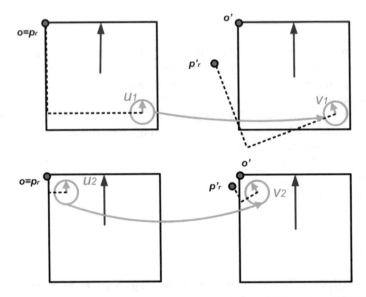

Fig. 9.6 Deviation in reference point prediction due to error in keypoint orientation (20°), when $\mathbf{p}_r$ coincides with an origin of coordinate system **o**. The match (u_2, v_2), where u_2 is situated near the reference point **p**, provides much better estimation of the reference $\mathbf{p}_r'$ than the match (u_1, v_1)

to detectors instability and the fact that the similarity transform is only an approximation. That means that more accurate estimations of $\mathbf{p}_r'$ are provided by the matches (u, v), where u is situated near the reference point. Therefore, these matches have higher prior probability to agree within given error bounds for $\mathbf{p}_r'$. Equation 9.24 is a special case of Eq. 9.26,[12] and choosing a good $\mathbf{p}_r$ has the same effect as choosing a proper origin of the coordinate system in Eq. 9.24. By default, we usually assume that the origin of an image is upper left pixel. In this case, the clusters near bottom right corner are likely to be faded away. The effect is illustrated in Fig. 9.6.

Keypoints in average are distributed uniformly around the centre of the image, that is why a reference point can be chosen to coincide with the centre of the first image. Another good position for $\mathbf{p}_r$ is a "mean" of keypoint coordinates in the image (Eq. 9.29).

$$\mathbf{p}_r = \left(\frac{1}{|F_1|} \sum_{u \in F_1} x_u \quad \frac{1}{|F_1|} \sum_{u \in F_1} y_u \right) \tag{9.29}$$

According to the experiments, both "fair" reference points provide almost the same result in average but still they are not ideal because the dependence on the

---

[12]By definition $\mathbf{t} = (t_x, t_y)$ is a position of the origin of coordinate system after the transformation, i.e. $\mathbf{t} = \mathbf{T}(\mathbf{p}_r - \mathbf{p}_u) + \mathbf{p}_v$, when $\mathbf{p}_r = (0, 0)$.

location of $\mathbf{p}_u$ remains (though it is less in average). That knowledge can be incorporated in the verification procedure (see the next Sect. 9.6.2).

Finding clusters in the set of tuples $\{(S^i, \theta^i, x'^i_r, y'^i_r), i = 1, \ldots, |M_{prime}|\}$ is the next step. K-means or ISODATA algorithms can be applied but the Hough transform is much faster and error bounds can be determined explicitly. In this case, a Hough accumulator (Hough table) is 4D parameter space and every bin is responsible for the specific rotation, scaling, and model position.

To avoid a problem of the boundary effect, each match votes for the two closest bins in each dimension, giving total of 16 entries for each match hypothesis [11]. An alternative solution [7] is to organize several Hough accumulators that are displaced against each other in each dimension by half of a bin size and increment appropriate bins in every table. Figure 9.7 illustrates a voting procedure with two Hough accumulators for 1D and 2D parameter spaces.

Note that the parameter space is 4D and the boundary effect remains until we use 16 displaced accumulators. Nevertheless, the number of the splits is reduced even if we use only few displaced tables. According to the experiments, a boundary effect has no significant impact on the result, when only two displaced accumulators are used (only 0.3% difference in number of correctly matched images in dataset db3). The advantage of two-accumulator approach (relatively to 16-closest bins voting) is that the overall number of entries to the accumulators decreases almost by a magnitude. This leads to a magnitude decrease of the number of clusters found (see Fig. 9.8) and, hence, accelerates the program.

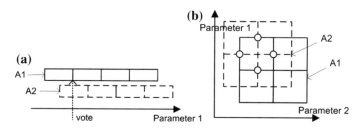

Fig. 9.7 Use of two Hough accumulators A1 and A2 to avoid a boundary effect: **a** 1D parameter space, **b** 2D parameter space (*circles* mark the points in parameter space, where clusters can be split)

Fig. 9.8 Number of bins with more than two votes versus total number of matches. This number is significantly reduced by partially ignoring boundary effect using two Hough accumulators. The plot was generated by matching hundreds of images having a non-overlapping content

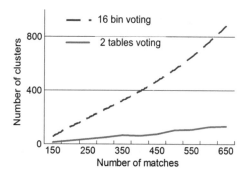

In the case of two displaced accumulators, twice larger bins should be used to guarantee an agreement in the same range (comparing to voting in neighbouring bins). The actual bin size choice depends on the number of clustered features. Complex coordinate transform can be approximated either by one similarity transform with large error bounds or by several similarity transforms with smaller error bounds. In this work, we recommend the following bin sizes as a compromise for different images, detectors, and descriptors: 30° for rotation, a factor of 2 for scale, and 0.25 of the maximal image resolution for a reference point location.

Using the bin sizes, Eqs. 9.18–9.19, and Eqs. 9.27–9.28 we can denote a function $g(u, v)$ that maps an arbitrary match to two 4D indexes in the Hough accumulators. It can be represented by Eq. 9.30, where $\mathbf{a}^t = \left(a_1^t, a_2^t, a_3^t, a_4^t\right)$, $t = \{0, 1\}$ are indexes along S, θ, x'_r, y'_r dimensions, respectively.

$$g : (u, v) = \left\{ (\mathbf{a}^1), (\mathbf{a}^2) \right\} \tag{9.30}$$

Bin size for reference point location is chosen with respect to scale of the matched features[13] [11]. Therefore, $a_3^1 = x'_r / \left(XY_{bin} \cdot S_{bin}^{a_1^1} \right)$ and $a_4^1 = y'_r / \left(XY_{bin} \cdot S_{bin}^{a_1^1} \right)$, where XY_{bin} and S_{bin} are the default sizes of bin accumulator for reference point position and scale, respectively.

Note that the Hough accumulator itself is immensely big but sparse. Effective implementation assumes the use of hash-tables [11]. A 4D bin number is a key. It is converted to 1D index in a hash-table, according to Eq. 9.31, where h is a hash-function, k is an index.

$$h : \left(a_1^i, a_2^i, a_3^i, a_4^i\right) \to k^i \tag{9.31}$$

Usually, there is an opportunity to assign some weight to every match, which relates to its quality (for example, using the nearest neighbour ratio). Therefore, in general, every match (u, v) increments a cell of one of Hough accumulators A^1 or A^2 according to Eq. 9.32, where k^i is a hash-index of the match (u, v) and $w(u, v)$ is a weighting function.

$$A^i(k^i) = A^i(k^i) + w(u, v) \tag{9.32}$$

Assigning the weights is usually task specific and, in practice, it is hard to relate a weight with real probability of a match to be correct. Therefore, $w(u, v) = 1$ usually works well but as shown in Sect. 9.6.2, there are cases, where use of w $(u, v) \neq 1$ is justified.

---

[13]We want smaller bins if the projected outline of the query image is small.

9.6.2   Verification of Cluster Hypotheses

The next goal of the CSVA is to reject accidental "false" clusters and to remove incorrect matches from the "correct" ones. The procedure relies on the mutual parameters of matched features within the detected clusters as this information is ignored during clustering. This is done by the RANSAC[14] that fits a plane model to a set of keypoint coordinates.

If we consider dynamic images, then different moving objects vote for different Hough bins. To deal with it, all the clusters larger than the predefined threshold (1% of total matches weights in M_{prime}) are verified independently and consequently (larger ones first). If some cluster hypothesis is verified, then all verified matches are removed from the rest of the clusters. The latter potentially enables fitting several models to one Hough peak as there is a high probability that a "twin" cluster (with almost the same set of matches) arises in the second Hough table.

Choosing a proper plane model is important because the correct matches within the peaks contain the keypoints from different surfaces. The RANSAC is known to fail, when several model instances need to be found. There is the PEaRL [57] that is more appropriate for multi-model cases but the number of inliers in peaks is not sufficient to describe several plane instances correctly and the number of planes is unknown a priori. (The PEaRL combines a model sampling from data points as in the RANSAC with iterative re-estimation of inliers and models the parameters based on a global regularization functional.) Hence, a single plane model should be used to loosely approximate all surfaces. Experiments with datasets db2 and db3 have shown that more general models still cannot fit the correct cluster data precisely. At the same time, they provide a greater probability to fit incorrect data. By these reasons, the similarity transform works better than the homography or the affine transform. The author of [38] came to almost the same conclusion for the task of 3D objects recognition (though, he used the least-squares instead of the RANSAC).

An individual verification of clusters imposes a lot of calculations. Assume that we verify a cluster with a set of matches M_{cl}. According to classical RANSAC, two matches $\sigma_1 = (u_1, v_1)$ and $\sigma_2 = (u_2, v_2)$ are randomly picked from M_{cl} and form a sample subset $\{\sigma_1, \sigma_2\}$, which is used to construct a hypothesis about similarity transform parameters $T^* = \{S^*, \theta^*, t_x^*, t_y^*\}$. Instead of using the parameters of individual matches (as it was done for clustering), we build a transformation hypothesis by solving a system of equations (Eq. 9.33).

---

[14]Note that it is possible to use the PROSAC [37] or the index-cloning [56] instead of the RANSAC to utilize information about matches quality (when weights $w(u, v) \neq 1$). Nevertheless, in most cases the incorporation of matches' quality does not give significant advantage over the RANSAC.

$$\begin{cases} x_{v_1} = S^* x_{u_1} \cos \theta^* - S^* y_{u_1} \sin \theta^* + t_x^* \\ x_{v_2} = S^* x_{u_2} \cos \theta^* - S^* y_{u_2} \sin \theta^* + t_x^* \\ y_{v_1} = S^* x_{u_1} \sin \theta^* + S^* y_{u_1} \cos \theta^* + t_y^* \\ y_{v_2} = S^* x_{u_2} \sin \theta^* + S^* y_{u_2} \cos \theta^* + t_y^* \end{cases} \qquad (9.33)$$

The final equations for the similarity transform parameters are given in Eqs. 9.34–9.37.

$$S^* = \sqrt{\left[(x_{v_1} - x_{v_2})^2 + (y_{v_1} - y_{v_2})^2 \right] / \left[(x_{u_1} - x_{u_2})^2 + (y_{u_1} - y_{u_2})^2 \right]} \qquad (9.34)$$

$$\theta^* = \arctan \left[\frac{(y_{v_1} - y_{v_2}) \cdot (x_{u_1} - x_{u_2}) - (x_{v_1} - x_{v_2}) \cdot (y_{u_1} - y_{u_2})}{(x_{u_1} - x_{u_2}) \cdot (x_{v_1} - x_{v_2}) + (y_{v_1} - y_{v_2}) \cdot (y_{u_1} - y_{u_2})} \right] \qquad (9.35)$$

$$t_x^* = x_{v_2} - S^* x_{v_1} \cos \theta^* + S^* y_{v_1} \sin \theta^* \qquad (9.36)$$

$$t_y^* = y_{v_2} - S^* x_{v_1} \sin \theta^* - S^* y_{v_1} \cos \theta^* \qquad (9.37)$$

Classical RANSAC iteratively constructs and compares hypotheses by counting the number of inliers that support $\{S^*, \theta^*, t_x^*, t_y^*\}$. The best hypothesis composes T^+. The probability P that T^+ after k trials is correct can be estimated by Eq. 9.38, where $f = p^2$ is a probability of taking an error-free pair $\{\sigma_1, \sigma_2\}$, P is a probability that σ_i is an inlier.

$$P = 1 - [1 - f]^k \qquad (9.38)$$

From Eq. 9.38, we can define Eq. 9.39 for the minimal number of trials to get a desired probability greater or equal to P.

$$k = \lceil \frac{\ln (1 - P)}{\ln [1 - f]} \rceil \qquad (9.39)$$

If $P = 0.5$ (50% is often considered to be the minimal number of inliers, when a least square solution works), then from Eq. 9.39 for $P = 0.95$, we get $k = 11$. That means that 11 iterations of the RANSAC is enough to converge to a correct result with the probability greater than 0.95. If $f = 0.3$ (30% is a chosen threshold), then $k = 32$.

There is a chance that the similarity transform parameters from single matches do not agree with $T^* = \{S^*, \theta^*, t_x^*, t_y^*\}$ in considered error bounds. In this case, we can skip the inliers counting step. Hence, a number of computationally heavy trials can be reduced. Moreover, there is a probability that parameters of two individual matches σ_1 and σ_2 do not agree in a range of half bin size (the guaranteed range of cluster detection). Let $\{S^i, \theta^i, p_r^i\}$ be the parameters that are recovered by the match

σ_i according to Eqs. 9.22–9.23 and Eqs. 9.27–9.28. Then the overall list of tests that have to be passed by $\{\sigma_1, \sigma_2\}$ are:

1. Scale agreement: $(S^1 \approx S^2)$ AND $(S^i \approx S^*)$.
2. Rotation agreement: $(\theta^I \approx \theta^2)$ AND $(\theta^i \approx \theta^*)$.
3. Reference point agreement: $\left(\mathbf{p}'_r{}^1 \approx \mathbf{p}'_r{}^2\right)$ AND $\left(\mathbf{p}'_r{}^i \approx \mathbf{p}_r^*\right)$.

The sign "$\approx$" means "agrees within error bounds" and $\mathbf{p}'_r{}^*$ is the position of a reference point obtained by applying T^* to $\mathbf{p}_r$. The three tests are not independent: when one test is passed, the probability of passing some other of the tests increases.

Let us find the upper bound of probability that an erroneous sample passes the agreement tests. Consider an error-prone set $\{\sigma_1, \sigma_2\}$ and two events: event A, when the condition $\mathbf{p}'_r{}^1 \approx \mathbf{p}'_r{}^2$ is satisfied, and event B, when the condition $\theta^i \approx \theta$ is satisfied.

Probability $Pr(A) < 1$ because two matches may not agree in reference position within the range of half bin size. The disagreement is possible along OX and OY axes. Therefore, $Pr(A)$ can be calculated by Eq. 9.40.

$$Pr(A) = Pr\left(x'_{ref}{}^1 \approx x'_{ref}{}^2\right) \cdot Pr\left(y'_{ref}{}^1 \approx y'_{ref}{}^2\right) \tag{9.40}$$

Rotation angle θ varies in the range $[0, 360]$ and does not depend on rotation of the single features θ^1 and θ^2. Therefore, the probability that θ agrees with θ^i within the range Δ is calculated via Eq. 9.41.

$$Pr(B) = 2\Delta/360 \tag{9.41}$$

Since A and B are independent, Eq. 9.42 is satisfied.

$$Pr(A, B) = Pr(A) \cdot Pr(B) \tag{9.42}$$

The probability of event C that random pair of matches $\{\sigma_i, \sigma_j\}$ is error-free equals the ratio of error-free pairs to the total number of the pairs in the set. If N is a number of all possible pairs, then $p^2 N$ of them are error-free and $(1 - p^2) \cdot N$ are errorneous. No more than $Pr(A, B)$ of last ones are able to pass the suggested tests, hence, we can estimate f in Eq. 9.39 via Eq. 9.43. Finally, Eq. 9.44 gives an estimation of the needed number of trials in the "RANSAC with tests"

$$f = Pr(C|A, B) \geq p^2 / [p^2 + Pr(A, B) \cdot (1 - p^2)] \tag{9.43}$$

$$k \leq \left\lceil \frac{\ln(1 - P)}{\ln[1 - p^2 / [p^2 + Pr(A, B) \cdot (1 - p^2)]]} \right\rceil \tag{9.44}$$

If we use the sizes of the Hough accumulator bins suggested in the previous subsection, we get[15] $Pr(A, B) = 3/64$ and $k \leq 2$ for $p = 0.5$, and $k \leq 3$ for $p = 0.3$. However, in our experiments on maximizing the objective function of the whole pipeline in dataset db3, $k = 1$ provides approximately the same number of correctly matched image pairs as any $k > 1$. That can be explained both by the "twin" effect in the Hough accumulator and the fact that Eq. 9.43 gives the upper bound of f, which actually is smaller. That means that the first sample satisfying the tests can be used to recover the transformation parameters.

Parameters T^+ are refined by the least squares. This is done only if there are more than 30% of inliers after RANSAC, otherwise the whole cluster is rejected without refining.

The least squares solution can be found in closed forms [38] by constructing the system of normal equations for the matches (u, v) that support T^+. The system can be written in a matrix form by Eq. 9.45, where $k, l, m = S \cos(\theta), n = S \sin(\theta)$ are unknowns.

$$
\begin{bmatrix} x_u & -y_u & 1 & 0 \\ y_u & x_u & 0 & 1 \\ & \cdots & & \end{bmatrix} \begin{bmatrix} m \\ n \\ k \\ l \end{bmatrix} = \begin{bmatrix} x_v \\ y_v \\ \cdot \\ \cdot \\ \cdot \end{bmatrix} \tag{9.45}
$$

Equation 9.40 can be written as $\mathbf{Ax} = \mathbf{b}$, and $\mathbf{x} = [\mathbf{A}^T\mathbf{A}]^{-1}\mathbf{A}^T\mathbf{b}$ is the least squares solution. Then we can compose the final set of parameters according to Eq. 9.46.

$$
T^{++} = \left\{ \theta^{++} = \text{acrtg}\,(n/m), \ S^{++} = n / \sin \theta^{++}, \ t_x^{++} = k, \ t_y^{++} = l \right\} \tag{9.46}
$$

The final outlier elimination checks that every match agrees with T^{++}. The bounds are $\sqrt{2}$ factor for scale and $15°$ for orientation. Translation is considered by comparing the predicted locations of the reference point in the test image. The two locations are obtained using the keypoint match parameters and recovered parameters T^{++}. The reference point location in the query image p_r is calculated according to Eq. 9.47, where Λ is a set of matches that support T^+. In other words, p_r is situated between the average position of keypoints in the cluster and verified keypoint location p_u.

---

[15]For two random variables $x_1, x_2 \sim U(0, 1)$, the probability to agree within 0.5 is $Pr(A) = Pr(|x_1-x_2| < 0.5) = 3/4$.

$$p_r = \left(\frac{1}{|\Lambda|} \left(\sum_{(w,l) \in M} p_w \right) + p_u \right) / 2 \qquad (9.47)$$

The threshold on the acceptable distance between the predicted locations of the reference point is also individual for different keypoints and is determined by Eq. 9.48, where α_1 and α_2 are predefined coefficients.

$$\left\| p_r \begin{bmatrix} S^i \cos \theta^i - m & -S^i \sin \theta^i + n \\ S^i \sin \theta^i - n & S^i \cos \theta^i - m \end{bmatrix} + \begin{bmatrix} t_x^i - k \\ t_y^i - l \end{bmatrix} \right\| < \alpha_1 \cdot \| p_r - p_{u^i} \| + \alpha_2 \qquad (9.48)$$

In our experiments $\alpha_1 = 0.3$, and α_2 equals 2% of test image diagonal length. The last two equations are useful because a transform fits better surfaces, where the keypoints of a cluster are situated, therefore the peripheral keypoints should have a discount. All matches that satisfy Eq. 9.48 in every verified cluster compose M_{final}.

9.7 Confidence Estimation

There is a probability that incorrect matches can accidently form a cluster that is not recognized as "false" using verification algorithm. Hence, it is very important to estimate the confidence level that the found cluster is correct. Naturally, large clusters resulted from few initial matches refer to a higher confidence. The combinatorics theory is applied to estimate probability $P(m \,|\, f)$ of the query image presence in pose m if the set f of k features has been matched between the query and test images [38]. The approach is applicable to different tasks and can be improved to consider the quality of the individual matches. Below there is an overview of the improved method.

The probability of getting k geometrically verified matches from n trials by an accident is given by the cumulative binomial distribution according to Eq. 9.49, where p is a probability of match accidently agreed with T^{++}.

$$P(f|\bar{m}) = \sum_{j=k}^{n} \binom{n}{j} p^j (1-p)^{n-j} \qquad (9.49)$$

Equation 9.49 can be efficiently calculated by many open source libraries (for example, Boost C++ libraries). An outline of the query image is projected onto the coordinate system of the test image using T^{++}. Number of keypoints that lie inside the projected outline gives n. That helps to consider that more textured surfaces are more likely to give accident false clusters [38].

Fig. 9.9 The nearest
neighbour ratio versus the
ratio of correct matches for
the BRISK and the SIFT
descriptors, respectively.
Experiments are carried out
with dataset db2

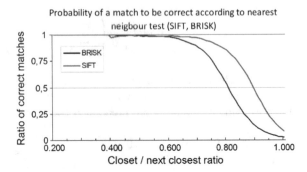

In work [38], $p = p_a$ is chosen according to Eq. 9.50, where l, r, s are the probabilities for a match to accidently satisfy the location, orientation, and scale constraints respectively.[16]

$$p_a = l\,r\,s \tag{9.50}$$

The values of l and r are related to the size of bins of the Hough accumulator, $s = 0.5$. For more information see [38]. Probability $P(f \mid m)$ can be calculated by Bayes rule (Eq. 9.51).

$$P(m|f) = \frac{P(f|m)\,P(m)}{P(f)} = \frac{P(f|m)\,P(m)}{P(f|m)\,P(m) + P(f|\bar{m})\,P(\bar{m})} \tag{9.51}$$

Equation 9.51 can be simplified to Eq. 9.52 by some task-invariant heuristics [38].

$$P(m|f) = \frac{0.01}{0.01 + P(f|\bar{m})} \tag{9.52}$$

In our experiments with dataset db2 according to Eq. 9.50 $p_a = 0.00256$. However, this equation does not take into account several aspects. We did brute force of p_a that maximizes recall given the false positive rate on dataset db2. Best p_a for SURF + SIFT (0.004) and SURF + BIRSK (0.001) is different. It can be explained by the fact that the BRISK descriptors are more distinctive, while the SIFT ones are more robust. Therefore, the SIFT is better description of the rough image correlations (such as similar configurations of contours) but that results in more false clusters. This effect is agreed with mutual keypoint relation. For example, a road and a river in the aerospace images can have approximately the same configuration that can result in a cluster of false matches that passes the verification procedure.

---

[16]Here we omit the probability of a keypoint to be matched to a different image from a dataset as we are concentrated on image-vs-image matching task instead of image-vs-database.

Figure 9.9 shows that there is a strong correlation of the nearest neighbour ratio and the probability of a match to be correct. This knowledge can be incorporated in Eq. 9.52 to reduce the number of false negatives.

If a solution is correct, then after the geometrical verification the average nearest neighbour ratio (nnr) must decrease. We calculate the mean nnr λ_m of the matches that remain after verification of the cluster. Then we estimate the probability of accidently selecting a match from set M_{prime} with nnr equal or greater than λ_m. This equals a ratio of matches with $\lambda^i > \lambda_m$ in the set (Eq. 9.53).

$$p_\lambda = \left| M_{NNR}\left(\lambda_m, M_{prime}\right) \right| / \left| M_{prime} \right| \qquad (9.53)$$

When a cluster is accidental, the parameters p_λ[17] and p_a are independent. Hence, we can multiply the two probabilities but according to our experiments a better result can be achieved by leveraging the impact of p_a relatively to p_λ (Eq. 9.54).

$$p = p_a \sqrt{2 p_\lambda} \qquad (9.54)$$

Parameter p_λ is doubled in Eq. 9.54 because its expected value for a false cluster is 0.5. Use of Eq. 9.54 instead of Eq. 9.50 reduces a number of false negatives by 25% with the same rate of false positives (dataset db2). This holds for different combinations of detectors and descriptors.

Threshold on $P(f \mid m)$ varies from task to task. In [38], solutions with $P(f \mid m) < 0.95$ are rejected, while in [11] a probability $P(f \mid m)$ should be greater than 0.98. In our experiments with dataset db2, the solutions with $P(f \mid m) > 0.9$ have the desired properties.

9.8   Application for Practical Tasks

The CSVA has been applied for the following practical tasks with minor adaptation. The matching 3D indoor scenes under strong viewpoint changes is discussed in Sect. 9.8.1. Section 9.8.2 provides the image retrieval for images of 3D scenes with the use of Bag of Words (BoWs) concept. The matching aerial and cosmic photographs with strong appearance changes caused by season, day-time, and viewpoint variation is considered in Sect. 9.8.3.

---

[17]Parameter p_λ is only approximation of probability to accidently select subset of matches with given nearest neighbor ratios.

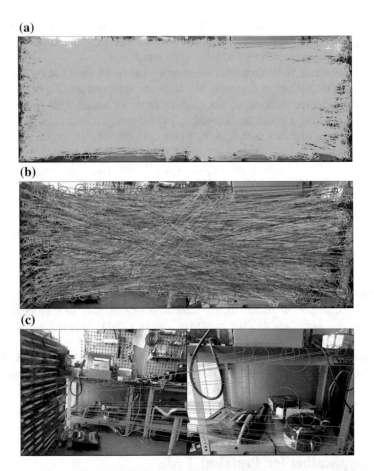

Fig. 9.10 Matching an image pair with the CSVA: **a** the result of the nearest neighbour search (set *M*) has 2.7% of correct matches, **b** set of matches M_{prime} after primary elimination has 5.7% of inliers, **c** after the CSVA there are 96.9% of inliers (In this case, a recall is only 0.34 but the dominant part of false negatives appears on the primary elimination step.)

9.8.1   Matching of Images of 3D Scenes

The CSVA as it is described in the previous sections was applied for the task of matching images of 3D scenes. According to the experiments with dataset db1, 5% of inliers is enough to accurately extract them from an initial match set. Figure 9.10 shows an example of matching an image pair with the SURF algorithm.

Dataset db3 was used for broader tests and comparison of the CSVA with the baseline methods. The results are represented in Table 9.3 (parameter Δ is an average distance of a keypoint in the test image from the etalon epipolar line).

Proprietary ERSP library was also used as a baseline (probably it implements a full pipeline version of the SIFT [54]). Noteworthy, different detectors and

Table 9.3 Results of matching of images from dataset db3 by different methods

Detector	Descriptor	Geometrical constrains	Number of image pairs with more than 50% of inliers	Ratio of correct matches, $|G(M)|/|M|$, %	Δ, pixel	Average number of matches, $|M_{final}|$
SURF	SIFT	Epipolar	205	53.8	106.9	285
ERSP Library			286	79	43.7	55
SIFT	SIFT	CSVA	300	82.9	32.7	56
SURF	SURF	CSVA	309	83.1	33.1	92
SURF	BRISK	CSVA	309	82.8	36.4	93
DENSE GRID	SURF	CSVA	301	81.8	36.8	485

descriptors give approximately the same results. This emphasizes the fact that the outlier elimination is a key factor for a successful matching. Results with the use of dense grid [58] show that the CSVA is applicable to dense detectors.[18] Figure 9.11 shows the successfully matched pairs.

The CSVA is applicable for non-static images that can be useful for tracking several objects at a time. Figure 9.12 demonstrates an example of non-static image matching.

9.8.2   Image Retrieval Using Bag of Words

The BoW approach is often applied in image retrieval tasks. The BoWs are the histograms of visual words in a view of the quantized descriptors obtained through a learning procedure. The compact representation of images allows savings in Random Access Memory (RAM) and Central Processing Unit (CPU) time. The major drawback of BoWs is that they do not support spatial information. Philbin et al. [59] proposed the "spatial re-ranking" of search results obtained by the straightforward BoW method. The approach was modified in [56] in order to improve a computational efficiency and provide more general spatial constraints. This was achieved by the application of the CSVA with minor adaptations. A description of the algorithm is below.

First of all, the BoWs are precomputed for every image in a database. The retrieval pipeline takes the following four steps:

1. Compute the BoW histogram for a query image.
2. Find N BoW histograms from the database closest to the query histogram.

---

[18]Interestingly, there are cases, when other detectors fail but dense grid works.

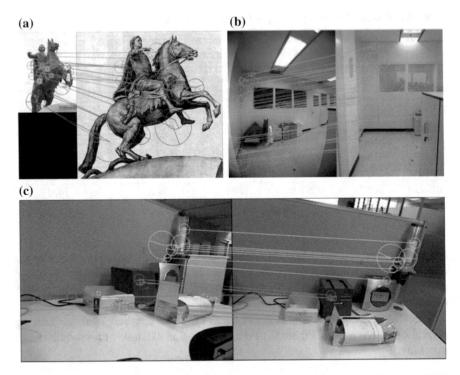

Fig. 9.11 Results of the CSVA application with different detectors and descriptors: **a** SIFT, **b** DENSE SURF, **c** SURF

3. For every ith of N best candidates, generate keypoint-to-keypoint matches of ith and the query image and eliminate outliers using the CSVA.
4. Re-rank search results according to the number of geometrically consistent matches.

Local matches are generated by interpreting corresponding histogram cells as a set of keypoint matches. Figure 9.13 illustrates this simple algorithm.

There are some adaptations of the CSVA (relatively to one used for 3D scene matching): no primary elimination algorithm is used ($M_{prime} = M$) and weighting scheme is applied. The latter is possible because the matches that arise from the histogram cells with multiple keypoints are objectively less reliable than those derived from single feature cells. We can assign $w(u, v)$ to every match according to Eq. 9.55, where n_1 and n_2 are the numbers of features assigned to a specific visual word in the query and test images, respectively.

$$w(u, v) = \min(n_1, n_2)/(n_1 \cdot n_2) \tag{9.55}$$

If we ignore the fact that corresponding features can be assigned to different visual words, then $w(u, v)$ equals the probability of match (u, v) to be correct. The

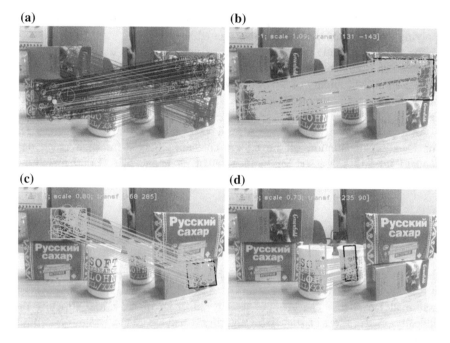

Fig. 9.12 Application of the CSVA to dynamic scene matching: **a** result of matching, **b** largest verified cluster, **c** second largest cluster, **d** third largest cluster

Fig. 9.13 Generation of local matches from two BoWs (histograms)

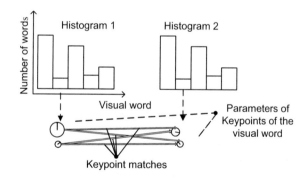

Table 9.4 Improvement bag of words with core algorithm	Method	Precision	Recall
	BoW	0.5	0.5
	BoW & CSVA, $N = 1$	0.95	0.48
	BoW & CSVA, $N = 20$	0.94	0.68

latter allows for the use of the improved version of the RANSAC [56],[19] as well as applying the weighted voting in the Hough table.

The tests were made with indoor image dataset db3. The query was always a picture taken from a different view angle and by a different camera (with strong lens distortions). A vocabulary of only 2000 visual words was used. Table 9.4 contains the results for the straightforward BoW method and improved method with the CSVA.

An important quality of the CSVA for a retrieval task is that it is especially fast. The C++ implementation of the CSVA consumes only 3 ms in average on a standard Intel i7 processor for verification of an image pair (1,000 matches in average).

9.8.3 Registration of Aerospace Images

The main challenge in aerospace image matching is the alterability of the Earth surface appearance, which is the result of seasonal, day-time, and anthropogenous (man-made) changes. The changes usually affect an image micro structure dramatically, hence, most of successful algorithms use the coarse image descriptions (e.g. set of lines, blobs, and angles) [60]. The keypoint-based approach is applicable to the task [40] because of the CSVA's ability to reject a huge number of outliers and rely on the features that are robust to the mentioned changes. The Earth surface shape can be approximated with a single similarity transform much more accurately than an arbitrary 3D scene. Therefore, only the biggest cluster in the Hough accumulator is verified without taking its actual size into account. We also use $M_{prime} = M_{ex} \cap M_{NNR}(0.98)$.

The conducted experiments with different combinations of detectors and descriptors led to dramatically different number of inliers. The SURF detector and the SIFT descriptor are the most appropriate to the aerospace matching task. The CSVA is robust in strong appearance changes caused by a season, day-time, and viewpoint variation (Fig. 9.14).

The anthropogenous (man-made) changes were also considered by matching the old photographs of city landscapes with modern ones. Figure 9.15 gives an interesting illustration of the CSVA properties.

Moreover, Synthetic Aperture Radar (SAR) and Infra-Red (IR) images were successfully matched with ones formed by optical sensors (Fig. 9.16).

The CSVA does not imply a constraint on the flatness of a surface and can register the dominant plane of 3D scenes. It is also able to match an aerial image against a map. Figure 9.17 illustrates some other matching examples for aerial images.

---

[19]In this case, the probability of picking a match equals $w(u, v)$. This can be efficiently implemented by cloning indexes that corresponds to more probable matches.

(a)   **(b)**   **(c)**

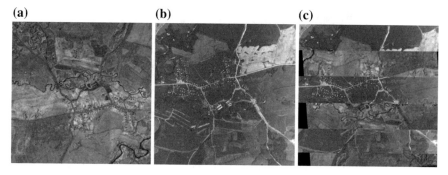

Fig. 9.14 Result of cosmic image registration under strong season changes: **a, b** image pair, **c** registration result

(a)

(b)

Fig. 9.15 Matching aerial images of New York of 1925 and 2014. Man-made changes have even affected the appearance of the coastline but the configuration of the bridges remains the same: **a** the inliers according to the CSVA describe the bridges, **b** registration based upon the inliers

We have calculated the recall and precision of the keypoint matches produced by the CSVA in dataset db2. The final precision is 0.99 and the final recall is 0.56. Figure 9.18 depicts the measures on different stages of the algorithm.

The CSVA is able to converge to a correct solution relying on a small subset from M (in some cases based on less than 0.1% of the whole multitude of matches).

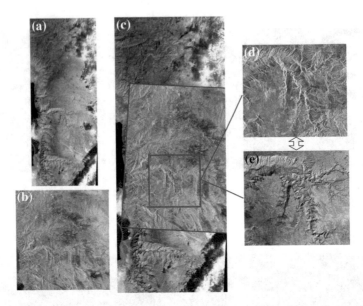

Fig. 9.16 Matching aerospace images of different sensors: **a** optical range image, **b** SAR image, **c** registration result, **d** and **e** human recognizable region of images

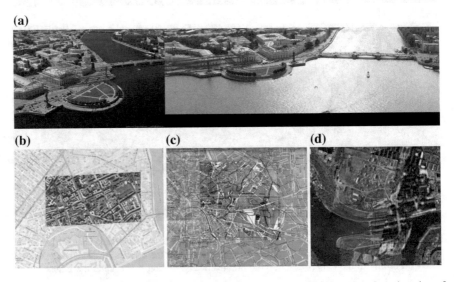

Fig. 9.17 Matching examples: **a** matching images taken from low-flying vehicle, **b** registration of cosmic image into the coordinate system of a map, **c** matching two maps of Moscow of different years (1961 and 1988), **d** consecutive matching aerial and cosmic images of Moscow taken in 1942, 1979, and 2014, respectively

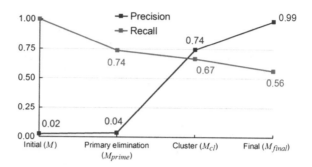

Fig. 9.18 Recall and precision on different stages of the CSVA

According to the experiments with dataset db2, 1% of inliers allows a successful registration with probability 0.7 (probability 0.95 is achieved with 3.5% of inliers). The CSVA fails in registering images, when knowledge about the objects is needed but works well, when a human can be confused by an immensely large number of details.

9.9   Conclusions

This chapter has introduced the CSVA as an algorithm for outlier elimination that can be used by developers and researches in different practical tasks with minor adaptation. The following applications are considered in the chapter: the matching of 3D indoor scenes (including non-static), retrieval images of 3D scenes, and matching aerial and cosmic photographs with strong appearance changes caused by season, day-time, and sensor variation. The CSVA is robust, computationally efficient and can be combined with different detectors and descriptors. Noteworthy, the algorithm is effective with binary descriptors (like the BRISK or the FREAK) that make it applicable for embedded systems with low computational resources. The CSVA allows successful matching images of 3D scenes with only 5% of inliers and matching aerial images with less than 1% of inliers. It can be recommended as a basic algorithm for various tasks.

Acknowledgements   This work was supported by the Ministry of Education and Science of the Russian Federation and the Government of the Russian Federation, Grant 074-U01. Special thanks to Shounan An and Vadim Lutsiv for useful comments and suggestions.

References

1. Lowe, D.: Automatic panoramic image stitching using invariant features. Int. J. Comput. Vis. **74**(1), 59–73 (2007)
2. Harris, C., Stephens, M.: A combined corner and edge detector. In: 4th Alvey Vision Conference, pp. 147–151 (1988)

3. Rosten, E., Drummond, T.: Machine learning for high-speed corner detection. In: Leonardis, A., Bischof, H., Pinz, A. (eds.) Computer Vision—ECCV 2006. LNCS, vol. 3951, pp. 430–443. Springer, Berlin, Heidelberg (2006)
4. Shi, J., Tomasi, C.: Good features to track. In: IEEE International Conference on Computer Vision and Pattern Recognition (CCVPR'1994), pp. 593–600 (1994)
5. Lowe, D., Matas, J., Chum, O., Urban, M., Pajdla, T.: Robust wide baseline stereo from maximally stable extremal regions. Image Vis. Comput. **22**(10), 761–767 (2004)
6. Agrawal, M., Konolige, K., Blas, M.R.: CenSurE. Center surround extremas for realtime feature detection and matching. In: 10th European Conference on Computer Vision (ECCV'2008), pp. 102–115 (2008)
7. Tuytelaars, T., Van Gool, L.: Wide baseline stereo based on local, affinely invariant regions. In: British Machine Vision Conference (BMVC'2000), pp. 412–422 (2000)
8. Rosten, E., Porter, R., Drummond, T.: Faster and better: a machine learning approach to corner detection. IEEE Trans. Pattern Anal. Mach. Intell. **32**(1), 105–119 (2010)
9. Hauagge, D., Snavely, N.: Image matching using local symmetry features. In: IEEE Conference on Computer Vision and Pattern Recognition (CVPR'2012), pp. 206–213 (2012)
10. Delponte, E., Noceti, N., Odone, F., Verri, A.: Spatio-temporal constraints for matching view-based descriptions of 3D objects. In: 8th International Workshop on Image Analysis for Multimedia Interactive Services (WIAMIS'2007), pp. 91–116 (2007)
11. Lowe, D.: Distinctive image features from scale-invariant keypoints. Int. J. Comput. Vis. **60**(2), 91–110 (2004)
12. Bay, H., Tuytelaars, T., Van Gool, L.: SURF: Speeded up robust features. J. Comput. Vis. Image Underst. **110**(3), 346–359 (2008)
13. Viola, P., Jones, M.: Rapid object detection using a boosted cascade of simple features. In: IEEE Computer Society Conference on Computer Vision and Pattern Recognition (CVPR'2001), pp. I-501–I-518 (2001)
14. Lutsiv, V., Malyshev, I., Pepelka, V.: Automatic fusion of the multiple sensor and multiple season images. In: Proceedings of the SPIE, vol. 4380, pp. 174–183 (2001)
15. Malashin, R., Peterson, M., Lutsiv, V.: Application of structural methods for stereo depth map improvement. In: AIP Conference Proceedings 1537, vol. 1, pp. 27–33 (2013)
16. Calonder, M., Lepetit, V., Strecha, C., Fua, P.: Brief: Binary robust independent elementary features. In: 11th European Conference on Computer Vision (ECCV'2010), Part IV, pp 778–792 (2010)
17. Rublee, E., Rabaud, V., Konolige, K., Bradski, G.: ORB: An efficient alternative to SIFT or SURF. In: International Conference on Computer Vision (ICCV'2011), pp. 2564–2571 (2011)
18. Leutenegger, S., Chli, M., Siegwart, R.: BRISK: Binary robust invariant scalable keypoints. In: International Conference on Computer Vision (ICCV'2011), pp. 2548–2555 (2011)
19. Ortiz, R., Vandergheynst, P.: FREAK. Fast retina keypoint. In: IEEE Conference on Computer Vision and Pattern Recognition (CVPR'2012), pp. 510–517 (2012)
20. Ke, Y., Sukthankar, R.: PCA-SIFT: A more distinctive representation for local image descriptors. In: IEEE Computer Society Conference on Computer Vision and Pattern Recognition (CVPR'2004), pp. 506–513 (2004)
21. Ojala, T., Pietikäinen, M., Harwood, D.: Performance evaluation of texture measures with classification based on Kullback discrimination of distributions. In: 12th IAPR International Conference on Pattern Recognition, vol. 1, pp. 582–585 (1994)
22. Ojala, T., Pietikäinen, M., Harwood, D.A.: Comparative study of texture measures with classification based on feature distributions. Pattern Recognit. **29**, 51–59 (1996)
23. Hafiane, A., Seetharaman, G., Zavidovique, B.: Median binary pattern for texture classification. In: 4th International Conference on Image Analysis and Recognition (ICIAR'2007), pp. 387–398 (2007)
24. He, C., Ahonen, T., Pietikäinen, M.: A Bayesian local binary pattern texture descriptor. In: 19th International Conference on Pattern Recognition (ICPR'2008), pp. 1–4 (2008)

25. Laptev, I., Lindeberg, T.: Local descriptors for spatio-temporal recognition. In: 1st International Conference on Spatial Coherence for Visual Motion Analysis (SCVMA'2004), pp. 91–103 (2004)
26. Laptev, I., Caputo, B., Schuldt, C., Lindeberg, T.: Local velocity-adapted motion events for spatio-temporal recognition. Comput. Vis. Image Underst. **108**(3), 207–229 (2007)
27. Koeser, K., Koch, R.: Perspectively invariant normal features. In: IEEE 11th International Conference on Computer Vision (ICCV'2007), pp. 1–8 (2007)
28. Wu, C., Clipp, B., Li, X., Frahm, J., Pollefeys, M.: 3D model matching with viewpoint-invariant patches (VIP). In: IEEE Computer Society Conference on Computer Vision and Pattern Recognition (CVPR'2008), pp. 1–8 (2008)
29. Mikolajczyk, K., Schmid, C.: Scale and affine invariant interest point detectors. Int. J. Comput. Vis. **60**(1), 63–86 (2004)
30. Averkin, A., Potapov, A., Lutsiv, V.: Construction of systems of local invariant image indicators based on the Fourier-Mellin transform. J. Opt. Technol. **77**(1), 28–32 (2010)
31. Cover, T., Hart, P.: Nearest neighbor pattern classification. IEEE Trans. Inf. Theory **13**, 21–27 (1967)
32. Muja, M., Lowe, D.: Fast approximate nearest neighbors with automatic algorithm configuration. In: International Conference on Computer Vision Theory and Applications (VISAPP'2009), pp. 331–340 (2009)
33. Muja, M., Lowe, D.: Fast matching of binary features. In: 19th Conference on Computer Robot Vision (CRV'2012), pp. 404–410 (2012)
34. Lutsiv, V., Potapov, A., Novikova, T., Lapina, N.: Hierarchical 3D structural matching in the aerospace photographs and indoor scenes. In: Proceedings of the SPIE, vol. 5807, pp. 455–466 (2005)
35. Peterson, M.: Clustering of a set of identified points on images of dynamic scenes, based on the principle of minimum description length. J. Opt. Technol. **77**, 701–706 (2010)
36. Fischler, M., Bolles, R.: Random sample consensus: a paradigm for model fitting with applications to image analysis and automated cartography. Commun. ACM **24**(6), 381–395 (1981)
37. Chum, O., Matas, J.: Matching with PROSAC—progressive sample consensus. In: IEEE Conference on Computer Vision and Pattern Recognition (CVPR'2005), pp. 1063–6919 (2005)
38. Lowe, D.: Local feature view clustering for 3D object recognition. In: IEEE Conference on Computer Vision and Pattern Recognition (CVPR'2001), pp. 682–688 (2001)
39. Malashin, R.: Methods of structural analysis of images of 3D scenes. Ph.D. thesis, National Research University of Information Technology (2014) (in Russian)
40. Malashin, R.: Matching of aerospace photographs with the use of local features. J. Phys.: Conf. Ser. **536**(1), 12–18 (2014)
41. Malashin, R.: Correlating images of three-dimensional scenes by clusterizing the correlated local attributes, using the Hough transform. J. Opt. Technol. **81**(6), 327–333 (2014)
42. Lowe, D.: Object recognition from local scale-invariant features. In: International Conference on Computer Vision (ICCV'1999), vol. 2, pp. 1150–1157 (1999)
43. Gordon, I., Lowe, D.: What and where: 3D object recognition with accurate pose. Toward Category-Level Object Recognit. **4270**, 67–82 (2006)
44. Brown, M., Lowe, D.: Recognising panoramas. In: International Conference on Computer Vision (ICCV'2003), pp. 1218–1225 (2003)
45. Li, Y., Wang, Y., Huang, W., Zhang, Z.: Automatic image stitching using SIFT. In: International Conference on Audio, Language and Image Proceedings (ICALIP'2008), pp. 568–571 (2008)
46. Scovanner, P., Ali, S., Shah, M.: A 3-dimensional sift descriptor and its application to action recognition. In: 15th ACM International Conference on Multimedia (MM'2007), pp. 357–360 (2007)
47. Niebles, J., Wang, H., Li, F.: Unsupervised learning of human action categories using spatial-temporal words. Int. J. Comput. Vis. **79**(3), 299–318 (2006)

48. Bicego, M., Lagorio, A., Grosso, E., Tistarelli, M.: On the use of SIFT features for face authentication. In: Conference on Computer Vision and Pattern Recognition Workshop (CVPRW'2006), pp. 35–35 (2006)
49. Luo, J., Ma, Y., Takikawa, E., Lao, S., Kawade, M., Lu, B.-L.: Person-specific SIFT features for face recognition. In: IEEE International Conference on Acoustics, Speech and Signal Processing (ICASSP'2007), pp. 593–596 (2007)
50. Park, U., Pankanti, S., Jain, A.: Fingerprint verification using SIFT features. In: SPIE Defense and Security Symposium, vol. 6944, no. 1, 69440K-1–69440K-9 (2008)
51. Shuai, X., Zhang, C., Hao, P.: Fingerprint indexing based on composite set of reduced SIFT features. In: 19th International Conference on Pattern Recognition (ICPR'2008), pp. 1–4 (2008)
52. Lutsiv, V.: Automatic Image Analysis: An Object-Independent Structural Approach. Lambert Academic Publishing, Saarbrücken, Germany (2011) (in Russian)
53. Morel, J., Yu, G.: ASIFT: A new framework for fully affine invariant image comparison. SIAM J. Image Sci. 438–469 (2009)
54. Brown, M., Lowe, D.: Invariant features from interest point groups. In: British Machine Vision Conference (BMVC'2002), pp. 656–665 (2002)
55. Ballard, D.H.: Generalizing the Hough transform to detect arbitrary shapes. Pattern Recognit. 13(2), 111–122 (1981)
56. Malashin, R.: Image retrieval with the use of bag of words and structural analysis. J. Phys: Conf. Ser. 735(1), 12–16 (2016)
57. Isack, H., Boykov, Y.: Energy-based geometric multi-model fitting. Int. J. Comput. Vis. 97 (2), 23–147 (2012)
58. Fei-Fei, L.: Bayesian hierarchical model for learning natural scene categories. In: IEEE Computer Society Conference on Computer Vision and Pattern Recognition (CVPR'2005), pp. 524–521 (2005)
59. Philbin, J., Chum, O., Isard, M., Sivic, J., Zisserman, A.: Object retrieval with large vocabularies and fast spatial matching. In: IEEE Conference on Computer Vision and Pattern Recognition (CVPR'2007), pp. 351–362 (2007)
60. Lutsiv, V., Malyshev, I., Potapov, A.: Hierarchical structural matching algorithms for registration of aerospace images. In: Proceedings of SPIE, vol. 5238, pp. 164–175 (2003)

Chapter 10
Strip-Invariants of Double-Sided Matrix Transformation of Images

Leonid Mironovsky and Valery Slaev

Abstract In this chapter, a strip-method suitable for reducing pulse interference in communication channels, cryptography, steganography, and other applications is considered. The invariants to fragmentation and double-sided matrix transformation of images provide the noise immunity and transmission security. The chapter contains new definitions of invariants, as well as invariant images of the first and second types. Moreover, tasks of analyzing and synthesizing both invariants and corresponding transformation matrices are set forth too. The criteria of their existences are derived and methods for creation of invariant images using eigenvectors of transforming matrices are proposed. Some cases of complex and multiple eigenvalues of a direct transformation matrix are considered. It was proposed to solve the problem of finding the matrices of direct and inverse transformations by means of a given set of invariant images. The solution of the task of arraying the matrix of double-sided transformation according to a given set of invariant images is suggested.

Keywords Scale · Rotatory · Negative (inverse) strip-invariants
Double-sided matrices image transformation · Analysis and synthesis
of invariant images

L. Mironovsky
St. Petersburg State University of Aerocosmic Instrumentation,
67, Bol. Morskaya Street, St. Petersburg 190000, Russian Federation
e-mail: miron@aanet.ru

V. Slaev (✉)
D.I. Mendeleyev Research Institute for Metrology, 19, Moskovsky Avenue,
St. Petersburg 190005, Russian Federation
e-mail: v.a.slaev@vniim.ru

© Springer International Publishing AG 2018
M.N. Favorskaya and L.C. Jain (eds.), *Computer Vision in Control Systems-3*,
Intelligent Systems Reference Library 135, https://doi.org/10.1007/978-3-319-67516-9_10

10.1   Introduction

It is reasonable that the strip-method is merely one of the methods used for
increasing the accuracy of signal and image transmission over communication
channels. A great number of publications are devoted to issues of raising the
noise-resistance of information transmission systems [1–35]. It is also necessary to
mention some works in the adjacent fields of activities, such as the researches in the
cluster systems of message transmission and linear pre-distortion of signals made
by Russian researches Ageyev, Babanov, Ignat'ev, Lebedev, Marigodov, Suslonov,
Tsibakov, Yaroslavsky, and others, methods of redundant variables, image
denoising, and linear transformation and block coding of signals and images, which
were done by American researches Abreu, Boyle, Buades, Costas, Chan, Clauset,
Dabov, DiGesu, Garnett, Huang, Kam, Lang, Leith, Najeer, Newman, Pierce,
Upatnieks [3, 4, 9, 10, 18, 24–35], and others.

Thus, the noise control based on introduction of pre-distortions at the stage of
signal transmission and optimal processing at the stage of signal reception is widely
used in information transmission systems. However, the majority of works deal
with the pre-distortion methods and correction using a root-mean-square criterion,
whereas the methods satisfying the requirements for optimizing the information
transmission systems with the help of a minimax criterion have been developed to a
significantly lesser degree. Therefore, perhaps, it would be more useful to develop
and study novel methods for blanking the pulse interference, which are supported
by use of the minimax criterion and modern computer processing for images.

The author contribution deals with the following research of a strip-method of
transforming images and signals, which was invented in 1980s [13], and also
careful examination of its properties, including a solution of the invariants task.

The issues how to increase the interference immunity, transmission speed as well
as to provide for the quality of a transmitted image are appeared during the image
transmission over a telecommunication channel. With that goal in a view at the
transmitting end of the channel, the image is subjected to some transformation
(pre-emphasis, coding and encoding, change of a frequency range, filtration, etc.)
and at the receiving end the reverse transformation is carried out. An appropriate
procedure is shown in Fig. 10.1, where the image transmitted before and after
transformation is indicated by X and Y, respectively, and the received image

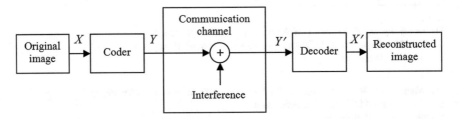

Fig. 10.1 Transmission of images over a communication channel

distorted within the process of transmission and then result of its restoration is denoted by Y' and X', respectively.

In the communication theory, there are a great number of versions of direct and inverse transformations. In current research, the double-sided matrix transformation described called as a strip-transformation is considered. In accordance with this, the original image is divided into the similar rectangular fragments (just as it is made in riddles of the puzzle type), which then are subjected to mixing and linear combining. Formally, the procedure of encoding is performed by the way of double-sided multiplication of the rectangular matrix, comprised by the image fragments, by numerical matrices (for example, the Hadamard matrices). As a result, each from the fragments of the image transformed carries information about all fragments of the original image.

At the receiving end, the reconstruction of an image is realized by the way of applying the inverse transformation that consists in the double-sided multiplication of the received image by the inverse matrices. Similar transformations can be useful for reducing pulse interferences, acting in the communication channels, crypto graphy, steganography, and other applications.

From the theoretical point of view, the issue, concerning the invariants of such transformation, i.e. the images that do not change their appearance as a consequence of the double-sided matrix transformation, is the subject of interest. Moreover, this issue is important in practice for the interference immunity (the presence of interferences, when they do not decay in the process of image reconstruction) and signal hiding.

Results in this field, which are described in researches [12–17], are related to a particular case of the double-sided matrix transformation, when in respect of its matrices some serious restrictions were applied (orthogonality, symmetry). In this chapter, a more general case, for which criteria of invariant image existence are established, is analyzed. The scale, rotation, and negative (inverse) strip-invariants are investigated and the methods of finding them are proposed. Moreover, the task of arraying the double-sided transformation on the basis of a given set of invariant images is solved.

The rest of the chapter is organized as follows. In Sect. 10.2, the double-sided matrix transformation of images is described in detail. Invariant images are maintained in Sect. 10.3 as a problem statement. The detailed analysis of strip-invariant images is carried out in Sect. 10.4. Section 10.5 provides a getting of invariant images using the eigenvectors of matrices, while Sect. 10.6 includes a synthesis of matrices by given invariant images. Section 10.7 is conclusions.

10.2 Double-Sided Matrix Transformation of Images

Let us to describe the procedure of the double-sided matrix transformation of images following the instructions from researches [1, 13, 15]. At the stage of encoding (see Fig. 10.1), the original image is divided into $m \times n$ similar rectangular fragments. As a result of this division, a block $m \times n$ matrix X is formed. Elements of this block matrix are the fragments of the original image. Then the matrix X is subjected to linear transformation (encoding) according to Eq. 10.1, where A and B are the nonsingular square matrices of dimensions $n \times n$ and m m, respectively.

$$Y = BXA \qquad (10.1)$$

Multiplication in Eq. 10.1 is performed according to the usual rule, i.e. "a line by a column". This action will result in a block matrix Y, every fragment of which represents a linear combination of the matrix X fragments. At multiplication of the fragment by a number, each of its pixels is multiplied by this number and when two fragments are summed up, the elementwise addition of their pixels takes place. It should be noticed that the double-sided matrix transformation of the image X does not accompanied by introducing the redundancy because the image Y will have the same dimensions just as the image X has. Moreover, if the matrices A and B are orthogonal, then Eq. 10.1 will be isometric, i.e. transformation preserving energy. Then the image Y is transmitted over the communication channel, where it can be distorted by interferences. The image obtained at the receiving end of the channel is subjected to the inverse transformation (decoding) provided by Eq. 10.2.

$$X' = B^{-1}Y'A^{-1} \qquad (10.2)$$

An error of the reconstructed image is determined:

$$X' - X = B^{-1}(Y' - Y)A^{-1}.$$

From this it follows that pulse interferences of the communication channel, which distort one of the fragments of the image Y being transmitted will be distributed over the whole reconstructed image X'. If the matrix elements A^{-1} and B^{-1} are close by their nominal absolute values, then this will bring about reducing the interferences due to spread them over the whole display area of the image.

The simplest version of transformation indicated by Eq. 10.1 is obtained using permutation matrices A and B. Then the multiplication XA will result in the permutation n of vertical bands of the original image and the multiplication BX will provide for the permutation m of horizontal bands. The two-sided multiplication BXA will result in some permutation mn of the image fragments. Something similar to this takes place in riddles of the puzzle type. However, the total number of possible permutations is $(mn)!$ And in the case under consideration it is only $m!n!$, which is noticeably less.

A convenient way from the various points of view is the use of symmetric normalized Hadamard matrices as the matrices A and B [1]. In this case $A^{-1} = A$ and $B^{-1} = B$, each fragment of the matrix Y represents a sum of the matrix X fragments taken with the sign "plus" or "minus". The amplitude of any pulse interference that distorts one of the fragments of the image Y being transmitted will be reduced by $\sqrt{mn}$. The similar reduction, though to a lesser extent, will take place for multiple pulse interferences.

A simple particular case can be obtained by taking $m = n$. At that, the matrix X will be quadratic and the matrices A and B will be of the same dimensions. Furthermore, in case of $A = B$ the transformation described by Eq. 10.1 will take the form $Y = AXA$. If the matrix A is symmetric and orthogonal, then the inverse transformation will be described by a similar formula $Y = AYA$.

Example 1 Let the image of the letter "T" be taken as an original image (Fig. 10.2a). To perform the double-sided matrix transformation at $m = n = 4$ the image needs to be divided into four parts along the vertical line and into four parts along the horizontal line. Let the obtained block matrix X be subjected to the double-sided strip-transformation with the help of the Hadamard matrix of the fourth order, taking $A = B$:

$$Y = AXA/4 \quad A = \begin{bmatrix} -1 & 1 & 1 & 1 \\ 1 & -1 & 1 & 1 \\ 1 & 1 & -1 & 1 \\ 1 & 1 & 1 & -1 \end{bmatrix}.$$

The transformed image Y is shown in Fig. 10.2b. Each of its 16 fragments carries information about all fragments of the original image.

Let us consider the case, when at the time of transmission the failure of the first fragment of the image transformed took place (instead of it the zero signal was obtained). The reconstructed image calculated by the formula $X' = AY'A/4$ is shown in Fig. 10.3a. It is seen that the distortion is evenly distributed over the whole image having made its quality somewhat worse. Figure 10.3b illustrates the form that the image would have obtained at the receiving end of the channel in the presence of a similar interference if the strip-method had not been applied (a user would have obtained the letter "Γ" instead of the letter "T").

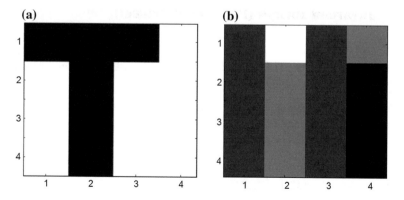

Fig. 10.2 Images of the letter "T": **a** original image, **b** transformed image

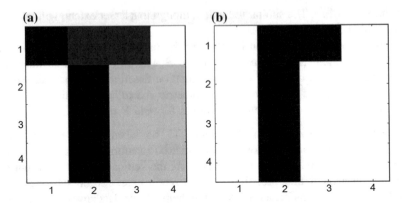

Fig. 10.3 Examples of image processing: **a** image obtained by the strip-transformation, **b** image without the strip-transformation

Example 2 Let a black-and-white photo of a bunch of flowers be used as the original image (Fig. 10.4a) and after dividing it into $4 \times 4 = 16$ portions the two-sided strip-transformation with the Hadamard matrix of the 4th order (Fig. 10.4b) be performed.

Let one-multiple interference in the form of a black square with dimensions 30×30 pixels (it is located in the second diagonal fragment in Fig. 10.4b) be added to the above. After inverse transformation, this interference became evenly distributed over all fragments and almost invisible on the reconstructed image (Fig. 10.4c). In Fig. 10.4d, it is possible to see the form that the image would have obtained at the receiving end of the channel if the strip-method had not been applied in the presence of the same interference.

The given examples show that the double-sided matrix transformation of images is sufficiently effective even if the transformation matrix is of a small order.

10.3   Invariant Images (Problem Statement)

Let the original image be divided into $m \times n$ equal rectangle fragments and double-sided matrix transformation of this image be performed in accordance with Eq. 10.1.

Definition 1 The image X is called as scale-invariant for the given matrix A and B provided it satisfies the relationship of the type:

$$BXA = \lambda X, \tag{10.3}$$

where $\lambda \in R$ is some scale factor.

(a)   (b)

(c)   (d)

Fig. 10.4 Examples of image processing: **a** image before transformation, **b** image transmitted, **c** image obtained with transformation, **d** image received without transformation

Let such images be called *invariant images of the first type*.

Thus, the scale-invariant image is transferred by double-sided matrix transformation into a similar image that differs only by brightness. Finding the invariant images for the given matrices A and B is an important task. If the interference in the communication channel coincides with the invariant image of the transformation used, then at inverse transformation it becomes unchangeable and it is not possible to achieve the effect of "smearing" the interference over the image. However, if the original image coincides with the invariant image, then at transferring it the effect of transmission security will not be achieved tough it can be useful, e.g. in solving problems of cryptography. On the other hand, the invariant images can appear to be useful for the purposes of steganography, when they are used as "containers" for hidden message transmission. Bellow there is an example of finding invariant images.

Example 3 Let $m = 2$, $n = 3$, i.e. the image is divided into 6 parts as it is shown in Fig. 10.5.

It is required to find invariant image X provided matrices A and B of the strip-transformation have the form:

Fig. 10.5 Fragmentation of images

x_1	x_2	x_3
x_4	x_5	x_6

$$A = \begin{bmatrix} -1 & 2 & 2 \\ 2 & -1 & 2 \\ 2 & 2 & -1 \end{bmatrix} \quad B = \begin{bmatrix} 1 & 1 \\ 1 & -1 \end{bmatrix}.$$

Notice that both matrices are symmetric and orthogonal. It is possible to normalize them dividing into $\|A\| = 3$ and $\|B\| = \sqrt{2}$, respectively. In order to find invariant images X, it is necessary to solve the following equation: $BXA = 3\sqrt{2}X$ (let $\lambda = \|A\| \cdot \|B\| = 3\sqrt{2}$ taken). Disclosing this equation, a homogeneous system of linear equations is obtained.

$$2x_1 - x_2 + 2x_3 - 2x_4 + x_5 - 2x_6 - 3\sqrt{2}x_5 = 0$$
$$2x_1 - x_2 + 2x_3 + 2x_4 - x_5 + 2x_6 - 3\sqrt{2}x_2 = 0$$
$$2x_1 + 2x_2 - x_3 - 2x_4 - 2x_5 + x_6 - 3\sqrt{2}x_6 = 0$$
$$2x_1 + 2x_2 - x_3 + 2x_4 + 2x_5 - x_6 - 3\sqrt{2}x_3 = 0$$
$$2x_2 - x_1 + 2x_3 - x_4 + 2x_5 + 2x_6 - 3\sqrt{2}x_1 = 0$$
$$2x_2 - x_1 + 2x_3 + x_4 - 2x_5 - 2x_6 - 3\sqrt{2}x_4 = 0$$

Let us write out the Jacobean of this system.

$$J = \begin{bmatrix} -1-3\sqrt{2} & 2 & 2 & -1 & 2 & 2 \\ 2 & -1-3\sqrt{2} & -1 & 2 & -1 & 2 \\ 2 & 2 & -1-3\sqrt{2} & 2 & 2 & -1 \\ -1 & 2 & 2 & 1-3\sqrt{2} & -2 & -2 \\ 2 & -1 & 2 & -2 & 1-3\sqrt{2} & -2 \\ 2 & 2 & -1 & -2 & -2 & 1-3\sqrt{2} \end{bmatrix}$$

The rank of this matrix is equal to 3. Therefore, its solution will contain three free parameters. For example, choosing them as x_2, x_3, x_6, the following solution is obtained.

$$x_1 = -x_2 + \frac{1}{2}x_3 + \frac{3}{4}\sqrt{2}x_6 \quad x_2 = -x_2 + \frac{1}{2}x_3 + \frac{3}{4}\sqrt{2}x_6$$

$$x_4 = -x_2 - x_3 - \frac{1}{2}x_6 + \sqrt{2}x_2 - \frac{1}{4}\sqrt{2}x_3 - \frac{3}{4}\sqrt{2}x_6$$

$$x_5 = -x_2 + x_3 + x_6 - \sqrt{2}x_2 + \sqrt{2}x_3$$

Thus, the set of root images is three-dimensional. Particularly, assuming that $x_2 = x_3 = x_6 = 1$, we get the following results.

$$x_1 = -\tfrac{1}{2} + \tfrac{3\sqrt{2}}{2} \qquad x_4 = -\tfrac{1}{2} \qquad x_5 = 1$$
$$x_1 = 1.621320343 \qquad x_4 = -0.5 \qquad x_5 = 1$$

The matrix of the corresponding root image has the form:

$$X_n = \begin{bmatrix} 1.6213 & 1 & 1 \\ -0.5 & 1 & 1 \end{bmatrix}.$$

This image is shown in Fig. 10.6.

Example 4 Now let us slightly change the matrix B by rearrangement of its rows:

$$B_2 = \begin{bmatrix} 1 & -1 \\ 1 & 1 \end{bmatrix}.$$

As a result, the matrix ceased to be symmetric and its proper numbers became complex: $\lambda \pm i$. Now the strip-transformation with the matrices A and B does not contain any root images at any k. Let us check it up solving the matrix equation $BXA = kX$ relative to matrix X.

$$B * X * A - k * X = 0$$

$$2x_1 - x_2 + 2x_3 - 2x_4 + x_5 - 2x_6 - kx_2 = 0$$
$$2x_1 - x_2 + 2x_3 + 2x_4 - x_5 + 2x_6 - kx_5 = 0$$
$$2x_1 + 2x_2 - x_3 - 2x_4 - 2x_5 + x_6 - kx_3 = 0$$
$$2x_1 + 2x_2 - x_3 + 2x_4 + 2x_5 - x_6 - kx_6 = 0$$
$$2x_2 - x_1 + 2x_3 - x_4 + 2x_5 + 2x_6 - kx_4 = 0$$
$$2x_2 - x_1 + 2x_3 + x_4 - 2x_5 - 2x_6 - kx_1 = 0$$

Let us write out the Jacobean of this system that has the following view.

Fig. 10.6 Fragmentation of images

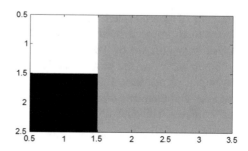

$$J = \begin{bmatrix} -1-k & 2 & 2 & 1 & -2 & -2 \\ 2 & -1-k & 2 & -2 & 1 & -2 \\ 2 & 2 & -1-k & -2 & -2 & 1 \\ -1 & 2 & 2 & -1-k & 2 & 2 \\ 2 & -1 & 2 & 2 & -1-k & 2 \\ 2 & 2 & -1 & 2 & 2 & -1-k \end{bmatrix}$$

Making its determinant equal to zero, a characteristic equation for k is obtained:

$$k^6 + 6k^5 + 18k^4 + 324k^2 + 1944k + 5832 = 0.$$

It is factorized as:

$$\left(k^2 - 6k + 18\right)\left(k^2 + 6k + 18\right)^2 = 0.$$

Let us drop the square and make the transformation:

$$\left(k^2 + 6k + 18\right)\left(k^2 - 6k + 18\right) = 324 + k^4 = 0.$$

It is possible to see that the equation obtained has only complex solutions:

$$-3+3i \quad -3-3i \quad 3+3i \quad 3-3i.$$

This is a root equation. Thus, it is possible to get it as a characteristic equation of a Kronecker product of matrices $\bar{A} = A^T \otimes B$. The concept of invariant images can be expanded if along with scaling of the image the latter is allowed to rotate by 180° in the plane of the picture.

Definition 2 Image X is called as a *rotationally invariant* one for the given pair of matrices A and B if such image satisfies the relation (ratio, correlation) of the type:

$$BXA = \lambda R(P_m X P_n), \tag{10.4}$$

where R is an operator of elementwise rotation of fragments by 180°, P_m and P_n designates anti-diagonal permutation matrices that are obtained by plane rotation of unity matrices by 90°.

Such images will be called the *invariant images of the second type*. Thus, with the help of double-sided transformation the rotationally invariant image is converted to an image that after rotation of each fragment by 180° coincides with the scaled image rotated by 180°.

When analyzing the problems relating to the existence of invariant images for a given pair of matrices A and B and constructing invariant images, it is possible to consider that elements of matrix X are not the fragments of the original image but real numbers. This fact simplifies the analysis and makes it possible to speak about invariant matrices X in Definitions 1 and 2 but not about invariant images. At that, Eq. 10.5 is replaced with a simpler one:

$$BXA = \lambda P_m X P_n. \tag{10.5}$$

The procedure is equivalent to matrix X rotation by $180°$, in packet MATLAB it can be executed with help of command rot90(X, 2).

As to the invariant matrices let the following tasks be set:

- *Task* 1 (criteria of existence). There are given matrices A and B. It is required to find out whether they have scale-invariant or rotationally-invariant matrices X.
- *Task* 2 (invariant matrices finding). For matrices A and B given, it is required to find all scale-invariant and rotationally-invariant matrices X, satisfying Eqs. 10.4–10.5.
- *Task* 3. Construction of matrices A and B, having a set of scale-invariant or rotationally-invariant matrices X.
- *Task* 4. Development of methods for constructing matrices A and B that have no scale-invariant or rotationally-invariant matrices X.

The first two tasks are connected with an analysis of invariant properties of double-sided matrix transformation of images. The remaining tasks are tasks of the synthesis of double-sided matrix transformation with properties given.

Let us start to consider the tasks of the simplest case when $m = n$. If it is assumed that $A = B$, then Eq. 10.1 takes a form $Y = AXA$. If matrix A is symmetric and orthogonal, then the inverse transformation will be described by the similar formula.

Comments. In researches [12–14], devoting to the two-sided matrix strip transformation, two versions of one-sided strip-transformation (the left and right ones) are considered. They are described as:

$$Z = BX \quad \text{and} \quad Z = XA,$$

where X is a result of column or line vectorization of image X, B, and A are the square matrix of the strip transformation being used.

In this case, the search for scale invariant images is reduced to the solution of the matrix equation $BX = kX$ or $XA = kX$, where k is a constant. We come to the well-known algebraic problem of eigenvalues, the solution of which is the left and right eigenvectors of the matrices B and A. In the case of two-sided matrix strip-transformation, we obtain a much more complex task that has separate features common with both the generalized problem of eigenvalues and theory of solvability of the Sylvester matrix equations.

10.4 Analysis of Strip-Invariant Images

In this section, the existence criteria of strip-invariant images for case $m = n$, $A = B$, renunciation of matrices equality condition $A = B$, and abandonment of the conditions of symmetry and orthogonality of matrices A and B are discussed in Sects. 10.4.1–10.4.3, respectively.

10.4.1 Existence Criteria of Strip-Invariant Images for Case m = n, A = B

Let the original image X be divided into $n \times n$ similar segments. Let us perform the double-sided transformation of this image according to Eq. 10.6, where A is an orthogonal matrix, e.g. the normalized Hadamard matrix.

$$Y = AXA \qquad (10.6)$$

Further raise the task to find the images, which are invariant with respect to transformation (Eq. 10.6), i.e. which turns them into the same image (with the accuracy to a constant factor), provided by Eq. 10.7.

$$AXA = \lambda A \qquad (10.7)$$

Such images will be called root images of transformation (Eq. 10.6). If the noise in the communication channel coincides with a root image of the transformation applied, then in the inverse transformation it will remain unchanged and attain the effect of "smearing" the noise over the image.

Root matrix problem: find all matrices X and those of number λ, satisfying relationship (Eq. 10.8) for a given non-singular matrix A.

Such a formulation resembles the algebra problem of finding eigenvalues of the matrix A. Therefore, the pair (λ, X) is called the *root value* and *root matrix* of the matrix A. Below the decision of the set problem for the orthogonal matrices, a matrix A is given. At first, determine the root value λ. In Eq. 10.8, let us move to the determinants:

$$|A| \cdot |X| \cdot |A| = \lambda^n |X| \Rightarrow \lambda^n = |A|^2.$$

Provided the matrix A is orthogonal, then $|A| = \pm 1$ and $\lambda^n = 1$. Thus, we have n root numbers $\lambda_1, \ldots, \lambda_n$ equal to the roots of unity. All of them are located on a unit circle in a plane of complex numbers. Among them only two numbers are real: $\lambda = 1$ (the image does not change) and $\lambda = -1$ (the image is replaced by a negative). Any other real roots do not exist. Moreover, if the matrix A is orthogonal and symmetrical (below the matrices of such a type will be considered), then the whole set of root numbers become exhausted by the value $\lambda = \pm 1$. This follows from the following theorem [8].

Theorem 1 *The minimal polynomial of an orthogonal symmetrical matrix $A \neq \pm I$, irrespective of its dimension, has the form $\lambda^2 - 1$.*

Proof To prove this theorem, it is sufficient to note that all eigenvalues are located on the unit circle and all eigenvalues of the symmetrical matrix are real.

Taken together all these conditions mean that a part of the eigenvalues of the matrix A are equal to 1, the remaining ones are equal to -1. Since the symmetrical matrices are diagonalizable, the minimal polynomial is equal to $(\lambda - 1)(\lambda + 1)$ with the exception of the identity matrix I (all its eigenvalues are equal to 1) and matrix I (all its eigenvalues are equal to -1). The minimal polynomials of these two matrices are equal to $\lambda - 1$ and $\lambda + 1$, respectively.

It should also be said that the amount of positive and negative eigenvalues in each of the symmetrical Hadamard matrices is the same and equal to $n/2$. The eigenvectors corresponding to them are formed two orthogonal complimentary subspace of the dimension $n/2$ each.

Let us move to find the root matrices of transformation (Eq. 10.6). From the above proved it follows that it is sufficient to consider two cases, following from Eq. 10.7 at $\lambda = +1$ and at $\lambda = -1$:

$$AXA = X \quad AXA = -X.$$

Let us consider each of them, beginning from the positive root matrices of transformation.

Case $\lambda = 1$. If the matrix A is orthonormal and symmetrical, then $A = A^{-1}$ and Eq. 10.1 take a form $AX = XA$. This means that the matrices A and X are commutative. Hence, their eigenvectors coincide, and the matrix X can be represented in the form of the polynomial from the matrix A provided by Eq. 10.8, where c_i is the arbitrary real numbers, m is an order of the minimal polynomial of the matrix A.

$$X = c_1 I + c_2 A + c_3 A^2 + \cdots + c_m A^{m-1} \qquad (10.8)$$

According to Theorem 1, for the orthogonal matrices we have $m = 2$, therefore, for such matrices Eq. 10.8 takes a form of Eq. 10.9.

$$X = c_1 I + c_2 A \qquad (10.9)$$

The corresponding root matrices represent by themselves a linear combination of the identity matrix and matrix A.

Example 5 Let A be the Hadamard matrix of the 4th order

$$A = \begin{bmatrix} -1 & 1 & 1 & 1 \\ 1 & -1 & 1 & 1 \\ 1 & 1 & -1 & 1 \\ 1 & 1 & 1 & -1 \end{bmatrix}.$$

Then in accordance with Eq. 10.9 we have

$$X = c_1 \begin{bmatrix} 1 & 0 & 0 & 0 \\ 0 & 1 & 0 & 0 \\ 0 & 0 & 1 & 0 \\ 0 & 0 & 0 & 1 \end{bmatrix} + c_2 \begin{bmatrix} -1 & 1 & 1 & 1 \\ 1 & -1 & 1 & 1 \\ 1 & 1 & -1 & 1 \\ 1 & 1 & 1 & -1 \end{bmatrix} = c_2 \begin{bmatrix} c & 1 & 1 & 1 \\ 1 & c & 1 & 1 \\ 1 & 1 & c & 1 \\ 1 & 1 & 1 & c \end{bmatrix}$$

$$c = \frac{c_1 - c_2}{c_2}.$$

For another version of the Hadamard matrix of the 4th order

$$A = \begin{bmatrix} -1 & 1 & 1 & 1 \\ 1 & 1 & -1 & -1 \\ 1 & -1 & -1 & 1 \\ 1 & -1 & 1 & -1 \end{bmatrix}$$

we obtain

$$X = c_1 E + c_2 \begin{bmatrix} -1 & 1 & 1 & 1 \\ 1 & 1 & -1 & -1 \\ 1 & -1 & -1 & 1 \\ 1 & -1 & 1 & -1 \end{bmatrix} = \begin{bmatrix} c_1 + c_2 & c_2 & c_2 & c_2 \\ c_2 & c_1 + c_2 & -c_2 & -c_2 \\ c_2 & -c_2 & c_1 - c_2 & -c_2 \\ c_2 & -c_2 & c_2 & c_1 - c_2 \end{bmatrix}.$$

Both formulae describe a two-parameter set of root matrices invariant in respect to transformation (Eq. 10.7). Below there are given two more examples of signals and noises, which are invariant in respect to the double-sided strip-transformation.

Example 6 In Fig. 10.7a, the "diagonal" noise is shown that distorts the diagonal blocks of the image transmitted over the communication channel. The identity matrix of the 16th order corresponds to this noise. After the double-sided transformation of the noise at the receiving end, same noise will be obtained (see Fig. 10.7b).

Example 7 In Fig. 10.8, the image is shown, the structure of which corresponds to the Hadamard matrix of the 8th order:

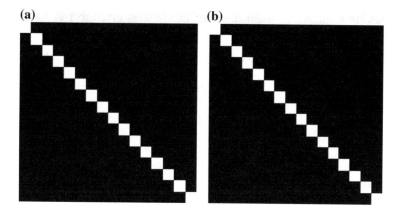

Fig. 10.7 Example of distorted matrix: **a** "diagonal" noise, **b** result of its transformation

Fig. 10.8 The example of the root noise for $n = 8$

$$A = \begin{bmatrix} 1 & 1 & 1 & 1 & 1 & 1 & 1 & 1 \\ 1 & -1 & 1 & -1 & 1 & -1 & 1 & -1 \\ 1 & 1 & -1 & -1 & 1 & 1 & -1 & -1 \\ 1 & -1 & -1 & 1 & 1 & -1 & -1 & 1 \\ 1 & 1 & 1 & 1 & -1 & -1 & -1 & -1 \\ 1 & -1 & 1 & -1 & -1 & 1 & -1 & 1 \\ 1 & 1 & -1 & -1 & -1 & -1 & 1 & 1 \\ 1 & -1 & -1 & 1 & -1 & 1 & 1 & -1 \end{bmatrix},$$

as well as the result of its double-sided strip-transformation with the help of the same matrix.

It is seen that after transformation, the image does not change, i.e. it is the root one. Equation 10.10 describes only a part of the root matrices. This is evident from the following example, in which the root matrix X is given that is not a linear combination of the identity matrix and matrix A.

Example 8 Let us take the matrix X of the form

$$X = \begin{bmatrix} 0 & 0 & 0 & 1 \\ 0 & 0 & -1 & 0 \\ 0 & -1 & 0 & 0 \\ 1 & 0 & 0 & 0 \end{bmatrix}$$

and transform it with the help of the Hadamard matrix of the fourth order $Y = AXA/4$:

$$A = \begin{bmatrix} -1 & 1 & 1 & 1 \\ 1 & -1 & 1 & 1 \\ 1 & 1 & -1 & 1 \\ 1 & 1 & 1 & -1 \end{bmatrix}$$

As the result we get

$$Y = AXA/4 = \begin{bmatrix} 0 & 0 & 0 & 1 \\ 0 & 0 & -1 & 0 \\ 0 & -1 & 0 & 0 \\ 1 & 0 & 0 & 0 \end{bmatrix}.$$

This shows that after transformation we get the same matrix, i.e. X is the root matrix with a root value $\lambda = 1$. The corresponding root image obtained in the MATLAB packet with the help of the command imagesc(X) is shown in Fig. 10.9.

It is evident that after transforming a noise at the receiving end is the same. This example shows that Eq. 10.10 describes not all decisions of the system $AX = XA$ but only a certain part of them. The general decision is provided by the following theorem.

Fig. 10.9 The root matrix X, which is not linear combination of matrices A and I

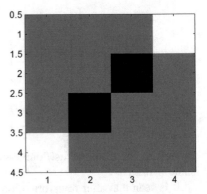

Theorem 2 *Let A be the orthogonal symmetrical matrix of the order n, m eigen-values of which are equal to 1, and H is the orthonormalized matrix of its eigen-vectors. Then, the general decision of a matrix equation $AXA = X$ contains $m^2 + (n - m)^2$ arbitrary constants and has the form of Eq. 10.10, where C_1 and C_2 are arbitrary square matrices of the order $m \cdot m$ and $(n - m) \cdot (n - m)$, respectively.*

$$X = H\tilde{X}H^T \quad \tilde{X} = \begin{bmatrix} C_1 & 0 \\ 0 & C_2 \end{bmatrix} \tag{10.10}$$

Proof Let the form of an original matrix equation have the form $AX = XA$. This is a particular case of the Sylvester Eqs. 10.8–10.9. To find its general decision, let us reduce the matrix A by a similarity transformation to the diagonal form:

$$H^TAH = E_0 \quad A = HE_0H^T \quad E_0 = \begin{bmatrix} E_m & 0 \\ 0 & -E_{n-m} \end{bmatrix},$$

where H is an orthonormal matrix of eigenvectors of the matrix A, E_m is a unity matrix of the order m.

It permits to write the equation $AX = XA$ in the form:

$$HE_0H^TX = XHE_0H^T.$$

Multiplying both parts by H^T from the left and by H from the right, as well as by denoting $\tilde{X} = H^TXH$, Eq. 10.11 is obtained.

$$E_0\tilde{X} = \tilde{X}E_0 \tag{10.11}$$

Let the matrix $\tilde{X}$ be represented in the block form $\tilde{X} = \begin{bmatrix} \tilde{X}_1 & \tilde{X}_2 \\ \tilde{X}_3 & \tilde{X}_4 \end{bmatrix}$, where the dimensions of the diagonal blocks are equal to m and $n - m$. Performing the multiplication and equaling blocks of the same name from the left and right, we find that $\tilde{X}_2 = 0$, $\tilde{X}_3 = 0$, i.e.

$$\tilde{X} = \begin{bmatrix} \tilde{X}_1 & 0 \\ 0 & \tilde{X}_4 \end{bmatrix}.$$

Returning to the original basis and denoting $C_1 = \tilde{X}_1$, $C_2 = \tilde{X}_4$, we obtain Eq. 10.10. Therefore, it is found the set of root matrices X corresponding to root value $\lambda = 1$.

Example 9 Let the root matrix X be determined for the Hadamard matrix of the 4th order

$$A = \begin{bmatrix} 1 & 1 & 1 & 1 \\ 1 & -1 & 1 & -1 \\ 1 & 1 & -1 & -1 \\ 1 & -1 & -1 & 1 \end{bmatrix} \qquad (10.12)$$

for the given matrix

$$\tilde{X} = \begin{bmatrix} 1 & 0 & 0 & 0 \\ 0 & 1 & 0 & 0 \\ 0 & 0 & 1 & 0 \\ 0 & 0 & 0 & 0 \end{bmatrix}.$$

Finding eigenvectors of the matrix A by using formula $X = H\tilde{X}H^{\mathrm{T}}$, we get

$$X = \begin{bmatrix} 1 & -1 & -1 & 0 \\ -1 & 2.5 & -0.5 & 0 \\ -1 & -0.5 & 2.5 & 0 \\ 0 & 0 & 0 & 3 \end{bmatrix}.$$

The corresponding root image is shown in Fig. 10.10.

Now let us move to considering negative (inverse) root images.

Case $\lambda = -1$. The whole set of the root images is described in this case by the following theorem.

Theorem 3 *Let A is the orthogonal symmetrical matrix of the order n, m eigenvalues of which are equal to 1, and H is the orthogonal matrix of its eigenvectors. The general decision of the matrix equation $AXA = -X$ contains $m^2 + (n - m)$ arbitrary constants and has the form of Eq. 10.13, where C_1 and C_2 are the arbitrary square matrices of the dimensions $m \cdot m$ and $(n - m) \cdot (n - m)$ correspondingly.*

Fig. 10.10 The root image (Example 9)

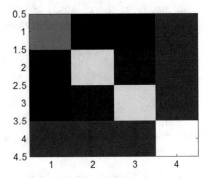

$$X = H\tilde{X}H^{\mathrm{T}} \quad \tilde{X} = \begin{bmatrix} 0 & C_1 \\ C_2 & 0 \end{bmatrix} \tag{10.13}$$

Proof The proof is performed in the same manner as it is done for Theorem 2, with the exception of the fact that in the right part of Eq. 10.11 the minus sign appears.

$$E_0\tilde{X} = -\tilde{X}E_0 \tag{10.14}$$

Due to this, the diagonal blocks of the matrix $\tilde{X}$ become equal to 0, and the matrix $\tilde{X} = \begin{bmatrix} 0 & \tilde{X}_2 \\ \tilde{X}_3 & 0 \end{bmatrix}$ takes the form:

$$\tilde{X} = \begin{bmatrix} 0 & \tilde{X}_2 \\ \tilde{X}_3 & 0 \end{bmatrix} = \begin{bmatrix} 0 & 0 & a & b \\ 0 & 0 & c & d \\ e & f & 0 & 0 \\ g & h & 0 & 0 \end{bmatrix}.$$

Thus, the general form of the matrix $\tilde{X}$ for the case $\lambda = -1$ has been obtained. As in the previous case it contains $m^2 + (n - m)^2$ arbitrary parameters.

Among all possible versions the permutation matrix will be noted. It has units on the lateral diagonal:

$$\tilde{X} = \begin{bmatrix} 0 & & 1 \\ & \cdot^{\cdot^{\cdot}} & \\ 1 & & 0 \end{bmatrix}.$$

It will comply with the root matrix

$$X = H\tilde{X}H^{\mathrm{T}} = [H_n \ \ldots \ H_1] \cdot [H_1 \ \ldots \ H_n]^{\mathrm{T}},$$

where H_i is the ith eigenvector of the matrix A.

Example 10 Let us find all "negative" or "inverse" root images for the Hadamard matrix of the 4th order. In accordance with Eq. 10.13, the matrix $\tilde{X}$ has the form:

$$\tilde{X} = \begin{bmatrix} 0 & \tilde{X}_2 \\ \tilde{X}_3 & 0 \end{bmatrix} = \begin{bmatrix} 0 & 0 & a & b \\ 0 & 0 & c & d \\ e & f & 0 & 0 \\ g & h & 0 & 0 \end{bmatrix},$$

where

$$X_1 = \begin{bmatrix} \left(-2f + 4\sqrt{2}h - 2c + 4\sqrt{2}d\right)\sqrt{2} \\ 2\sqrt{2}f + 4h - 4a + 2\sqrt{2}c + 8\sqrt{2}b - 8d \\ 2\sqrt{2}f + 4h + 4a + 2\sqrt{2}c - 8\sqrt{2}b - 8d \\ \left(-6f + 2c - 4\sqrt{2}d\right)\sqrt{2} \end{bmatrix}$$

$$X_2 = \begin{bmatrix} \left(-2\sqrt{2}e + 8g + 2f - 4\sqrt{2}h + 2c + 2\sqrt{2}d\right)\sqrt{2} \\ 4e + 4\sqrt{2}g - 2\sqrt{2}f - 4h + 4a - 2\sqrt{2}c + 4\sqrt{2}b - 4d \\ 4e + 4\sqrt{2}g - 2\sqrt{2}f - 4h - 4a - 2\sqrt{2}c - 4\sqrt{2}b - 4d \\ \left(-6\sqrt{2}e + 6f - 2c - 2\sqrt{2}d\right)\sqrt{2} \end{bmatrix}$$

$$X_3 = \begin{bmatrix} \left(-2\sqrt{2}e - 8g + 2f - 4\sqrt{2}h + 2c + 2\sqrt{2}d\right)\sqrt{2} \\ -4e - 4\sqrt{2}g - 2\sqrt{2}f - 4h + 4a - 2\sqrt{2}c + 4\sqrt{2}b - 4d \\ -4e - 4\sqrt{2}g - 2\sqrt{2}f - 4h - 4a - 2\sqrt{2}c - 4\sqrt{2}b - 4d \\ \left(6\sqrt{2}e + 6f - 2c - 2\sqrt{2}d\right)\sqrt{2} \end{bmatrix}$$

$$X_4 = \begin{bmatrix} \left(2f - 4\sqrt{2}h - 6c\right)\sqrt{2} \\ -2\sqrt{2}f - 4h - 12a + 6\sqrt{2}c \\ -2\sqrt{2}f - 4h + 12a + 6\sqrt{2}c \\ \left(6f + 6c\right)\sqrt{2} \end{bmatrix}$$

For example, let us take $a = 1/4$, while the remaining seven parameters will be equal to 0. Then, the root matrix, in which the first and the last rows are equal to 0, can be obtained:

$$X = \begin{bmatrix} 1 & 0 & 0 & 0 \\ -1 & 1 & 1 & -3 \\ 1 & -1 & -1 & 3 \\ 0 & 0 & 0 & 0 \end{bmatrix} \quad Y = -X.$$

The corresponding image is shown in Fig. 10.11.

Fig. 10.11 The root image (Example 10)

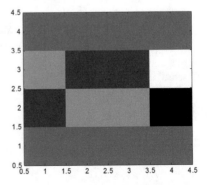

Thus, both the positive and negative root images of the two-sided strip-transformation with the symmetrical Hadamard matrix A are determined by formula $X = H\tilde{X}H^T$, where H is the orthonormal matrix of eigenvectors of the matrix A.

At that, the matrix A has the form:

$$\tilde{X} = \begin{bmatrix} \tilde{X}_1 & 0 \\ 0 & \tilde{X}_4 \end{bmatrix} \text{ if } \lambda = 1 \text{ or } \tilde{X} = \begin{bmatrix} 0 & \tilde{X}_2 \\ \tilde{X}_3 & 0 \end{bmatrix} \text{ if } \lambda = -1.$$

In the first case, the images remain unchanged after transformation and, in the second case, they became inverted. The performed analysis of root matrices shows that the double-sided strip-transformation with the symmetric Hadamard matrix possesses a great set of root images. This is due to a high symmetry of Hadamard matrices. In most cases the root images inherit this symmetry.

To decrease the power of a variety of root images (for example, taking into account requirements of cryptography, steganography, and information protection), the asymmetrical versions of the Hadamard matrices should be applied. Moreover, the conditions $A = B$ and $m = n$ accepted from considerations of calculation resources should be declined.

10.4.2 Renunciation of Matrices Equality Condition A = B

Let original image X be divided into $m \times n$ equal parts. Let fulfill its strip-transformation in accordance with Eq. 10.1 provided by Eq. 10.15, where A and B are the non-singular square matrices of dimensions $n \times n$ and $m \times m$, respectively.

$$Y = BXA \tag{10.15}$$

In this case, the task to find scale-invariant image X:

$$BXA = \lambda X$$

at λ given is reduced to solution of Sylvester's equation of the type

$$XA_1 = BX, \tag{10.16}$$

where X is $n \times m$ matrix, $A_1 = \lambda^{-1}$.

This is the Sylvester homogeneous equation. The result given in [13, 15] is known.

Theorem 4 *If matrices A_1 and B have no common eigenvalues (i.e. characteristic polynomials $P_{A_1} = |\mu I - A_1|$ and $P_B = |\mu I - B|$ are mutually simple), then the Sylvester Eq. 10.16 has only trivial solution $X = 0$.*

From this theorem, a number of conclusions can be made. First, this theorem gives a simple algebraic criterion for checking up the presence of root images with root number λ in the given pair of square matrices A and B. It is reduced to calculation of resultant $R(P_A, P_B)$ of their characteristic polynomials $P_{A_1} = |\mu I - A_1|$ and $P_B = |\mu I - B|$. If the result obtained does not equal to 0, then there is no root images. Second, if matrices A and B are orthogonal and symmetric, then they have common eigenvalues of the form ± 1. Third, the advantage of Eq. 10.11 is compared with simplified version $Y = AXA$ considered above.

Example 11 Let us consider the case with $m = 4$, $n = 6$, when the original image is divided into 6 parts vertically and into 4 parts horizontally, i.e. it is cut into 24 small rectangles (matrix X will have dimensions 4×6).

$$
A = \frac{1}{\sqrt{5}}
\begin{bmatrix}
0 & 1 & 1 & 1 & 1 & 1 \\
1 & 0 & 1 & -1 & -1 & 1 \\
1 & 1 & 0 & 1 & -1 & -1 \\
1 & -1 & 1 & 0 & 1 & -1 \\
1 & -1 & -1 & 1 & 0 & 1 \\
1 & 1 & -1 & -1 & 1 & 0
\end{bmatrix}
\qquad
B = \frac{1}{2}
\begin{bmatrix}
-1 & 1 & 1 & 1 \\
1 & -1 & 1 & 1 \\
1 & 1 & -1 & 1 \\
1 & 1 & 1 & -1
\end{bmatrix}
$$

Here matrix A is the normalized C-matrix of the 6th order, and matrix B is the normalized Hadamard matrix of the 4th order. Both matrices are symmetric and orthogonal, thus they have to possess root images. Let us find them.

In the case given, matrix equality $BXA - X = 0$ represents 24 equations with 24 unknown $x_1, x_2, \ldots, x_{24},$ i.e. elements of matrix X. The rank of the matrix of this system turned out to be equal to 12. Thus, the general solution contains 12 arbitrary parameters. Supposing, for example, that all elements x_i with odd indices Eq. 10.1, the following matrix of the root image can be obtained:

$$
X =
\begin{bmatrix}
1 & -1 & 1 & \sqrt{5} & 1 & -1 \\
1 & -1 & 1 & \sqrt{5} & 1 & -1 \\
1 & -1 & 1 & \sqrt{5} & 1 & -1 \\
1 & -1 & 1 & \sqrt{5} & 1 & -1
\end{bmatrix}.
$$

The corresponding image is shown in Fig. 10.12. Other version of the root image is shown in Fig. 10.13. It meets the matrix

$$
X =
\begin{bmatrix}
0 & 1 & -0.618 & 0.618 & -1 \\
0 & 0 & 0 & 0 & 0 \\
0 & 0 & 0 & 0 & 0 \\
0 & -1 & 0.618 & -0.618 & 1
\end{bmatrix}
$$

and satisfies equality $BXA = X$ too.

Fig. 10.12 The root image (Example 11)

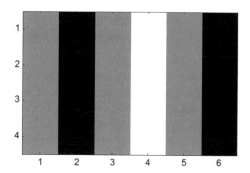

Fig. 10.13 Other version of the root image (Example 11)

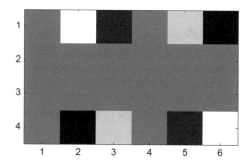

Notice, that both matrices X have rank 1.

Example 12 To increase the interference immunity as well as to get rid of root images, let us take an asymmetric version of matrix B:

$$A = \frac{1}{\sqrt{5}} \begin{bmatrix} 0 & 1 & 1 & 1 & 1 & 1 \\ 1 & 0 & 1 & -1 & -1 & 1 \\ 1 & 1 & 0 & 1 & -1 & -1 \\ 1 & -1 & 1 & 0 & 1 & -1 \\ 1 & -1 & -1 & 1 & 0 & 1 \\ 1 & 1 & -1 & -1 & 1 & 0 \end{bmatrix} \quad B = \frac{1}{2} \begin{bmatrix} 1 & 1 & 1 & 1 \\ -1 & 1 & -1 & 1 \\ 1 & 1 & -1 & -1 \\ -1 & 1 & 1 & -1 \end{bmatrix}.$$

With complex eigenvalues $\frac{\sqrt{2}}{2}(\pm 1 \pm i)$. Let matrix A be remained the same. Writing out matrix equation $BXA - X = 0$, we ascertain that now the matrix of this system is not a degenerated one (its rank is equal to 24) and, therefore, the system has only the zero solution, and there is no root images. By the above, a possibility of complete abandonment of root images in applying the strip-transformation with orthogonal matrices A and B is shown.

10.4.3 Abandonment of Conditions of Symmetry and Orthogonality of Matrices A and B

In the case of arbitrary non-singular matrices A and B, the problem of root image existence is reduced to an analysis of solvability of the linear system of Eq. 10.7 $BXA = \lambda X$ at some real number λ. Let us rewrite this system in an equivalent form provided by Eq. 10.17, where $\bar{X}$ is a vector with mn components, which was obtained as a result of "extending" into a column of matrix X, $\bar{A}$ is a square matrix of order $m\,n$.

$$\bar{A} = \lambda \bar{X} \tag{10.17}$$

Equations 10.17 will have a non-trivial solution only in that case if there exists a real number λ, at which matrix $\bar{A} - \lambda I$ is degenerated. This represents the existence criterion of the root image (I is the identity matrix of the $m\,n$ order).

Thus, in order to get a reply to the question about the presence of a root image in the given pair of matrices A and B, it is necessary to write out a characteristic polynomial of matrix $\bar{A}$ of Eq. 10.17 and to check up the presence of real roots in it. Each of such roots λ will corresponds to root image X_i that can be found solving Eq. 10.17 at $\lambda = \lambda_i$.

Example 13 Provided $m = n = 2$, i.e. the original image is divided into 4 parts. Let us to make its strip-transformation by formula $Y = BXA$, where matrices A and B have the form:

$$A = \begin{bmatrix} -11 & -30 \\ 6 & 16 \end{bmatrix} \quad B = \begin{bmatrix} -22 & -50 \\ 10 & 23 \end{bmatrix}.$$

Notice, that both of them are not orthogonal and not symmetric. Write out matrix equality in order to check up the existence of root images.

$$\begin{bmatrix} -22 & -50 \\ 10 & 23 \end{bmatrix} \begin{bmatrix} x_1 & x_3 \\ x_2 & x_4 \end{bmatrix} \begin{bmatrix} -11 & -30 \\ 6 & 16 \end{bmatrix} = \lambda \begin{bmatrix} x_1 & x_3 \\ x_2 & x_4 \end{bmatrix}.$$

This matrix equality is in accordance with the homogeneous system of equations.

$$\begin{aligned} 242x_1 + 550x_2 - 132x_3 - 300x_4 &= \lambda x_1 \\ -110x_1 - 253x_2 + 60x_3 + 138x_4 &= \lambda x_2 \\ 660x_1 + 1500x_2 - 352x_3 - 800x_4 &= \lambda x_3 \\ -300x_1 - 690x_2 + 160x_3 + 368x_4 &= \lambda x_4 \end{aligned} \tag{10.18}$$

It will have the zero solution if the determinant of the system is equal to 0 $\det(\bar{A} - \lambda I) = 0$, where

$$\bar{A} = \begin{bmatrix} 242 & 550 & -132 & -300 \\ -110 & -253 & 60 & 138 \\ 660 & 1500 & -352 & -800 \\ -300 & -690 & 160 & 368 \end{bmatrix}.$$

Opening the determinant results in the equation for λ:

$$\lambda^4 - 5\lambda^3 - 98\lambda^2 + 120\lambda + 576 = 0.$$

It is factorized and reduced to the form:

$$(\lambda - 3)(\lambda + 8)(\lambda - 12)(\lambda + 2) = 0$$

This equation has 4 real roots. Four matrices of root images correspond to these real roots.

$$X_1 = \begin{bmatrix} -5 & -25/2 \\ 2 & 5 \end{bmatrix} \quad X_2 = \begin{bmatrix} -5 & -10 \\ 2 & 4 \end{bmatrix} \quad X_3 = \begin{bmatrix} -4 & -10 \\ 2 & 5 \end{bmatrix}$$
$$X_4 = \begin{bmatrix} -2 & -4 \\ 1 & 2 \end{bmatrix}.$$

Two of these matrices are of the first rank, two others are of the second rank. Thus, the criterion of existence of invariant images of the first order has been gotten.

Criterion 1. The necessary and sufficient condition of root image existence is the presence of a real eigenvalue of matrix $\bar{A} = A^T \otimes B$.

The other formulation of this criterion can take place in the case, when among the products of eigenvalues $\lambda_i \mu_j$ of matrices A and B there is at least one real number. This concerns any double-sided strip-transformation with a pair of matrices A and B that have not obligatory to be orthogonal and symmetric.

Let us consider a more complicated task, i.e. finding matrices A and B, for which there is no both versions of invariant images (usual and "turned up"). Determine the Criterion 1 of the invariant image existence. Let the two-sided strip-transformation be described by Eq. 10.1 $BXA = \lambda X$, where B and A are the non-singular square matrices of the m and n orders. The next task will consist in finding the images that are turned by such transformation into an image of the same type with accuracy up to rotation by $180°$ within the plane of the picture and constant factor:

$$BXA = \lambda PXP.$$

Here P is the anti-diagonal permutation matrix. Let this equation be rewritten in the form of

$$PBXAP = \lambda X \quad \text{or} \quad B_1 X A_1 = \lambda X.$$

Applying Criterion 1 to the last equation, we obtain the next result.

Criterion 2. A necessary and sufficient condition of existence of an invariant image of the second type is the presence of a real eigenvalue of matrix $\bar{\tilde{A}} = \tilde{A}^{\mathrm{T}} \otimes \tilde{B}$, *where matrix* $\tilde{B}$ *has been obtained from B by permutation of rows in the inverse order and matrix* $\tilde{A}$ *has been obtained from A by permutation of columns inversely.*

Just as the root numbers of the first type are equal to pair-wise products of eigenvalues of matrices A and B, the root numbers of the second type, i.e. the eigenvalues of matrix $\bar{\tilde{A}}$, are equal to various pair-wise products of the eigenvalues of matrices $\tilde{A}$ and $\tilde{B}$. From this it follows that there exists some other formulation of Criterion 2, which is realized via the presence of at least one real number among the products of eigenvalues $\lambda_i \mu_j$ of matrices $\tilde{A}$ and $\tilde{B}$. To get rid of both versions of invariant images (the usual and "inverted" ones), it needs that neither in matrix $\bar{A} = A^{\mathrm{T}} \otimes B$ nor in "inverted" matrix $\bar{A}$, i.e. $P\bar{A}$, where P is the anti-diagonal permutation matrix, there would exist real eigenvalues.

10.5   Getting Invariant Images Using Eigenvectors of Matrices

An effective way of finding invariant images at given matrices A and B is based on the use of eigenvectors of matrices provided by Eq. 10.19, where B and A are the given non-singular matrices of dimensions $m \times m$ and $n \times n$.

$$BXA = kX \tag{10.19}$$

It is obvious that a trivial decision $X = 0$ is out of interest. Therefore, the conditions, under which the decisions with matrix X of rank 1 exist, is not equal to 0. Let us begin with the case of matrices X of the unit rank: $rankX = 1$. It is known that such matrices (dyads) can be presented in the form of a product provided by Eq. 10.20, where $h \in R^m$ and $g \in R^n$ are the vectors-columns.

$$X = hg^{\mathrm{T}} \tag{10.20}$$

Let $h_1, \ldots, h_m$ are the eigenvectors of matrix B satisfying its eigenvalues $\mu_1, \ldots, \mu_m$, and $g_1, \ldots, g_m$ are the eigenvectors of matrix A^{T}, satisfying its eigenvalues $\lambda_1, \ldots, \lambda_n$.

The following result takes place.

Theorem 5 *All real decisions X of rank 1 of matrix Eq. 10.19 can be obtained by substitution of real eigenvectors* (h_i, g_j) *into Eq. 10.20:*

$$X_{ij} = h_i g_j^T \qquad i = \overline{1, m} \qquad j = \overline{1, n}. \tag{10.21}$$

At that, constant k in Eq. 10.19 is equal to the product of corresponding eigenvalues $k_{ij} = \mu_i \lambda_j$.

The proof follows from the chain of equalities:

$$BX_{ij}A = (Bh_i)\left(A^T g_j\right)^T = \mu_i h_i \lambda_j g_j^T = \mu_i \lambda_j X_{ij}.$$

From Theorem 5 it follows that for existence of invariant images of rank 1 there is some other criterion, i.e. the *rank criterion*.

Criterion 3. The necessary and sufficient condition for the scale-invariant images of rank 1 to exist is the presence of at least one pair of real eigenvalues (μ_i, λ_j) in matrices A and B.

The total number of invariant images of rank 1 will be defined by the number of such pairs. If all $m + n$ eigenvalues of matrices A and B are real and different, then the total number of invariant images will be equal to product *mn*.

Example 14 Let we have $m = 2$, $n = 3$ and given matrices A, B:

$$A = \begin{bmatrix} -1 & 2 & 2 \\ 2 & -1 & 2 \\ 2 & 2 & -1 \end{bmatrix} \quad B = \begin{bmatrix} 1 & 1 \\ 1 & -1 \end{bmatrix}.$$

In this case, Eq. 10.19 has the form:

$$\begin{bmatrix} 1 & 1 \\ 1 & -1 \end{bmatrix} \begin{bmatrix} x_{11} & x_{12} & x_{13} \\ x_{21} & x_{22} & x_{23} \end{bmatrix} \begin{bmatrix} -1 & 2 & 2 \\ 2 & -1 & 2 \\ 2 & 2 & -1 \end{bmatrix} = k \begin{bmatrix} x_{11} & x_{12} & x_{13} \\ x_{21} & x_{22} & x_{23} \end{bmatrix}.$$

For applying Criterion 3, let the eigenvalues and eigenvectors of matrices A and B be found.

Then the characteristic polynomials of matrices A and B are written out:

$$P_A = \lambda^3 + 3\lambda^2 - 9\lambda - 27 \quad P_B = \mu^2 - 2.$$

Their eigenvalues will be:

$$\lambda_1 = 3 \quad \lambda_2 = \lambda_3 = -3 \quad \mu_1 = -\sqrt{2} \quad \mu_2 = \sqrt{2}.$$

The eigenvectors given below correspond to the above eigenvalues:

$$g_1 = \begin{bmatrix} 1 \\ 1 \\ 1 \end{bmatrix} \quad g_2 = \begin{bmatrix} -1 \\ 1 \\ 0 \end{bmatrix} \quad g_3 = \begin{bmatrix} -1 \\ 0 \\ 1 \end{bmatrix} \quad h_1 = \begin{bmatrix} 1 - \sqrt{2} \\ 1 \end{bmatrix} \quad h_1 = \begin{bmatrix} 1 \\ 1 + \sqrt{2} \end{bmatrix}.$$

All eigenvalues and eigenvectors are real numbers. Therefore, six invariant images of rank 1 have to exist. Let them be found with the help of Eq. 10.21.

$$X_{11} = h_1 g_1^T = \begin{bmatrix} 1-\sqrt{2} & 1-\sqrt{2} & 1-\sqrt{2} \\ 1 & 1 & 1 \end{bmatrix}$$

$$X_{12} = h_1 g_2^T = \begin{bmatrix} -1+\sqrt{2} & 1-\sqrt{2} & 0 \\ -1 & 1 & 1 \end{bmatrix}$$

$$X_{13} = h_1 g_3^T = \begin{bmatrix} \sqrt{2}-1 & -\sqrt{2}+1 & 0 \\ -1 & 1 & 0 \end{bmatrix}$$

$$X_{21} = h_2 g_1^T = \begin{bmatrix} 1 & 1 & 1 \\ 1-\sqrt{2} & 1-\sqrt{2} & 1-\sqrt{2} \end{bmatrix}$$

$$X_{22} = h_2 g_2^T = \begin{bmatrix} -1 & 1 & 0 \\ 1-\sqrt{2} & \sqrt{2}-1 & 0 \end{bmatrix}$$

$$X_{23} = h_2 g_3^T = \begin{bmatrix} -1 & 0 & 0 \\ 1-\sqrt{2} & 0 & \sqrt{2}-1 \end{bmatrix}$$

Three of these invariant images are shown in Fig. 10.14.

Let us consider a particular case, when matrices A and B coincide: $A = B$. In this case, the invariant images correspond not only to the real eigenvalues of matrix A but also to the complex ones. Really, let matrix A have complex eigenvalue $\lambda_1 = a + ib$ and corresponding complex eigenvector g_1. Then matrix A^T has to contain conjugate eigenvalue $\mu_1 = a - ib$ and conjugate eigenvector h_1. Product $\mu_1 \lambda_1$ will be real and this is a sign of the invariant image existence. For this image to be found it is sufficient to take the real and imaginary parts of complex invariant image $X = h_1 g_1^T$. The rank of these images will be equal to two.

Example 15 Consider transformation $Y = AXA$ with the orthogonal matrix:

$$A = \begin{bmatrix} 2 & -1 & 2 \\ 2 & 2 & -1 \\ -1 & 2 & 2 \end{bmatrix}.$$

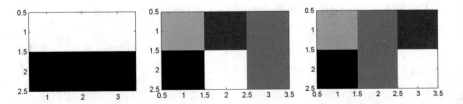

Fig. 10.14 Invariant images of rank 1 (Example 14)

One of the eigenvalues of the matrix is real $\lambda_1 = 3$, other two are complex $\lambda_{2,3} = \frac{2}{3}\left(1 \pm i\sqrt{3}\right)$.

Eigenvectors corresponding to them are:

$$h_1 = \begin{bmatrix} 1 \\ 1 \\ 1 \end{bmatrix} \qquad h_2 = \begin{bmatrix} -1+i\sqrt{3} \\ -1-i\sqrt{3} \\ 2 \end{bmatrix} \qquad h_3 = \begin{bmatrix} -1-i\sqrt{3} \\ -1+i\sqrt{3} \\ 2 \end{bmatrix}.$$

The eigenvectors of matrix A^{T} have the form:

$$g_1 = h_1 \qquad g_2 = h_3 \qquad g_3 = h_2.$$

The invariant image of rank 1 corresponds to the real eigenvalue:

$$X_1 = h_1 g_1^{\mathrm{T}} = \begin{bmatrix} 1 & 1 & 1 \\ 1 & 1 & 1 \\ 1 & 1 & 1 \end{bmatrix}.$$

The invariant images of rank 2, which correspond to the complex eigenvalues, are obtained by the following formulae.

$$X_2 = h_2 g_3^{\mathrm{T}} + h_3 g_2^{\mathrm{T}} = \begin{bmatrix} -1 & 1 & 1 \\ 0 & 1 & -1 \\ 1 & -1 & 1 \end{bmatrix}$$

$$X_3 = h_2 g_3^{\mathrm{T}} - h_3 g_2^{\mathrm{T}} = \begin{bmatrix} -1 & 2 & -1 \\ 2 & -1 & -1 \\ -1 & -1 & 2 \end{bmatrix}$$

The corresponding invariant images are shown in Fig. 10.15.

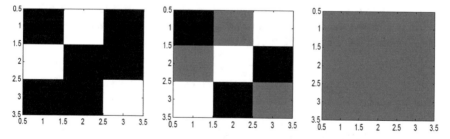

Fig. 10.15 Invariant images (Example 15)

10.6   Synthesis of Matrices by Given Invariant Images

Above we were solving the images X at the given matrices of double-sided strip-transformation A and B. Undoubtedly, the inverse problem is also of interest, when invariant image X is given and it is required to find transformation matrices A and B, for which it will be invariant. The applied significance of this problem consists in the fact that it is possible to take image X, add a "secret" text or picture to it and after subjecting it to the strip-transformation to transmit the message obtained over the communication channel. For the inverse problem to be solved, let us present the given invariant image in the form of a sum of matrices of the unit rank (dyads):

$$X = P_1 + \cdots + P_n.$$

This can be done, for example, with the help of spectral decomposition, when the summands have the form:

$$P_i = \lambda_i h_i g_i^T,$$

where h_i and g_i are the right and left eigenvectors of matrix X, which correspond to eigenvalue λ_i.

Earlier it was shown that the invariant images are formed by the way of multiplying various pairs of eigenvectors of matrices A and B. Thereby, we get a problem of reconstructing matrices A and B by the known matrices of their right and left eigenvectors H_A, G_A, H_B, and G_B. Let us explain the corresponding procedure by a simple example.

Example 16 Let the matrix of an invariant image of the 3rd order be given

$$X = \begin{bmatrix} 2 & 2 & 2 \\ 4 & 2 & 4 \\ 4 & 2 & 4 \end{bmatrix}. \tag{10.22}$$

It is required to find matrices A and B of strip-transformation $Y = BXA$, for which it is invariant, $Y = X$. In the case given, matrix X has rank 2 and can be represented in the form of the sum of two matrices of rank 1:

$$X = P_1 + P_2 = \begin{bmatrix} 1 & 2 & 1 \\ 1 & 2 & 1 \\ 1 & 2 & 1 \end{bmatrix} + \begin{bmatrix} 1 & 0 & 1 \\ 3 & 0 & 3 \\ 3 & 0 & 3 \end{bmatrix}$$

The images corresponding to matrices P_1, P_2, X, are shown in Fig. 10.16.

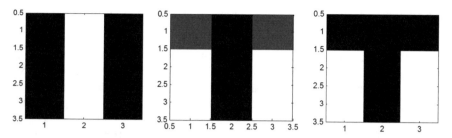

Fig. 10.16 Invariant images (Example 16)

Now let us form the matrices of eigenvectors.

$$H_B = \begin{bmatrix} 1 & 1 & 1 \\ 1 & 3 & 2 \\ 1 & 3 & 3 \end{bmatrix} \quad G_B = \begin{bmatrix} 1.5 & 0 & -0.5 \\ -0.5 & 1 & -0.5 \\ 0 & -1 & 1 \end{bmatrix}$$

$$H_A = \begin{bmatrix} 0.5 & 1 & -0.5 \\ 0.5 & -0.5 & 0 \\ -0.5 & 0 & 0.5 \end{bmatrix} \quad G_A = \begin{bmatrix} 1 & 2 & 1 \\ 1 & 0 & 1 \\ 1 & 2 & 3 \end{bmatrix}$$

After that we use the formula of spectral decomposition.

$$B = k_1 h_{B1} g_{B1}^{\mathrm{T}} + k_2 h_{B2} g_{B2}^{\mathrm{T}} + k_3 h_{B3} g_{B3}^{\mathrm{T}}$$
$$A = k_1 h_{A1} g_{A1}^{\mathrm{T}} + k_2 h_{A2} g_{A2}^{\mathrm{T}} + k_3 h_{A3} g_{A3}^{\mathrm{T}}$$

Substituting

$$k_1 = 2 \quad k_2 = 4 \quad k_3 = 2 \quad h_3 = \begin{bmatrix} 1 \\ 2 \\ 3 \end{bmatrix} \quad g_3 = \begin{bmatrix} 1 \\ 2 \\ 3 \end{bmatrix},$$

we obtain

$$A = \begin{bmatrix} 4 & 0 & 2 \\ -1 & 2 & -1 \\ 0 & 0 & 2 \end{bmatrix} \quad B = \begin{bmatrix} 1 & 2 & -1 \\ -3 & 8 & -3 \\ -3 & 6 & -1 \end{bmatrix}.$$

Performing the verification, we make certain that given matrix X (Eq. 10.22) is invariant with respect to transformation BXA.

10.7 Conclusions

The chapter given contains the description of the strip-method based on the double-sided matrix transformation of two-dimensional images, as well as examples of its use. Such transformations can be useful for decreasing pulse interferences, taking place in communication channels. They are also useful for the purposes of cryptography, steganography, and other applications. The authors introduced the concepts of invariant images that do not change their form because of the double-sided matrix transformation, including scale, rotational and negative (inverse) invariants. The criteria of existence of invariant images for the given transformation matrices and methods to find them are described. The procedure developed for constructing transformation matrices without any invariant images is presented. A consideration is given to the problem of transformation matrices synthesis by invariant images.

References

1. Mironovsky, L., Slaev, V.: Implementation of Hadamard matrices for image processing. In: Favorskaya, M.N., Jain, L.C. (eds.) Computer Vision in Control Systems-1, ISRL, vol. 73, pp. 311–349. Springer International Publishing, Switzerland (2015)
2. Andrews, H.: Computer Techniques in Image Processing. Academic Press, New York (1970)
3. Costas, J.: Coding with linear systems. PIRE **40**(9), 1101–1103 (1952)
4. Crowther, W., Rader, C.: Efficient coding of vocoder channel signals using linear trans-formation. Proc. IEEE **54**(11), 1594–1595 (1966)
5. Gantmacher, F. The Theory of Matrices. AMS Chelsea Publishing. Reprinted by American Mathematical Society (2000)
6. Golub, G., van Loan, C.: Matrix Computations, 3rd edn. John Hopkins University Press, Baltimore, MD, USA (1996)
7. Kramer, H.: The covariance matrix of vocoder speech. Proc. IEEE **55**(3), 439–440 (1967)
8. Kramer, H., Mathews, M.: A linear coding for transmitting a set of correlated signals. IRE Trans. Inf. Theory IT **2**, 41–46 (1956)
9. Lang, G.: Rotational transformation of signals. IEEE Trans. Inf. Theory IT **9**(3),191–198 (1963)
10. Leith, E., Upatnieks, J.: Reconstructed wavefronts and communication theory. J. Opt. Soc. Am. **52**(10), 112 (1962)
11. Mironovskii, L., Slaev, V.: Invariant relations between means of small samples. Meas. Tech. **57**(2), 145–152 (2014)
12. Mironowsky, L., Slaev, V.: Double-sided noise-immune strip transformation and its root images. Meas. Tech. **55**(10), 1120–1127 (2013)
13. Mironovsky, L., Slaev, V.: Strip-Method for Image and Signal Transformation. De Gruyter, Berlin (2011)
14. Mironovsky, L., Slaev, V.: Root images of two-sided noise immune strip-transformation. In: International Workshop on Physics and Mathematics (IWPM'2011) (2011)
15. Mironovsky, L., Slaev, V.: The strip method of noise-immune image transformation. Meas. Tech. **49**(8), 745–775 (2006)
16. Mironovsky, L., Slaev, V.: The strip method of transforming signals containing redundancy. Meas. Tech. **49**(7), 631–638 (2006)

17. Mironovsky, L., Slaev, V.: Optimal Chebishev pre-emphasis and filtering. Meas. Tech. **45**(2), 126–136 (2002)
18. Pierce, W.: Linear-real codes and coders. Bell Syst. Tech. J. **47**(6), 1067–1097 (1968)
19. Mironovsky, L., Slaev, V.: Optimal test signals as a solution of the generalized Bulgakov problem. Autom. Remote Control **63**(4), 568–577 (2002)
20. Mironovsky, L., Slaev, V.: Invariants in metrology and technical diagnostics. Meas. Tech. **39** (6), 577–593 (1996)
21. Slaev, V.: Metrological problems of information technology. Meas. Tech. **37**(11), 1209–1216 (1994)
22. Mironovsky, L., Slaev, V.: Technical diagnostics of dynamic systems on the basis of algebraic invariants. In: 3rd Symposium of the IMEKO, TC on Technical Diagnostics, Budapest, pp. 243–251 (1983)
23. Mironovsky, L., Slaev, V.: Equalization of the variance of a nonstationary signal. Telecommun. Radio Eng. **29–30**(5), 24–27 (1975)
24. Abreu, E., Lightstone, M., Mitra, S., Arakawa, S.: A new efficient approach for the removal of impulse noise from highly corrupted images. IEEE Trans. Image Proc. **5**(6), 1012–1025 (1996)
25. Boyle, R., Sonka, M., Hlavac, V.: Image Processing, Analysis, and Machine Vision, 1st edn. University Press, Cambridge (2008)
26. Buades, A., Morel, J.: A non-local algorithm for image denoising. In: IEEE Computer Society Conference on Computer Vision Pattern Recognition (CVPR'2005), vol. 2, pp. 60–65 (2005)
27. Chan, R., Ho, C., Nikolova, M.: Salt-and-pepper noise removal by median-type noise detectors and detail-preserving regularization. IEEE Trans. Image Proc. **14**(10), 1479–1485 (2005)
28. Clauset, A., Newman, M., Moore, C.: Finding community structure in very large networks. Phys. Rev. E **70**(6), 066111 (2004)
29. Dabov, K., Foi, A., Katkovnik, V., Egiazarian, K.: Image denoising by sparse 3-D transform-domain collaborative filtering. IEEE Trans. Image Proc. **16**(8), 2080–2095 (2007)
30. Di Gesu, V., Staravoitov, V.V.: Distance-based functions for image comparison. Pattern Recognit. Lett. **20**(2), 207–213 (1999)
31. Garnett, R., Huegerich, T., Chui, C., He, W.: A universal noise removal algorithm with an impulse detector. IEEE Trans. Image Proc. **14**(11), 1747–1754 (2005)
32. Huang, T.S. (ed.): Two-Dimensional Digital Signal Processing II: Transforms and Median Filters. Springer, Berlin, New York (1981)
33. Kam, H., Tan, W.: Noise detection fuzzy (NDF) filter for removing salt and pepper noise. In: Zaman, H.B., Robinson, P., Petrou, M., Olivier, P., Schröder, H., Shih, T.K. (eds.) Visual Informatics: Bridging Research and Practice, LNCS, vol. 5857, pp. 479–486. Springer, Berlin, Heidelberg (2009)
34. Najeer, A., Rajamani, V.: Design of hybrid filter for denoising images using fuzzy network and edge detecting. Am. J. Sci. Res. **3**, 5–14 (2009)
35. Newman, M.: Analysis of weighted networks. Phys. Rev. E **70**(5), 056131 (2004)

Chapter 11
The Fibonacci Numeral System for Computer Vision

Oleksiy A. Borysenko, Vyacheslav V. Kalashnikov,
Nataliya I. Kalashnykova and Svetlana M. Matsenko

Abstract One of the most important challenges when creating efficient systems of technical vision is the development of efficient methods for enhancing the speed and noise-resistance properties of the digital devices involved in the system. The devices composed of counters and decoders occupy a special niche among the system's digital tools. One of the most common ways of creating noise-proof devices is providing special coding tricks dealing with their informational redundancy. Various frameworks make that possible, but nowadays, an acute interest is attracted to noise-proof numeral systems, among which the Fibonacci system is the most famous. The latter helps generate the so-called Fibonacci codes, which can be effectively applied to the computer vision systems; in particular when developing counting devices based on Fibonacci counters, as well as the corresponding decoders. However, the Fibonacci counters usually pass from the minimal form of representation of Fibonacci numbers to their maximal form by recurring to the

O.A. Borysenko · V.V. Kalashnikov · S.M. Matsenko
Department of Computer Science, Sumy State University,
Rimsky-Korsakov st. 2, Sumy 40007, Ukraine
e-mail: 5352008@ukr.net

S.M. Matsenko
e-mail: slavkamx@mail.ru

V.V. Kalashnikov (✉)
Department of Industrial Engineering, School of Engineering and Sciences,
Tecnológico de Monterrey (ITESM), Campus Monterrey, Av. Eugenio Garza Sada
2501 Sur, 64849 Monterrey, Nuevo León, Mexico
e-mail: kalash@itesm.mx

V.V. Kalashnikov
Department of Experimental Economics, Central Economics
and Mathematics Institute (CEMI), Russian Academy of Sciences (RAS),
Nakhimovsky pr. 47, Moscow 117418, Russian Federation

N.I. Kalashnykova
Department of Physics & Mathematics (FCFM), Universidad Autónoma de Nuevo León
(UANL), Av. Universidad S/N, Ciudad Universitaria, 66455 San Nicolás de los Garza,
Nuevo León, Mexico
e-mail: nkalash2009@gmail.com

© Springer International Publishing AG 2018
M.N. Favorskaya and L.C. Jain (eds.), *Computer Vision in Control Systems-3*,
Intelligent Systems Reference Library 135, https://doi.org/10.1007/978-3-319-67516-9_11

special operations catamorphisms and anamorphisms (or "folds" and "unfolds"). The latter makes the counters quite complicated and time-consuming. In this chapter, we propose a new version of the Fibonacci counter that relies only on the minimal representation form of Fibonacci numerals and thus leads to the counter's faster calculation speed and a higher level of the noise-resistance. Based on the above-mentioned features, we also present the appropriate fast algorithm implementing the noise-proof computation and the corresponding fractal decoder. The first part of the chapter provides the estimates of the new method's noise-immunity, as well as that of its components. The second problem studied in the chapter concerns the efficiency of the existing algorithm of Fibonacci representation in the minimal form. Based on this examination, we propose a modernization of the existing algorithm aiming at increasing its calculation speed. The third object of the chapter is the comparative analysis of the Fibonacci decoders and the development of the fractal decoder of the latter.

Keywords Fibonacci numbers · Minimal presentation form · Digital devices Noise-immunity · Counting velocity

11.1   Introduction

Computer vision systems are nowadays spreading away and widely applied in recognition devices, controllers of technical and natural objects, robotic complexes, high-level techniques such as electronic microscopes, etc. These systems are compound structures involving different blocks such as information transmission, processing, and compression devices making use of counters. Various counters are also extensively used in measurement tools: timers, rangefinders, frequency counters, among others, which, in their turn, are often parts of many computer vision systems.

Especially often, the counters are elements of video cameras, which are indispensable attributes of any computer vision system, and the operation quality of which guarantee the high level of the system's performance. It is worthwhile to note that *not* all existing video cameras are apt for the computer vision systems. In a lot of instances, their structure needs a serious improvement, especially on the part of their scanning systems; the latter was noticed when the control and recognition procedures named "local windows" were applied to the surface control of piston rings [1, 2]. In this case, as in many others, the exploited video cameras needed very fast counters, the development of which may prove to be a great challenge (especially if they are subject to extremely high standards of the computation velocity and noise immunity).

Indeed, such video cameras boasting additional exterior counters required a special structure of the whole control system [1]. Namely, the system contained two video cameras that, with the aid of the counters, simultaneously scanned the surface of the controlled piston rings. The latter were pushed to the cameras with a robot device composed of a data buffer and an electromechanical system putting the rings under control and removing them afterward. Depending upon the existence of any defects on the surface of the piston rings (which were detected with the computer vision system), an appropriate mechanical device distributes the controlled rings to the corresponding poles. Even more, the counters with the buffer data memory composed a specialized calculator that analyzed the data and detected the defects. This calculator was based on a unique method of recognition of defects, namely, the local windows technology [2]. As a result, the above-described computer vision system proved to be extremely fast, highly reliable and required the minimum of device cost for its implementation. At the time when the system was used, its ratio "price/quality" was about ten times lower than the best known similar systems in the world. Exactly the use of the counters as the principal elements of the system's structure determined its moderate cost. And since the counters' technical parameters are mainly implied by the numeral system used, the choice of the latter affects drastically the counter's efficiency. Nowadays, the classical binary system is almost a must in the counters because that makes them sufficiently simple, reliable, and cheap. However, the counters based upon the binary system lack a natural informational redundancy, which prevents them from finding easily the errors occurring when they work. That is why various methods of enhancing the counters' noise immunity are often proposed, including those employed in the computer vision. Some of these methods apply an artificial noise-proof coding, doubling, and majority coding.

Nevertheless, the counters' noise immunity may also be elevated by making use of the natural informational redundancy inherent in the structure of the exploited numeral systems. Fibonacci numeral systems with the Fibonacci numbers serving as the weights in the coding words boast such a redundancy to a higher grade than other systems. In addition to their ability to detect errors arising when functioning, the counters based on Fibonacci numeral systems reveal the high computational speed as well. That is why our chapter is devoted mainly to such numeral systems.

Notwithstanding, apart from the Fibonacci systems, there exist other numeral systems with the property of the natural redundancy, which allows them to detect errors arising when exploiting the counters making use of those systems. One could mention, for example, the factorial and binomial numeral systems, among others. These systems select factorial and binomial expressions, respectively, for the weights in their code words. However, the factorial-based counters use several natural numbers in addition to 0 and 1, which makes it difficult to implement them within the binary digital devices. Moreover, the noise immunity of the factorial systems is poorer than that of the Fibonacci systems. As for the binomial counters, even though their level of noise immunity is higher than the Fibonacci-based ones, they require elevated hardware costs and demonstrate the operation speed much lower than the Fibonacci counters. Therefore, the Fibonacci-based digital devices are preferable for the computer vision systems, in which the operation speed is one of the primordial requirements.

The above-mentioned example of the rings control system [1] illustrates the importance of the fast performance and noise immunity of the digital devices (counters) exploited in the modern computer vision systems. One of the corner-stones of such quality of the counters is the coding system forming their base. As it was mentioned previously, the coding system's redundancy makes way to its reliability and noise immunity. The informational redundancy of the system can be provided not only by a redundant coding system but also by the noise-proof numeral systems distinct from the standard decimal one. The most well-known numeral systems of this kind are the binomial, factorial, and Fibonacci systems [3–8]. Those noise-proof numeral systems often use the binary codes, which makes them extremely easy in applications and enhance their effective performance.

One of the advantages of involving the noise-proof numeral systems in the digital devices is the following feature. Having introduced the redundancy on the initial stage of their creation, they permit one to control not only the processes of the information accumulation and transmission but also the performed logic and arithmetic operations. Moreover, such devices in the computer vision systems can exchange information coded with the same noise-proof rules. The latter makes the control of both the information processing and information transmission in-between the devices much easier and comfortable. Hence, a thorough control is possible to realize within the computer vision system at once. This makes the system greatly more efficient since it allows one not only to enhance the system's noise immunity but also to increase its velocity as well as decrease the device cost.

The noise-proof numeral systems often help also generate combinatorial objects of special kinds, such as for example, permutations and/or combinations. The latter can be of a great use when solving combinatorial optimization problems, such as the counting of combinatorial objects, recognition of images, compression and defense of information, etc., which are readily met in the complex computer vision systems.

Next, it is worthy to mention that the digital devices based on the non-standard numeral systems boast the following useful property. Namely, the code's redundancy that helps detect errors, is uniformly distributed along their schematic structure, which makes them technologically homogeneous and hence, easier implementable. In addition, the simplicity of coding in such systems makes them cheaper in the apparatus sense, as well as faster and more reliable than the coding systems based on traditional numeral systems (decimal, binary, etc.).

Among as yet few really working noise-proof numeral systems, the Fibonacci system is the simplest and hence, a universal one. Based on the Fibonacci numeral system, schemes of various noise-proof digital devices such as processors, counters, adders (summers), and even computers have been developed [4, 7]. However, when implementing those devices and circuits, as a rule, two distinct representation forms: minimal and maximal, were exploited, which allowed one, by making transitions from the minimal form to the maximal one, and vice versa, to do operations over the Fibonacci numbers.

In order to complete such transitions, special digital circuits are necessary that realize the operations of catamorphisms and anamorphisms (or "folds" and "unfolds"), which transform the Fibonacci numbers from the maximal form to the minimal form, and vice versa [4, 7, 9]. These operations, in the end, made the Fibonacci devices' (e.g. the Fibonacci counters' and counting circuits') performance more complicated and slow. This prohibited the thorough use of all potential capabilities of the Fibonacci codes as implemented in efficient Fibonacci counters. Therefore, an important and challenging problem arose that asked if it were possible to implement only the minimal presentation form of the Fibonacci numbers in the Fibonacci counters. This problem was solved in general in the paper [10], thus permitting to exclude the transitions in-between the maximal and minimal presentation forms for Fibonacci numbers.

However, up to date, the efficiency of the Fibonacci counters handling only the minimal presentation form for the Fibonacci numbers, in the part of their information transmission velocity and noise-immunity, hasn't been estimated. Exactly this lack of development is removed by the presented chapter, which provides all the necessary evaluations.

Next, among various components involved in the digital devices supporting the computer vision technique, the Fibonacci decoder is by no means less important than the Fibonacci counter. Its value is especially evident because it not only decodes but also detects errors of transmission, which makes it indispensable in order to develop efficient counting circuits. Therefore, another task fulfilled in this chapter is the description of Fibonacci decoders boasting the minimal apparatus costs. Summed together, these two particular topics solve the general problem of the development of fast and noise-proof counters for computer vision systems.

The remaining part of the chapter is organized as follows. In Sect. 11.2, we describe the Fibonacci numbers, while Sect. 11.3 introduces Fibonacci codes and reminds their main properties. Section 11.4 and its two Sects. 11.4.1 and 11.4.2 deal with the minimal (and maximal) representation forms for the Fibonacci numerals, estimate their noise immunity, and discuss the noise-proof Fibonacci counter, respectively. Finally, Sect. 11.5 considers the principles of Fibonacci fractal decoding devices, while Sect. 11.6 makes conclusions and outlines the possible future research. The chapter is finished by an Acknowledgment and the list of References.

11.2 Fibonacci Numbers

The theory of Fibonacci numbers has started attracting a vivid interest and theoretical efforts on part of eminent researchers since the beginning of the 1960s after the Russian mathematician Nikolai Vorobyov published in 1961 his seminal and popular booklet "Fibonacci Numbers" [5]. In 1969, the American mathematician Verner Hoggatt published his book "Fibonacci and Lucas Numbers", while in 1989, the book by Steven Vajda "Fibonacci & Lucas Numbers, and the Golden Section:

Theory and Applications" was published [8]. These three sources laid the foundation of the theory of Fibonacci numbers. An idea of developing the so-called "Fibonacci computers" was popular in the 1970s [4, 7]. However, at that time, this idea didn't find a great support owing to the overwhelming progress in computing hardware based on the binary numeral system. Nevertheless, the interest to Fibonacci computers has survived and is fueled by various sporadic publications worldwide.

Consider the minimal form of presentation of the Fibonacci codes, which is also a basic form for that. Indeed, it was this form that was used initially for Fibonacci numbers, namely, determined by the weights of the digits associated with the Fibonacci numbers. They form the elements of the sequence (series) 1, 1, 2, 3, 5, 8, 13, 21, 34, ..., F_n, ..., where each element with the index n greater or equal to 3 is found as the sum of the two previous ones [4, 5]. The latter condition is expressed by the (Fibonacci) recursion formula

$$F_n = F_{n-1} + F_{n-2}, \quad n = 3, 4, \ldots, \tag{11.1}$$

where $F_1 = F_2 = 1$.

Equation 11.1 is the weight function of the Fibonacci numeral system that determines the weights of all positions used in a Fibonacci number. This weight function is essentially different from the power weight functions involved in the natural systems, such as the binary, ternary, octal numeral systems, and others. According to the name of the recursion relation (Eq. 11.1), the numeral systems based on the latter are traditionally called as *Fibonacci systems*.

It is clear enough that other Fibonacci-type sequences and the respective weight functions can be generated, like, for instance, 1, 1, 1, 3, 5, 9, ..., where each element starting from the fourth is calculated by summing up the previous *three* elements. Taking that into account, one can discuss other Fibonacci-like numeral systems the number of which may be unlimited, just as is with the traditional power numeral systems. However, only the very first, original Fibonacci sequence 1, 1, 2, 3, 5, ..., and its particular subsequence 1, 2, 3, 5, ..., used by Fibonacci himself in order to solve the problem of the reproduction of rabbits, have been extensively applied in practice. Thus, the properties of only these two Fibonacci series are studied intensely by the researchers into the Fibonacci theory. Up to date, quite a number of those properties have been discovered, and many new interesting features continue to reveal more and more every year. In our chapter, we are mostly indulged into the study of only one of those properties, namely, the possibility of constructing a noise-proof Fibonacci numeral system.

11.3 Fibonacci Numbers and Fibonacci Codes

It is not difficult to establish that any natural number and zero can be represented by the Fibonacci numeral function [4]:

$$A = a_n F_n + a_{n-1} F_{n-1} + \cdots + a_i F_i + \cdots a_1 F_1. \tag{11.2}$$

The same can be written shorter as follows:

$$A = a_n a_{n-1} \ldots a_i \ldots a_1,$$

where $a_i \in \{0, 1\}$ is a binary character of the ith unit in the positional representation (Eq. 11.2), n is a number of units in each code word, F_i is a weight of the ith unit, which in our case coincides with the ith Fibonacci number.

The array of all Fibonacci presentations from a given range makes the *Fibonacci code*.

Fibonacci codes boast a series of important properties, namely:

- The Fibonacci word having 1's in all odd positions 1, 3, 5, ..., n represents the maximum Fibonacci number among all Fibonacci words of the *odd* length n and equals $F_{n+1} - 1$.
- The Fibonacci word having 1's in all even positions 2, 4, 6, ..., n presents the maximum Fibonacci number among all Fibonacci words of the *even* length n and equals $F_{n-1} - 1$.
- The range of Fibonacci words of length n is equal to

$$P = F_{n+1} = F_n + F_{n-1}. \tag{11.3}$$

- The Fibonacci words in the minimal form do not allow two 1's following one another, i.e. occupying two successive positions [4].

- The latter property is exactly what makes Fibonacci codes in the minimal form noise-proof. Fibonacci words permit the operations of catamorphisms and anamorphisms (or "folds" and "unfolds") applied to their units, which makes a transition from their minimal form to the maximum form. These operations leave their quantitative equivalent invariant.

The above-listed properties of Fibonacci words are illustrated in Table 11.1. There, for the Fibonacci sequence 1, 2, 3, 5, and 8, one can find Fibonacci words in the minimal form. If in any of them one adds the weights (that is, the Fibonacci numbers) corresponding to 1's, then the sums will equal their quantitative values, which follow each other in the increasing order with the step 1. As a result, Table 11.1 generates all natural numbers from 0 to 12. Exactly these numbers in the same order will be produced by a Fibonacci adding counter.

Table 11.1 Fibonacci code words for the Fibonacci series 1, 2, 3, 5, 8

Unit no	5	4	3	2	1	Unit no	5	4	3	2	1
Unit's weight	8	5	3	2	1	Unit's weight	8	5	3	2	1
Digit i	a_5	a_4	a_3	a_2	a_1	Digit i	a_5	a_4	a_3	a_2	a_1
0	0	0	0	0	0	7	0	1	0	1	0
1	0	0	0	0	1	8	1	0	0	0	0
2	0	0	0	1	0	9	1	0	0	0	1
3	0	0	1	0	0	10	1	0	0	1	0
4	0	0	1	0	1	11	1	0	1	0	0
5	0	1	0	0	0	12	1	0	1	0	1
6	0	1	0	0	1						

11.4 Minimal and Maximum Representation Forms for Fibonacci Numbers

Since the Fibonacci codes are represented in two forms, namely, the minimal (normal) and maximal ones, the operations of catamorphisms and anamorphisms (or "folds" and "unfolds") were developed in order to transform the minimal form of a Fibonacci code to its maximal form, and backward. In more detail, the minimal form is transformed to the maximal one making use of the anamorphism ("unfold") operation [7]. For example, the number "3" in its minimal form is coded as 100, and by unfolding it one comes to its maximal form 011. On the contrary, the maximal form 011 is reduced to its minimal form 100 with the aid of the converse operation of folding. As could be expected, the numerical equivalent of a Fibonacci code in the maximal form coincides with the same in the minimal form. In the above example, if the third unit of the Fibonacci code is 1 while its two previous units are 0, then after unfolding this unit 1 to the sequence 011 the total weight of the previous units is intact equaling 3.

Folding and unfolding operations simplify a lot carrying out the arithmetic operations over Fibonacci codes; hence they are widely used when developing the Fibonacci arithmetic. However, these operations are not too easy for implementing in the Fibonacci counters without increasing the complexity of the devices and decreasing their computation velocity. In order to avoid this inconvenience, in this chapter, we propose how to create the counters with the use of only the minimal form of the Fibonacci codes but not their maximal form. This way can lead to enhancing the new counters' velocity and noise immunity, and to the decrease of their hardware costs, as compared to the counters relying on the operations of folding and unfolding.

11.4.1   Estimate of the Noise Immunity of the Fibonacci Codes in Their Minimal Form

Fibonacci codes boast a natural noise immunity property. The latter makes them quite popular when designing various noise-proof digital devices, including computers [4]. In addition, a series of publications promote their use in information transmission devices, in particular, within the self-synchronizing transmission/reception systems [7]. The development of such procedures badly needs new estimates of the Fibonacci codes noise immunity as compared to the other existing noise-proof codes, both taking into account the transmission channel's properties and not. In the latter case, only the noise immunity of proper Fibonacci digital devices exploited for the preliminary information workout (before the transmission) will be examined. However, the real-life transmission/reception systems use to apply Fibonacci coding of both the digital devices and the transmission channels.

One of the remarkable (but not well studied as yet) attribute of the Fibonacci codes is the fact that the arithmetical operations fulfilled in these codes don't require transitions in-between the unit positions (in contrast to the decimal and other positional numeral systems). This genetic endowment of the Fibonacci codes may potentially lead to the development of extremely fast circuits conducting all arithmetical operations. Nevertheless, any increase in velocity is directly associated with faster transition processes in the micro-circuits, which inevitably brings up a higher number of errors. Again, the latter disadvantage may be coped with a noise-proof coding. Therefore, the Fibonacci codes being such noise-proof ones can efficiently reduce the above-mentioned defect generic for the fast digital devices by combining their two helpful features: being capable of both enhancing the devices' velocity and detecting the occurring errors.

One the other hand, one of the unpleasant minor points of the Fibonacci codes is that they can detect the erroneous transmission of 0 as 1 but not vice versa; that is, mistaken transitions of 1–0 aren't discovered. However, the number of digits 0 in the Fibonacci codes is much higher than that of digits 1; therefore, erroneous transmissions of 0–1 would considerably override the number of those of 1–0 even in symmetric channels. This allows one to conclude that the Fibonacci code in its minimal form is sufficiently noise-proof. Moreover, in the real-life digital circuits, the errors use to be non-symmetric hence the minimal form Fibonacci codes can detect the majority of transmission errors. The latter means that the Fibonacci codes are ideally adapted to catching out transmission errors arising in the digital circuits similar to non-symmetric transmission channels. Fibonacci codes cannot have several 1's in a row, so such clustered errors are traced very easily; nonetheless, isolated (separated) sole errors are also found out readily, which makes Fibonacci codes extremely reliable. Furthermore, isolated errors found between two 1's are also corrected immediately.

As is mentioned above, the minimal form of a Fibonacci code word doesn't permit two or more 1's in a row, which is equivalent to prohibiting zero digit(s) to appear between two 1's in the code's maximal form [4]. Is this rule is violated there must be errors in the Fibonacci code words. Such code words are called *prohibited combinations*, and their number, together with the allowed code words, equals $N = 2^n$.

Errors arising in Fibonacci code words, as well as in other coding systems, can be detected and corrected only when permitted words are transmitted as prohibited combinations. Thus, the greater is the number of prohibited combinations in a code, the higher is its noise immunity (the probability of catching out the arising transmission errors). Because of that, the Fibonacci counters handling the higher number of units in the code words will be more reliable than those with the lower number of units.

Table 11.2 provides the 19 prohibited combinations that complement the 13 allowed ones listed in the above Table 11.1. All these prohibited combinations boast at least one pair (or more) of adjacent 1's in a row. Such words if received mean that some errors have occurred. Since the total number of prohibited combinations is determined by the difference $2^n - P = 2^n - (F_n + F_{n-1})$, where n is the number of digital units in each word, Table 11.2 contains exactly $2^5 - (F_5 + F_4) = 32 - (8 + 5) = 19$ prohibited combinations.

Fibonacci numbers are widely used in noise-proof digital devices solving special problems such as timers, rangefinders, frequency counters, etc. because it helps enhance the noise resistance of not only those devices but also the possibility of development of noise-proof channels that transmit the information generated by them. In order to implement such a possibility, the idea of a throughout control with the aid of the same codes of both the digital devices and the transmission channels is exploited [7]. It is worthwhile to note here that the use of other, more powerful noise-proof codes for such hybrid tasks of information processing and transmission are usually not very efficient, as overhead costs involving the development of special coders and decoders arise. Those coders/decoders are necessary to fulfill

Table 11.2 Prohibited combinations of the Fibonacci code for the series 1, 2, 3, 5, 8

Unit no	5	4	3	2	1	Unit no	5	4	3	2	1
Unit's weight	8	5	3	2	1	Unit's weight	8	5	3	2	1
Digit i	a_5	a_4	a_3	a_2	a_1	Digit i	a_5	a_4	a_3	a_2	a_1
1	0	0	0	1	1	11	1	0	1	1	2
2	0	0	1	1	0	12	1	1	0	0	0
3	0	0	1	1	1	13	1	1	0	0	1
4	0	1	0	1	1	14	1	1	0	1	0
5	0	1	1	0	0	15	1	1	0	1	1
6	0	1	1	0	1	16	1	1	1	0	0
7	0	1	1	1	0	17	1	1	1	0	1
8	0	1	1	1	1	18	1	1	1	1	0
9	1	0	0	1	1	19	1	1	1	1	1
10	1	0	1	1	0						

transitions from the digital devices to the transmission channel and vice versa. Moreover, these coders/decoders affect negatively the information processing speed.

For Fibonacci codes, we must start determining the percentage of detected errors, and hence the proportion (probability) of undetectable errors, as well. In the first task, the desired estimate equals the ratio of the forbidden combinations to the whole number of possible combinations. As for the second task, we will calculate the fraction of the allowed combinations to their total number minus the probability of the undistorted transmission. Those two calculated ratios represent the probability of forbidden and allowed combinations obtained, respectively. If we assume that each combination can arise with the same probability ($1/N$), then we will characterize the properties of the code only, without taking into account the transmission channel's parameters. Therefore, the universal characteristics of the proper codes will be estimated. One also has to remember that the undetectable errors are perceived as the right messages. Moreover, they may arise suddenly and untimely thus causing unexpected damage. It is impossible to exclude them completely; however, their number can be considerably lowered by introducing additional noise-proof coding tools.

The proportion of detectable errors is found by the well-known formula

$$D = 1 - \frac{P}{N},$$
(11.4)

where P represents a number of allowed coding combinations, while N is a number of all possible combinations of length n.

Having substituted P with its expression from Eq. 11.3 and $N = 2^n$, the percentage of detectable errors in the Fibonacci code is readily obtained:

$$D = 1 - \frac{F_n + F_{n-1}}{2^n}.$$
(11.5)

Now apply Eq. 11.5 to calculate the estimate of the portion of detectable errors for Fibonacci codes with the combination length $n = 2, 3, \ldots, 32$, and compile this data to Table 11.3.

When $n = 1$, the Fibonacci numbers boast only allowed combinations 0 and 1, while no prohibited combination exists. Because of that, Table 11.3 starts with $n = 2$, since in this case, apart from the allowed Fibonacci combinations 00, 01, and 10, a single prohibited combination 11 (composed of two ones in a row) appears.

Next, make use of Eq. 11.5 and determine the proportions (probabilities) of detectable errors. The latter are listed in Table 11.4. Table 11.4 permits one to draw the graph of the probabilities of detectable errors depicted in Fig. 11.1 with the solid line. The graph in Fig. 11.1 shows that the proportion of detected errors grows along with the number of positions n and tends to 1.

Table 11.3 The number of prohibited Fibonacci combinations for $n = 2, 3, \ldots, 32$

n	The number of prohibited combinations	n	The number of prohibited combinations
2	1	18	2.554×10^5
3	3	19	5.133×10^5
4	8	20	1.031×10^6
5	19	21	2.068×10^6
6	43	22	4.148×10^6
7	94	23	8.314×10^6
8	201	24	1.666×10^7
9	423	25	3.336×10^7
10	881	26	6.679×10^7
11	1.816×10^3	27	1.337×10^8
12	3.719×10^3	28	2.676×10^8
13	7.582×10^3	29	5.355×10^8
14	1.54×10^4	30	1.072×10^9
15	3.117×10^4	31	2.144×10^9
16	6.295×10^4	32	4.289×10^9
17	1.269×10^5		

Table 11.4 Proportions of detectable errors in the Fibonacci codes

n	Proportions of detectable errors	n	Proportions of detectable errors
2	0.25	18	0.974
3	0.375	19	0.979
4	0.5	20	0.983
5	0.594	21	0.986
6	0.672	22	0.989
7	0.734	23	0.991
8	0.785	24	0.993
9	0.826	25	0.994
10	0.86	26	0.995
11	0.887	27	0.996
12	0.908	28	0.997
13	0.926	29	0.997
14	0.94	30	0.998
15	0.951	31	0.998
16	0.961	32	0.999
17	0.968		

Fig. 11.1 Graph of the portions of detectable errors as the function of the number of positions n in the Fibonacci codes

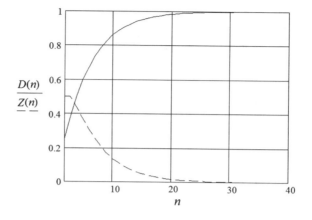

Now the proportions of *undetectable* errors are deduced as follows:

$$Z = \frac{P-1}{N}. \qquad (11.6)$$

Having inserted in Eq. 11.6 the number of permitted combinations by Eq. 11.3, the percentage of undetectable errors reduces to

$$Z = Z(n) = \frac{F_n + F_{n-1} - 1}{2^n}. \qquad (11.7)$$

Table 11.5 The number of permitted combinations in the Fibonacci code with $n = 2, 3,..., 32$

n	The number of permitted combinations	n	The number of permitted combinations
2	3	18	6.765×10^3
3	5	19	1.095×10^4
4	8	20	1.771×10^4
5	13	21	4.637×10^4
6	21	22	7.503×10^4
7	34	23	1.214×10^4
8	55	24	1.964×10^5
9	89	25	5.142×10^5
10	143	26	8.32×10^5
11	232	27	1.346×10^6
12	377	28	2.178×10^6
13	610	29	5.703×10^6
14	986	30	9.227×10^6
15	1.596×10^3	31	1.493×10^7
16	2.584×10^3	32	2.416×10^7
17	4.181×10^3		

Table 11.6 Proportions of undetectable errors of the Fibonacci codes

n	The proportions of undetectable errors of the Fibonacci codes	n	The proportions of undetectable errors of the Fibonacci codes
2	0.5	18	0.026
3	0.5	19	0.021
4	0.438	20	0.017
5	0.375	21	0.014
6	0.313	22	0.011
7	0.258	23	8.944×10^{-3}
8	0.211	24	7.235×10^{-3}
9	0.172	25	5.854×10^{-3}
10	0.139	26	4.736×10^{-3}
11	0.113	27	3.831×10^{-3}
12	0.092	28	3.1×10^{-3}
13	0.074	29	2.508×10^{-3}
14	0.06	30	2.029×10^{-3}
15	0.049	31	1.641×10^{-3}
16	0.039	32	1.328×10^{-3}
17	0.032		

Table 11.5 provides the number of permitted combinations in the Fibonacci code for each length $n = 2, 3, \ldots, 32$. Based upon Table 11.5 and Eq. 11.7 we construct Table 11.6 yielding the proportions of undetectable errors.

The dashed line in Fig. 11.1 illustrates the decreasing proportion of undetectable errors as the length n grows, which confirms the growth of the noise-immunity of the Fibonacci codes along with the combination length n. This graph allows one to estimate (for a given Fibonacci code combination length) the probabilities (proportions) of both detectable and non-detectable errors.

As it was already mentioned above, the proportion of detectable errors by the Fibonacci codes grows along with the combinations length tending in the limit to 1. Such codes are extremely efficient when implemented in devices with asymmetric errors because Fibonacci codes detect errors only in transitions of 0–1. Now since the modern digital gadgets mainly boast asymmetric errors, the definitely asymmetric nature of Fibonacci codes promotes their use in the devices.

However, in the above discussion, we supposed that any combination is generated with the (uniform) probability $1/N = 1/2^n$, which implies the equality $D + Z + 1/2^n = 1$. In the latter case, as the combination length n grows, the probability of any (correct or erroneous) combination grows to zero. Nevertheless, in real life coding and transmitting systems, this probability never drops below some positive threshold. Therefore, if the transitions of correct combinations to the permitted or prohibited ones are equiprobable, the probability of detecting the errors decreases. However, luckily, the proportions (probabilities) of detectable errors tend to grow anyway. If the transitions from 0 to 1 and 1 to 0 have unequal probabilities, proportions of detectable and non-detectable errors are calculated with other formulas (*cf.*, e.g. [7]). Equations 11.4 and 11.6 used above are special cases of the most general ones, in the case when the transitions of the correct combinations to prohibited and permitted are equiprobable.

11.4.2   Noise-Proof Fibonacci Counting

Even in the modern times, the task of obtaining highly reliable results of operations of digital devices is still extremely important. As we mentioned above, one of the promising ways to reach this aim is to apply the Fibonacci numeral systems. Based upon these systems, a noise-proof algorithm of running those digital gadgets was developed, which, however, required to transform the Fibonacci numbers represented in the minimal form to the same numbers but written in the maximal representation form. It is worthy to explain that the latter procedure consumed the additional running time, as well as implied a more complicated hardware of the corresponding counter. In order to avoid these shortcomings, a method of summarizing noise-immune Fibonacci calculus in the normal (minimal) form was developed. This method indeed boasted high enough velocity and noise immunity (*cf.*, [10]).

In more detail, this method consists in finding two subsequent 0's in the current minimal form representation of a Fibonacci number and then replacing the first (from the right) of them with 1. This transformation is accompanied by translating all the lower rank positions (that is, standing to the right) to zero. For example, consider the following Fibonacci number in its minimal (normal) form 01000101. Since the weights of its digits are (from the left to the right) 34, 21, 13, 8, 5, 3, 2, and 1, respectively, this Fibonacci number's value equals $21 + 3 + 1 = 25$. Therefore, according to the above-described algorithm, the next Fibonacci number is found as 01001000, which indeed produces $21 + 5 = 26$. This procedure of adding 1 can be repeated until the last Fibonacci number has no two subsequent 0's and boasts 1 standing in the highest (the extreme left) position. The latter Fibonacci number will be the maximum possible in the present range, and its value equals the sum of the weights of the highest and the second highest position minus 1. That is, in the present example, the maximum possible Fibonacci number is 10101010, whose value is $34 + 21 - 1 = 54$. Indeed, the same value is obtained by summing up the weights of the positions containing 1's: $34 + 13 + 5 + 2 = 54$.

Table 11.1 above demonstrates (in an increasing order) all possible normal Fibonacci numbers with the weights of 8, 5, 3, 2, and 1, from the corresponding range comprising $8 + 5 = 13$ numbers generated by the just described algorithm. One of the remarkable advantages of the method in question is that it needs no operations of catamorphisms and anamorphisms (or "folds" and "unfolds"), as was the case for the previous algorithms of Fibonacci counting [1, 2]. This advantage enhances the running speed and the robustness (reliability) of the counters, in which the Fibonacci numeral system is implemented.

The operation of adding 1 to Fibonacci numbers can be illustrated for the 5-digit minimal form in Fig. 11.2. All 5-digit Fibonacci numerals are listed in the increasing order in Table 11.2. The additions start from the number 00000 with its value of 0. According to the described algorithm, we look through its positions from the right to the left. As a result, we see that zero in the first position is followed by zero in the second position (these two sequential zero are included in a box). The method rules to replace zero in the first position by 1 thus getting the next Fibonacci number 00001 with its numerical value 1.

Now we have to analyze the structure of the new Fibonacci numeral. Because it still has two adjacent zero digits to the left from the newly obtained 1 in the first position, we replace the first (right) zero with 1 but simultaneously replacing the previously generated 1 with zero thus coming to the next Fibonacci numeral 00010 having value 2. It is easy to see that in the next step of the algorithm the next Fibonacci number 00100 will be found, and so on (see Fig. 11.2). The procedure stops after having reached the Fibonacci number 10101 free from two or more zero digits in a row. The latter is the maximal possible Fibonacci number in this range with its value of 12: summing the weights yields $8 + 3 + 1 = 12$.

Fig. 11.2 Adding 1 in Fibonacci calculus in the minimal form

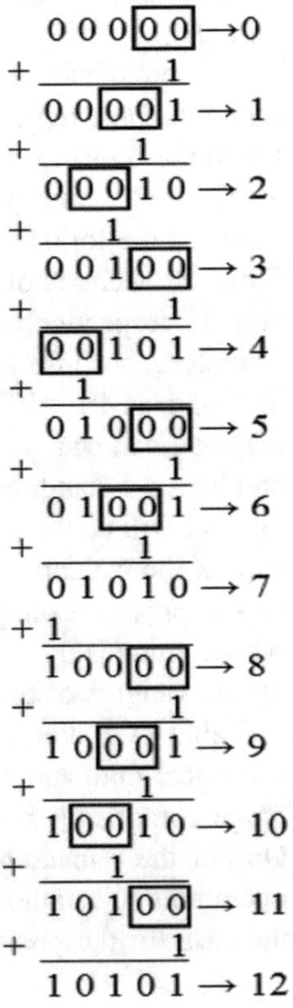

$$
\begin{array}{l}
0\ 0\ 0\ \boxed{0\ 0} \rightarrow 0 \\
+\quad\quad\quad\ 1 \\
\hline
0\ 0\ \boxed{0\ 0}\ 1 \rightarrow 1 \\
+\quad\quad\ 1 \\
\hline
0\ \boxed{0\ 0}\ 1\ 0 \rightarrow 2 \\
+\quad\ 1 \\
\hline
0\ 0\ 1\ \boxed{0\ 0} \rightarrow 3 \\
+\quad\quad\quad\ 1 \\
\hline
\boxed{0\ 0}\ 1\ 0\ 1 \rightarrow 4 \\
+\ 1 \\
\hline
0\ 1\ 0\ \boxed{0\ 0} \rightarrow 5 \\
+\quad\quad\quad\ 1 \\
\hline
0\ 1\ \boxed{0\ 0}\ 1 \rightarrow 6 \\
+\quad\quad\ 1 \\
\hline
0\ 1\ 0\ 1\ 0 \rightarrow 7 \\
+\ 1 \\
\hline
1\ 0\ 0\ \boxed{0\ 0} \rightarrow 8 \\
+\quad\quad\quad\ 1 \\
\hline
1\ 0\ \boxed{0\ 0}\ 1 \rightarrow 9 \\
+\quad\quad\ 1 \\
\hline
1\ \boxed{0\ 0}\ 1\ 0 \rightarrow 10 \\
+\quad\ 1 \\
\hline
1\ 0\ 1\ \boxed{0\ 0} \rightarrow 11 \\
+\quad\quad\quad\ 1 \\
\hline
1\ 0\ 1\ 0\ 1 \rightarrow 12
\end{array}
$$

The above-depicted algorithm can be formally described by the following steps:

Step 1. The search from the right to the left within a Fibonacci number finds (at least) two adjacent positions containing zero digits.

Step 2. The first (from the right) of those digits is replaced with 1 while all the lower (standing to the right) positions are set zero.

Step 3. Two adjacent 1's are prohibited (if such a pair appears that means that an error has occurred; then stop and verify all the previous steps).

Step 4. The procedure is repeated until at least one pair of adjacent zero digits exist.

Step 5. Stop.

The method under consideration can be represented as a combination of two elementary logical operations:

Step 1. The search from the right to the left within a Fibonacci number $A = a_n a_{n-1} \ldots a_i a_{i-1} \ldots a_1$ finds (at least) two adjacent positions with their logical sum (disjunction) equaling 0, that is, $a_i \vee a_{i-1} = 0$.

Step 2. If no pair with $a_i \vee a_{i-1} = 0$ is detected, stop: all the possible Fibonacci numbers in the given range have been generated.

Step 3. Otherwise, i.e. such a pair with $a_i \vee a_{i-1} = 0$ is run into, then $a_{i-1} := 1$ and $a_j := 0, j = i - 2, \ldots, 1$.

Step 4. Check all adjacent pairs: If $a_i \wedge a_{i-1} = 0$ for all $i = 2, 3, \ldots, n$, then the just generated Fibonacci number is correct. Go to Step 1.

Step 5. Otherwise, if there exists an adjacent pair of digits with $a_i \wedge a_{i-1} = 1$, report an error and return to Step 2.

Even though the above-described method of adding 1 in the minimal form seems to be very simple, its implementation in real-life Fibonacci counters required a development of new, previously unknown structures. With the aid of the latter, the Fibonacci counting algorithm has been used to construct original, fast, and noise-proof counters.

Every Fibonacci-based counter and the corresponding numerical device work within a finite range of the Fibonacci numbers. The latter range is determined by the sum of the weights of the first and the second (from the left) positions of the Fibonacci code. Thus, if the weights of first and the second positions are 8 and 5 (as in the above-mentioned example), then the total range of the represented Fibonacci numbers is 13. Namely, the minimal number is 00000 (representing zero) and the maximal number is 10101, which represents 12. Since the Fibonacci sequence is unbounded, then any desired range can be obtained by increasing the number of positions of the Fibonacci code words. For instance, if the Fibonacci position weights are selected as 1, 2, 3, 5, 8, 13, 21, and 34, then the range of the 8-position Fibonacci counter (or its calculation coefficient) equals 55. This means that the minimal Fibonacci number represented in the latter counter is 00000000 (corresponding to 0) while the maximal number is 10101010 that equals 54.

There is also another way of extending the range of the represented numbers. Namely, one can link sequentially several Fibonacci counters say with the range 13, thus making it possible to count the numbers in the new range obtained by multiplying the ranges of the connected smaller counters. For example, if the arranged link connects sequentially 10 Fibonacci counters with range 13 each, then the total range of this connection, that is, the aggregate number of the represented Fibonacci words will equal 13^{10}.

11.5   Decoding the Fibonacci Combinations in the Minimal Representation Form

As already mentioned above, digital counting gadgets comprise various digital devices, among which decoders are of special importance. Their task is to transform the counter's states into distinct signals, as for example, in control gadgets or image transmitters. Even counting circuits need some elements of deciphering in order to select certain states of the counter. Therefore, practically no digital device can function without solving the decoding problem, and hence without decoders.

The Fibonacci decoders are special due to the fact that they are to decipher only Fibonacci numerals as a subclass of binary ones. As was shown above, Fibonacci numbers cannot have two or more 1's in a row. Owing to that, on the one hand, any complete binary decoder is capable of deciphering Fibonacci numerals, which is one of the definite advantages of the Fibonacci codes. However, on the other hand, by making use of the general binary decoders, we miss some possibilities of decreasing the hardware cost, together with enhancing the counters' reliability, as well as diminishing their energy consuming. Therefore, on the basis of the special properties of the Fibonacci codes (such as the wider list of prohibited code words), one can apply the well-known and robust methods of minimization of the binary codes with the aim of saving on the hardware costs.

Since the collection of Fibonacci words is a subset of binary numbers, certain redundant information arises. Its numerical value is determined by the difference between the logarithm of the total quantity of binary words and the same of the total number of Fibonacci combinations. The information redundancy generic to the Fibonacci numbers is exactly what permits one to detect errors occurring when transmitting, storing, and processing the information. The same property allows one to generate Fibonacci decoders of a special Structure, which is called *fractal* or *reversal*.

The proposed method of fractal decoding helps cut essentially the hardware costs of the decoders as compared to the usual deciphering devices. This fact follows from the generic property of the Fibonacci numbers saying that every Fibonacci numeral except for the very initial 1 and 2 equals the sum of the two preceding Fibonacci numbers [4]. Moreover, the first of those two addends determines the total quantity of the Fibonacci numbers within the range starting with 0, and the (smaller) second term determines the quantity of the Fibonacci within the same range starting with 1.

As was demonstrated above, one of the remarkable properties of the Fibonacci numbers is their *fractal* (or exhibiting a repeating pattern) nature. Strictly speaking, this fractal property follows from the very definition of the Fibonacci numbers. Indeed, consider the first five Fibonacci numerals 1, 2, 3, 5, and 8. Every one of them is constructed according to the repeating pattern, that is, each is equal to the sum of the two preceding ones. For instance, $8 = 5 + 3$, $5 = 3 + 2$, $3 = 2 + 1$. These equalities are called the *Fibonacci fractals of rank 1*. (By the way, there exist also Fibonacci numbers of rank 2, rank 3, etc.).

Such a deciphering procedure (based on the similar, or fractal decoding the preceding digits) is readily applied to decoding the Fibonacci numbers. This approach permits one to essentially save on the circuits realizing logical products implemented in the real-life decoding hardware. Indeed, once done, the logical product need not being repeated twice. This repeating pattern (fractal property) of the Fibonacci numbers allows one to diminish the total number of Fibonacci words that need to be decoded. Nevertheless, there always exist a number of Fibonacci words (in the first half of their list ordered to increase and starting with 0) that lack this repeating pattern and thus need to be deciphered with an individual decoding procedure. For example, in Table 11.1, such are three ($8 - 5 = 3$) 5-positional Fibonacci numbers: 01000, 01001, and 01010, equaling 5, 6, and 7, respectively.

The fractal decoder repeats the deciphering procedure exported from another, more general decoder. Even this property is enough to guarantee the existence of high-technological structures to generate the decoders. Such a procedure may boast a large number of repeated stages. In this chapter, we restrict ourselves to only two stages. The summary quantity of elements needed to implement a decoder at each stage is lower than if a standard (without minimization) deciphering procedure would be applied.

In order to explain the principle of operating a fractal decoder, let us repeat once again all thirteen Fibonacci numbers from Table 11.1 in the new Table 11.7, with the first and the last 5 numbers written in bold. A crucial property of those selected numbers is the following: they differ from each other only in the highest, 5th position, while the lower 4 positions contain the same digits. This implies that if necessary we need to decipher *only one* of the two identical 4-positional combinations, whereas the senior (the 5th) position containing 0 or 1, serves to identify to which of two groups of 5-positional combinations the word in question belongs.

The remaining (not in bold) three Fibonacci numbers all have 0 in the senior position. The latter means that a standard decoding procedure is applicable only to those three intermediate 5-positional Fibonacci numbers. The other 10 Fibonacci words are decoded in pairs: indeed, by deciphering 4 positions of the first 5 Fibonacci numbers, we do the same with respect to the last 5 Fibonacci words, too.

Remark 1 Of course, switching the output device to the first 5 outcomes, when the senior position of the deciphered word would contain 0, and to the last 5 outcomes

Table 11.7 Fractal structure of Fibonacci numbers

No.	Combinations				
	$X_5 = 8$	$X_5 = 8$	$X_5 = 8$	$X_5 = 8$	$X_5 = 8$
0	0	0	0	0	0
1	0	0	0	0	1
2	0	0	0	1	0
3	0	0	1	0	0
4	0	0	1	0	1
5	0	1	0	0	0
6	0	1	0	0	1
7	0	1	0	1	0
8	1	0	0	0	0
9	1	0	0	0	1
10	1	0	0	1	0
11	1	0	1	0	0
12	1	0	1	0	1

otherwise (when the senior digit is 1), eats the running time and reduces the decoding speed to approximately one-half.

Remark 2 The three exclusive 5-positional Fibonacci words, namely: 01000, 01001, and 01010,—need a standard (non-fractal) procedure to decode them.

Similar relationships are true for the set of ordered Fibonacci numbers within an arbitrary range. The number of those having 0 in the senior position n equals the weight F_n of this position, while the number of those boasting 1 in the same position coincides with the weight F_{n-1} of position $n - 1$. Moreover, the Fibonacci words having 0 in the senior position run their absolute values from 0 to $F_n - 1$, whereas those with 1 in the senior position cover the range from F_n to $(F_n + F_{n-1}) - 1$. The latter numbers occupy the places at the very end of the increasing order containing altogether $F_n + F_{n-1}$ numerals. For example, in Table 11.7, they are the last 5 numbers among the total of $8 + 5 = 13$ numerals. Finally, the total of $F_n - F_{n-1}$ words is deciphered with a general (non-fractal) algorithm.

Fibonacci fractals of rank 2 are the recursions of rank 1, in which the greater Fibonacci number is replaced with the Fibonacci fractal of rank 1. For instance, the equality $13 = ((5 + 3) + 5)$ is a Fibonacci fractal of rank 2 deduced from the rank 1 Fibonacci fractal $13 = (8 + 5)$. Inductively expanding this rule, one can obtain Fibonacci fractals of rank 3, 4, and so on. Thus, for the same number 13, its rank 3 Fibonacci fractal is defined by the equality $13 = ((3 + 2) + 3) + (3 + 2)$ yielded from the rank 2 Fibonacci fractal $13 = ((5 + 3) + 5)$.

Therefore, the fractal method of decoding the Fibonacci words can be described in its most general form as follows:

Step 1. A table containing all the Fibonacci words from the given range is produced.

Step 2. The lower $n - 1$ (right) positions are analyzed, and the Fibonacci words boasting the same tails are split into two groups: Group 1 (constituent 1) contains the words with 0 in the senior position and Group 2 (constituent 2) comprises the Fibonacci code words with 1 in the senior position (position n).

Step 3. A Fibonacci word to be decoded having arrived, it is classified as belonging to one of the two constituents.

Step 4. A signal from the constituent containing the obtained word (which depends on the digit, 0 or 1, occupying the senior position) is transmitted to one of the two corresponding outputs.

Step 5. The Fibonacci words having 1 in the penultimate position (position $n - 1$) are decoded by a standard (non-fractal) deciphering procedure.

This deciphering algorithm can be implemented in the form of both a computer code and a flowchart called the *Fibonacci decoder*. To illustrate the latter scheme, let us develop the Fibonacci decoder flowchart for the thirteen Fibonacci 5-positional words listed in Table 11.7.

Figure 11.3 depicts the flowchart of the proposed Fibonacci decoder for the 5-positional Fibonacci words. Its description in more detail can be found in [11]. The flowchart is composed of two decoders: decoder 1 and decoder 2 boasting $n - 1$ and n outputs, respectively that decipher the Fibonacci combinations of length $n - 1$ and n. In addition, the circuit comprises two commutators to switch to the needed outputs the signals coming from decoder 1.

This Fibonacci decoder runs as follows. The binary values $X_1, X_2, \ldots, X_{n-1}$ of the signals obtained from positions 1 through $n - 1$ of the Fibonacci combination $X_1, X_2, \ldots, X_{n-1}, X_n$ (which has arrived at the inputs of decoder 2) are transmitted to the inputs of decoder 1, whose outputs are oriented to the corresponding inputs of commutators 1 and 2.

Fig. 11.3 Flowchart of the Fibonacci decoder

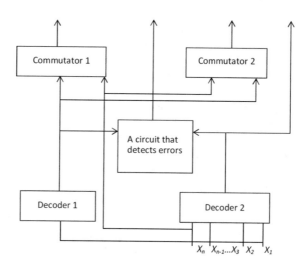

Meanwhile, the signal X_n from the senior position is transmitted to the inverse control input of commutator 1 and to the direct control input of commutator 2. If this signal is 0, then commutator 1 turns on; otherwise, i.e. if it is 1, then commutator 2 is activated. Next, the signal transmitted from the output of decoder 1 appears on one of the outputs of commutators 1 or 2. The aggregate number of outputs from the two commutators is $2F_{n-1}$. If no signal appears on the outputs of both commutators, then, if the word is a Fibonacci combination, a signal is read on one of the outputs from decoder 2.

Finally, if no signal is detected on the outputs from decoders 1 or 2, it means that the current combination is erroneous, which activates the circuit that detects errors. The latter reports the error on its output.

11.6 Conclusions and Future Research

One of the most difficult challenges when developing a competent system of technical vision is to create reliable methods to elevate the velocity and noise immunity of the digital devices composing the system. The gadgets combining counters and decoders are indispensable devices for such a system. In this chapter, we propose to develop such fast and noise-proof devices based upon special coding tools boasting informational redundancy. Various frameworks make that possible, but nowadays, an acute interest is attracted to noise-proof numeral systems, among which the Fibonacci system is extremely popular. The latter helps generate the so-called Fibonacci codes, which can be effectively applied to the computer vision systems; in particular when developing counting devices based on Fibonacci counters, as well as the corresponding decoders.

However, the Fibonacci counters usually pass from the minimal form of representation of Fibonacci numbers to their maximal form by recurring to the special operations catamorphisms and anamorphisms (or "folds" and "unfolds"). The latter makes the counters quite complicated and time-consuming. In this chapter, we propose a new version of the Fibonacci counter that relies only on the minimal representation form of Fibonacci numerals and, thus, leads to the counter's faster performance and a higher level of the noise-immunity. Based on the above-mentioned features, we also present the appropriate fast algorithm implementing the noise-proof counting and the corresponding fractal decoder.

In the first part of the chapter, we estimated the new method's noise-immunity, as well as that of its components. The second problem studied in the chapter concerns the efficiency of the existing algorithm of Fibonacci representation in the minimal form. Based on this examination, we proposed a modernization of the existing algorithm aiming at increasing its performance velocity. The third object of the chapter was to realize the comparative analysis of the Fibonacci decoders, as well as the development of the fractal Fibonacci decoders.

Since factorial, binomial, and golden section numeral systems are also promising in overcoming the standard obstacles when developing counters and decoders of high velocity and noise-immunity, we are planning to develop similar schemes with the use of the above-mentioned numeral systems in the future research.

Acknowledgements This research was partially supported by the CONACYT grant CB-2013-01-221676 (Mexico). The authors would also like to express their profound gratitude to Prof. Dr. Anatoliy Ya. Beletsky and Prof. Dr. Neale R. Smith Cornejo, whose critical remarks and suggestions have helped us a lot in the revision of the chapter as well as in improving its contents and presentation.

References

1. Borisenko, A.A., Solovey, V.A., Uskov, M.K., Chinaev, P.I.: A robotic-technical combined device for the automatic control of surface defects with an aid of computer vision. Probl. Mach. Build. Autom. **2**, 31–37 (1991) (in Russian)
2. Borisenko, A.A., Solovey, V.A.: A method for the automatic control of surface defects and its implementation. USSR Author Rights Certificate No. 1715047 (1991) (in Russian)
3. Borisenko, A.A., Kalashnikov, V.V., Protasova, T.A., Kalashnykova, N.I.: A new approach to the classification of positional numeral systems. In: Neves-Silva, R., Tshirintzis, G.A., Uskov, V., Howlett, R.J., Jain, L.C. (eds.) Frontiers in Artificial Intelligence and Applications (FAIA), vol. 262, pp. 44–450 (2014)
4. Stakhov, A.P.: Introduction to the Algorithmic Measurements Theory. Soviet Radio, Moscow (1997) (in Russian)
5. Vorobyov, N.N.: Fibonacci numbers. Nauka, Moscow (1969) (in Russian)
6. Goryachev, A.E., Degtyar, S.A.: A method for generation of permutations based on factorial numbers and complementary arrays. Trans. Sumy State Univ. Tech. Sci. **4**, 86–92 (2012) (in Russian)
7. Stakhov, A.P.: Fibonacci and Golden Proportion Codes as an Alternative to the Binary Numeral System. Part 1. Academic Publishing, Germany (2012)
8. Vajda, S.: Fibonacci and Lucas Numbers, and the Golden Section: Theory and Applications. Ellis Horwood Ltd., Chichester (1989)
9. Azarov, O.D., Chernyak, O.I., Murashenko, O.G.: A method of developing fast Fibonacci counters. Probl. Inf. Control **46**, 5–8 (2014) (in Ukrainian)
10. Borysenko, O.A., Matsenko, S.M., Polkovnikov, S.I.: A noise-proof Fibonacci counter. Trans. Natl. Technol. Univ. Kharkiv **18**, 77–81 (2013) (in Russian)
11. Borysenko, O.A., Stakhov, A.P., Matsenko, S.M.: A Fibonacci decoder. In: 5th International Conference on Information Technologies and Computer Engineering, Vasyl Stefanik Carpathian National University, Ivano-Frankivsk–Vynnytsya, pp. 279–281 (2015) (in Ukrainian)

Printed in the United States
By Bookmasters